Solid-state Electronic Amplifiers

Solid-state Electronic Amplifiers

Contributors

Antonio Agnesi and Federico Pirzio et al.

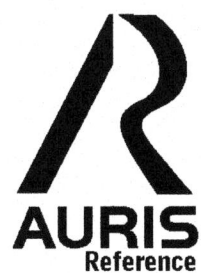

AURIS
Reference

www.aurisreference.com

Solid-state Electronic Amplifiers

Contributors: Antonio Agnesi and Federico Pirzio et al.

Published by Auris Reference Limited

www.aurisreference.com

United Kingdom

Solid-state Electronic Amplifiers

ISBN: 978-1-78154-920-9

British Library Cataloguing in Publication Data
A CIP record for this book is available from the British Library

Printed in the United Kingdom

Exclusively distributed by CBS Publishers & Distributors Pvt. Ltd.

Sales & Distribution Rights only for India, Pakistan, Bangladesh, Sri Lanka, Nepal and Bhutan.This book is not to be sold outside these territories.

Contents

List of Abbreviations

ASE	amplified spontaneous emission
B3G	Beyond the 3rd generation
CH	Cherry-Hooper
CML	collimation microlens
CMFB	common-mode feedback
CS	Current-steering
DPSS	Diode-pumped solid state
DBDP	Dual Band Digital Pre distortion
EM	Electromagnetic
HWP	half-wave plate
H-FWG	Helix-Folded Waveguide
HCS	Hybrid current-steering
LO	local oscillator
MP	memory polynomial
MT	Multiplier
OPG	optical parametric generation
OPO	Optical parametric oscillator
OFDM	orthogonal frequency-division multiplexing
PAE	power added efficiency
PA	power amplifier
PAE	Power-added efficiency
PVT	Process, voltage, and temperature
SHG	second harmonic generation
SDR	Software-defined radio
SSPAs	Solid state power amplifiers
TAS	Transadmittance stage
TIS	Transimpedance stage
TWT	Traveling wave tube
VGA	Variable-gain amplifier
WLAN	Wireless Local Area Network

List of Contributors

Antonio Agnesi
Laboratorio Sorgenti Laser Dipartimento di Elettronica dell'Università di Pavia Via Ferrata 1 - 27100 Pavia Italy

Federico Pirzio
Laboratorio Sorgenti Laser Dipartimento di Elettronica dell'Università di Pavia Via Ferrata 1 - 27100 Pavia Italy

Jiaxing Liu
Beijing National Laboratory of Condensed Matter Physics, the Institute of Physics, Chinese Academy of Science, No. 8, 3rd South Street, Zhongguancun, Haidian District, Beijing 100190, China

Wei Wang
Department of Physics, Capital Normal University, 105 West Third Ring Road North, Haidian District, Beijing 100048, China

Zhaohua Wang
Beijing National Laboratory of Condensed Matter Physics, the Institute of Physics, Chinese Academy of Science, No. 8, 3rd South Street, Zhongguancun, Haidian District, Beijing 100190, China

Zhiguo Lv
School of Physics and Optoelectronic Engineering, Xidian University, 266 Xinglong Section of Xifeng Road, Xi'an 710126, Shanxi, China

Zhiyuan Zhang
School of Science, China University of Mining & Technology, Ding No. 11 Xueyuan Road, Haidian District, Beijing 100083, China

Zhiyi Wei
Beijing National Laboratory of Condensed Matter Physics, the Institute of Physics, Chinese Academy of Science, No. 8, 3rd South Street, Zhongguancun, Haidian District, Beijing 100190, China
School of Physics and Optoelectronic Engineering, Xidian University, 266 Xinglong Section of Xifeng Road, Xi'an 710126, Shanxi, China

Tianxiang Zhuge
School of Physical Electronics, University of Electronic Science and Technology of China, Chengdu 610054, China

Yulu Hu
School of Physical Electronics, University of Electronic Science and Technology of China, Chengdu 610054, China

Mikhail Grishin
Institute of Physics & EKSPLA uab, Lithuania

Andrejus Michailovas
Institute of Physics & EKSPLA uab, Lithuania

Tsung-Sum Lee
National Yunlin University of Science and Technology, Taiwan (R.O.C.)

Alessandro Cidronali
Department of Electronics and Telecommunications, University of Firenze, Italy

Iacopo Magrini
Department of Electronics and Telecommunications, University of Firenze, Italy

Gianfranco Manes
Department of Electronics and Telecommunications, University of Firenze, Italy

Rajeev Kumar Ranjan
Department of Electronics Engineering, Indian School of Mines, Dhanbad, Jharkhand 826004, India

Surya Prasanna Yalla
Department of Electronics Engineering, Indian School of Mines, Dhanbad, Jharkhand 826004, India

Shubham Sorya,
Department of Electronics Engineering, Indian School of Mines, Dhanbad, Jharkhand 826004, India

Sajal K. Paul
Department of Electronics Engineering, Indian School of Mines, Dhanbad, Jharkhand 826004, India

Zhengyu Sun
Institute of Microelectronics of Chinese Academy of Sciences, Beijing 100029, China

Yuepeng Yan
Institute of Microelectronics of Chinese Academy of Sciences, Beijing 100029, China

Huiyong Li
School of Electronic Engineering, University of Electronic Science and Technology of China, Chengdu 611731, China

Xun Li
School of Electronic Engineering, University of Electronic Science and Technology of China, Chengdu 611731, China

Chen Wei
School of Electronic Engineering, University of Electronic Science and Technology of China, Chengdu 611731, China

Fernando Gregorio
Conicet Department of Electrical and Computer Engineering, Universidad Nacional del Sur, Av. Alem 1253, Bahía Blanca 8000, Argentina

Juan Cousseau
Conicet Department of Electrical and Computer Engineering, Universidad Nacional del Sur, Av. Alem 1253, Bahía Blanca 8000, Argentina

Stefan Werner
Aalto University School of Electrical Engineering P.O. Box 13000, FI-00076 Aalto, Finland

Taneli Riihonen
Aalto University School of Electrical Engineering P.O. Box 13000, FI-00076 Aalto, Finland

Risto Wichman
Aalto University School of Electrical Engineering P.O. Box 13000, FI-00076 Aalto, Finland

J. Jacob
ESRF, Grenoble, France

Preface

An amplifier or electronic amplifier is an electronic device that can increase the power of a signal. It does this by taking energy from a power supply and controlling the output to match the input signal shape but with a larger amplitude. In this sense, an amplifier modulates the output of the power supply to make the output signal stronger than the input signal. An amplifier is effectively the opposite of an attenuator: while an amplifier provides gain, an attenuator provides loss. An amplifier can either be a separate piece of equipment or an electrical circuit within another device. This book emphasizes on the solid-state electronic devices fundamentals, and amplifiers widely used in almost all electronic equipment. In first chapter, we review simple, yet effective, numerical models for grazing-incidence class of amplifiers, for several operating regimes such as cw and pulsed up to multi-kHz repetition rate. Second chapter presents on the high energy picosecond laser operating at a repetition rate of 1 kHz and the high average power picosecond laser running at 100 kHz based on bulk Nd-doped crystals. A design of a V-band Helix-Folded Waveguide (H-FWG) cascaded traveling wave tube (TWT) is presented in third chapter. Dynamics of continuously pumped solid-state regenerative amplifiers has been presented in fourth chapter. Fifth chapter describes the design of two 1V fully differential CMOS switched-capacitor amplifiers in a standard CMOS technology using improved bootstrapped switches. Flexible power amplifier architectures for spectrum efficient wireless applications has been presented in sixth chapter. In seventh chapter, a new analog comb filter based on notch filter is proposed. In eighth chapter, a VGA circuit is presented for high-speed data communication systems. The analysis of the performance of multi-beam forming in memory nonlinear power amplifier is presented in ninth chapter. Tenth chapter proposes a MIMO-PD system that linearizes the power amplifier response and compensates nonlinear crosstalk and IQ imbalance effects for each branch of the multiantenna system. The aim of last chapter is to introduce some important developments made in the generation of high radio frequency (RF) power by combining the power from hundreds of transistor amplifier modules.

Chapter 1

HIGH GAIN SOLID-STATE AMPLIFIERS FOR PICOSECOND PULSES

Antonio Agnesi and Federico Pirzio

Laboratorio Sorgenti Laser Dipartimento di Elettronica dell'Università di Pavia Via Ferrata 1 - 27100 Pavia Italy

INTRODUCTION

Picosecond solid-state lasers are attractive for many industrial and scientific applications, such as precision material processing (Dausinger et al., 2003; Breitling et al., 2004), nonlinear optics (McCarthy & Hanna 1993; Ruffling et al., 2001; Sun et al., 2007) and laser spectroscopy (Mani et al., 2001). In contrast with traditional laser processing performed with 10-100 ns multi-kHz sources, picosecond laser-matter interaction is basically non-thermal, relying on multi-photon ionisation and photo-ablation processes that allow cleaner and much higher spatial definition in laser marking, drilling and cutting. Femtosecond pulses would perform even better in principle, but at the expense of a significant increase in complexity of the laser system that most often is unwelcome in industrial environments.

Furthermore, the multi-kW peak-power levels allowed by cw mode-locked picosecond lasers with average power of at least few watts are already sufficient to produce efficient frequency conversion by harmonic, sum- or difference-mixing and parametric generation. Semiconductor saturable absorber mirrors (SESAMs) are widely employed for the passive mode-locking of picosecond solid-state lasers (Keller, 2003). SESAMs are very effective and highly reliable when used in low-power oscillators; however, when employed in highpower oscillators, they require a special design as their thermal management becomes a very important issue (Burns et al., 2000; Neuhaus et al., 2008). Indeed, the intense intracavity radiation of this particular operating regime may induce significant optical and thermomechanic stress effects, leading to rapid degradation of their performance. An alternative (and not new) approach to powerful cw picosecond sources is to use a master-oscillator power-amplifier

(MOPA) system, in which a seed from a low-power, robust picosecond laser is amplified to the required average power levels through extracavity diode-pumped amplifiers.

Some recent results have pointed out the great potential of this approach (Snell et al., 2000; Agnesi et al., 2006a; Nawata et al., 2006; Farrell & Damzen, 2007; McDonagh et al., 2007). To our knowledge, the most powerful cw picosecond source reported to date was a Nd:YVO$_4$ MOPA system longitudinally-pumped with 216 W at 888 nm, where a 60-W cw picosecond mode-locked laser was amplified to 111 W, with 53% amplifier extraction efficiency (McDonagh et al., 2007). Though the master oscillator of this example could be well classified as "high-power", the effectiveness and the convenience of the power amplification to reach very high power levels is clear. It is also worth noticing that fibre laser technology is rapidly approaching maturity also in the field of high-power ultrafast laser applications (Fermann & Hartl, 2009), though reliable large-area, photonic fibres sustaining picosecond pulses with energy >1 μJ are still subject of extensive research. Most likely, this will take some time before full commercial exploitation. Indeed, all-fibre amplification of femtosecond pulses, stretched to nanosecond or subnanosecond time duration, is definitely easier than direct amplification of pulses only few picosecond long. However, robust picosecond master oscillators, passively mode-locked by SESAMs (Okhotnikov et al., 2003) or other techniques (Porta et al., 1998), can be successfully realised with readily available fibre laser components. These considerations suggest an attractive approach to picosecond MOPAs consisting in a compact rugged picosecond fibre oscillator and a powerful diode-pumped solid-state amplifier with mode size properly scaled in order to avoid damage. Grazing-incidence side-pumped Nd:YVO$_4$ slabs allow efficient power extraction owing to the very high single-pass gain achievable in such conFigureuration (Bernard & Alcock, 1993; Damzen et al., 2001). This article reviews picosecond MOPA systems employing this particular amplification technique. Simple numerical models are presented and applied to the design of MOPAs, as well as for the interpretation of their performance and limitations. A four-pass amplification setup including a photo-refractive phase-conjugating mirror was first reported to yield 12.8 W, nearly diffraction-limited, 8.7-ps pulses (Ojima et al., 2005) starting from a 100-MHz, 290-mW commercial cw mode-locked Nd:YVO4 master oscillator. An improved and more powerful setup also employing phase-conjugating mirror delivered up to 25 W (Nawata et al., 2006). A simpler setup with either a single- or double-pass slab yielded as much as 8.4 W with 7.5- ps pulses at 150 MHz (30% amplifier extraction efficiency from 28-W pump power), using a 50-mW seed oscillator (Agnesi et al., 2006a).

Other applications require instead intense picosecond pulses from compact diode-pumped solid-state laser systems at lower repetition rates (P1 MHz). Usually, intra- or extra-cavity pulse-picking from a diode-pumped low-power picosecond oscillator at ~ 100-MHz repetition rate is used to seed a regenerative amplifier. By this means, the pulse energy is increased from nanojoules to within a range from few microjoules to few millijoules, depending on the operating frequency (Siebold et al., 2004; Kleinbauer et al., 2005; Killi et al., 2005). For this aim, too, grazing-incidence bounce amplifiers provide an interesting alternative, since their gain of $\approx$ 30 dB/pass allows efficient energy extraction in just two or three passes, thus avoiding the higher complexity of the former schemes and requiring only a versatile, extra-cavity acousto-optic pulse-picker. This needs to be appropriately synchronised to the mode-locked train, as well as to the amplifier pump pulse (if the amplifier is not cwpumped). In the first demonstration of such an amplifying conFigureuration, the high-frequency picosecond seeder output was sampled synchronously, and the selected pulse (with energy < 1 nJ) was injected into a two-stage amplifier, yielding an output energy up to 10 μJ at 100 Hz (Agnesi et al., 2006b). A remarkable 1-MHz high repetition frequency and pulse energy up to 76 μJ were later reported by Nawata et al., (Nawata et al., 2007) who employed a more complex cw-pumped phase-conjugation setup with a double-pass grazing-incidence slab. Repetition frequency as high as 4 MHz was also reported for an extra-cavity multi-pass.

Nd:YVO4 amplifier (Gerhard et al., 2008), yielding 80 μJ at low frequency and 1.8 μJ at 4 MHz. Effective material processing results were demonstrated with such a laser system. However, the highest pulse energy was achieved with a refined qcw-pumped single-pass two-slabs amplifier design (Agnesi et al., 2008a), yielding as much as 210 μJ, up to 1 kHz repetition rate, and 11-ps time duration pulses. Highly efficient harmonic conversion to 532 nm and 266 nm was readily observed, owing to the $\approx$ 20-MW peak power of the amplified picosecond pulses. Travelling-wave parametric generation spanning the ranges 770-1020 nm (signal), 1110-1720 nm (idler) was also demonstrated (Agnesi et al., 2006b). Particular applications such as photo catode injection (Will et al., 2005), low-threshold parametric generation (Agnesi et al., 1993; Butterworth et al., 1996) and implementation of pulse-format and wavelength typical of free-electron-lasers (Edwards et al., 2002) by allsolid-state laser technology require instead amplification of bunches of picosecond pulses. Again, side-pumped high-gain bounce amplifiers can be successfully employed to increase the pulse energy of pulse trains as long as ~ 1 μs. A remarkable example of such a laser source delivering trains of ~ 2500 pulses of 12-ps time duration, 5-GHz repetition rate at 1064 nm, and train energy of 250 mJ, was reported recently (Agnesi et al., 2008b). This review is organised as follows. Section 2 gives the theoretical background for understanding and modelling grazing-incidence slab amplifiers; Sections 3-5 review some representative experimental results achieved by our research group with slab amplifiers operated in several regimes, as well as interesting frequency conversion applications; in Section 6 we finally summarise our results and trace few conclusions.

NUMERICAL MODEL

In this Section we review simple, yet effective, numerical models for grazing-incidence class of amplifiers, for several operating regimes such as cw and pulsed up to multi-kHz repetition rate. Limitations due to the finite amplifying bandwidth are discussed. These models will be applied to the interpretation of experimental results reviewed in the next Sections. More generally, their use extends to a wider class of amplifier, including, for example, cascaded systems and bounce amplifiers based on different laser materials, provided the pump absorption depth is of the order of ~ 1 mm or less, and the integrated single-pass gain reaches useful levels.

Operation in cw Regime

The model is based on standard cw amplifier theory (Koechner, 2006). Let us introduce a reference system for bounce beam propagation in the active material slab shown in Figure. 1. The length and the thickness of the slab are L and W, respectively. A particularly useful transverse local frame for the seed beam is ξ–y. Assuming small grazing angles, i.e. $\theta \ll 1$ rad, we may approximate $x(s, \xi) = \theta L/2 + \xi - s\theta$. The propagation inside the amplifier occurs with a nearly-constant beam cross section in order to optimise the overlap efficiency, therefore the gain can be calculated along the longitudinal s coordinate as in a ray-tracing approximation:

$$\frac{dI(\xi, y)}{ds} = \sigma n(x, y) I(\xi, y)$$

(1)

described by coordinates (ξ, y). The saturated population inversion density is

$$n(x, y) = \frac{R_p(x, y)\tau}{1 + I(\xi, y) / I_s}$$

(2)

and the pump rate is given by

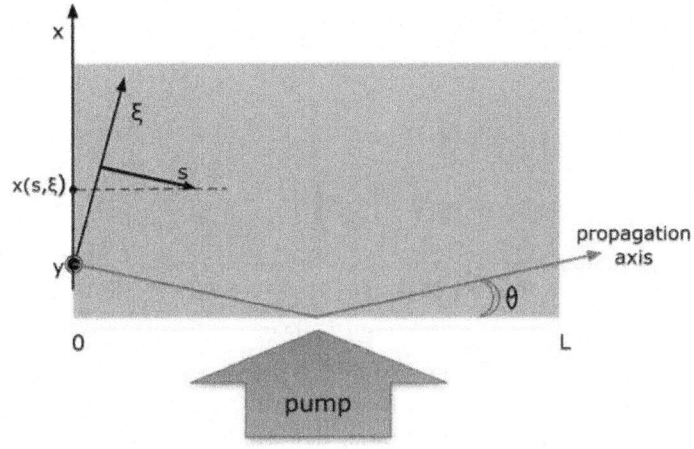

Figure. 1: Model of the slab amplifier (seen from above): the beam enters from the left side with a grazing angle θ. The beam cross section is most easily

$$R_p(x,y) = \frac{\lambda_p}{\lambda_L} \frac{\alpha_p}{WL} \frac{P_{inc}\Theta(y)e^{-\alpha_p x}}{hv}$$

$$(3)$$

The step function Θ is defined as: $\Theta(y) = 1$ for $|y| < W/2$, $\Theta(y) = 0$ for $|y| > W/2$; σ is the stimulated emission cross section at the laser frequency $v = c/\lambda_L$; τ is the fluorescence time; $I_s = hv/(\sigma\tau)$ is the laser saturation intensity; α_p is the saturated absorption coefficient at the pump wavelength λ_p, corresponding to the incident pump intensity $I_p = P_{inc}/(WL)$ (P_{inc} is the power transmitted through the pump face and absorbed by the crystal) $\alpha_P = \alpha_{P0}/(1 + I_P/I_{SP})$ and I_{SP} is the pump absorption saturation intensity (Bermudez et al., 2002). Inserting Eqs. (2) and (3) into Eq. (1) and integrating, one obtains

$$\ln\left[\frac{I_o(\xi,y)}{I_i(\xi,y)}\right] + \frac{I_o(\xi,y) - I_i(\xi,y)}{I_s} = \frac{\lambda_p}{\lambda_L} \frac{\alpha_p P_{inc}\Theta(y)}{WLI_s} \int_0^L ds \exp\left[-\alpha_p|x(\xi,s)|\right]$$

$$(4)$$

The double-pass model can be readily derived from Eqs. (1-3), requiring that the forward travelling intensity equals the backward-travelling intensity at s = L. We assume that, after the first pass, the beam retraces its path backward through the amplifier. Eventually, we end up with two equations, one for the single-pass (q = 1) and one for the double-pass amplifier (q = 1/2):

$$q \ln \left[\frac{I_o(\xi, y)}{I_i(\xi, y)} \right] + \frac{I_o(\xi, y) - I_i(\xi, y)}{I_s} = \frac{\lambda_p}{\lambda_L} \frac{\alpha_p P_{inc} \Theta(y)}{WLI_s} \psi(\xi)$$

(5)

Where

$$\psi(\xi) = \frac{2}{\alpha_p \theta} \left[1 - \cosh(\alpha_p \xi) \exp\left(-\frac{\alpha_p \theta L}{2} \right) \right]$$

(6)

Notice that the small-signal gain coefficient is proportional to the right-hand side of Eq. (5): it is assumed to have a flat-top profile in the vertical direction while it has a smooth symmetric profile in the horizontal ξ-direction, with a peak on axis. This is a consequence of the bounce occurring at the mid-point of the slab. Moreover, thermal distortions that usually accompany the pump deposition are averaged accordingly, yielding the same symmetry as for the gain distribution along the ξ axis. Thermal lensing retains the usual distribution along the y axis

As anticipated, an arbitrary seed beam profile can now be traced along the amplifier with Eqs. (5) and (6). A particularly handy approximation, that speeds up the calculations significantly, consists in assuming an average gain coefficient over the horizontal seed diameter:

$$\bar{\psi} = \frac{2}{w_s} \int_0^{w_s/2} \psi(\xi) d\xi = \frac{2}{\alpha_p \theta} \left[1 - \frac{2}{\alpha_p w_s} \sinh(\alpha_p w_s/2) \exp(-\alpha_p \theta L/2) \right]$$

(7)

Rather than the transverse intensity distribution of the rays with coordinates (ξ, y), the spatially-averaged Eq. (5) yields now a single output intensity, given the seed cross section $w_s \times W$ and the input intensity

$$q \ln \left[\frac{I_o}{I_i} \right] + \frac{I_o - I_i}{I_s} = \frac{\lambda_p}{\lambda_L} \frac{\alpha_p P_{inc}}{WLI_s} \bar{\psi}$$

(8)

It is worth noticing that the crystal end faces limit the effective width of the input seed. The beam width may be comparable with the distance between the centre of the beam and the edge of the slab input face, i.e. $\theta L/2$. On the other hand, when increasing the angle θ there is a possibility that the beam width becomes comparable with its distance from the other edge of the slab, of width D. Therefore, the effective beam diameter to be amplified is the minimum out of the quantities w_s, θL and $2(D - \theta L/2)$.

Amplification in pulsed, low-frequency regime $^{(f \ll 1/\tau)}$.

In pulsed amplification regime, usually a pulse-picker selects a single pulse from a cw mode-locked laser and sends it to the amplifier which was previously pumped to achieve an appropriate gain level. Assuming pump pulse duration T, and averaging the spatial gain distribution as in Eq. (7), the pump rate equation yields the small-signal exponential gain:

$$\frac{dn}{dt} = R_p - \frac{n}{\tau}$$

(9)

$$g_0 = \left\langle \int_0^L \sigma n \, ds \right\rangle = \frac{2}{w_s} \int_0^{w_s/2} d\xi \left[\int_0^L \sigma n(\xi, s) \, ds \right] = \sigma R_p \tau \left(1 - e^{-T/\tau} \right) \overline{\psi}$$

(10)

$$g_0 = \frac{\lambda_p}{\lambda_L} \frac{E_{inc}}{WLF_s} \frac{\tau}{T} \left(1 - e^{-T/\tau} \right) \overline{\psi}$$

(11)

We may now use the Franz-Nodvik amplifier model (Franz & Nodvik, 1963) to calculate the output fluence Fo, given the input fluence Fi and the amplifier saturation fluence $F_S = h\nu/\sigma$:

$$F_o = F_s \ln \left[1 + e^{g_0} \left(e^{F_i/F_s} - 1 \right) \right]$$

(12)

So far we have considered amplification in a single slab. However, the output fluence from the first slab may be used as the input for a second pass or to the next slab, and so on, allowing in general different beam sizes and grazing angles to be considered for each stage of the amplifier chain.

Amplification in pulsed, high-frequency regime $(f \gtrsim 1/\tau)$

This operating regime is mostly employed in laser systems for industrial applications, where high processing speed is extremely important, even at the expense of some pulse energy reduction. In this case a cw pump laser is chosen owing to duty-cycle limitations of more powerful qcw laser diode arrays employed at lower frequency (typically $^{< 1/\tau)}$. The amplifier gain (or the inversion population) between two amplified pulses is still restored according to Eq. (9), starting from the gain g_f just after the amplified pulse and reaching g_i just before the next seed pulse enters the amplifier. The Franz-Nodvik model can still be used, provided we use the right initial gain, which, in turns, depends on the pulse repetition frequency (Koechner, 2006):

$$F_o = F_s \ln\left[1 + \left(e^{F_i/F_s} - 1\right)e^{g_i}\right]$$

(13)

$$g_i = g_\infty - \left(g_\infty - g_f\right)e^{-\frac{1}{ft}}$$

(14)

$$F_o = F_i + \left(g_i - g_f\right)F_s$$

(15)

We notice that, according to Koechner's notation, $g_\infty = \lim_{f \to 0} g_i(f)$ hence we identify $g_\infty = g_0$ as given in Eq. (11). Again, Eqs. (13-15) apply to single-pass amplification, but sequential application of the same model readily accounts for multi-pass schemes or amplifier chains.

Bandwidth limitation and gain narrowing

The most important of limitations arising from the finite bandwidth of the amplifier concern single- or multi-pulse amplification in pulsed regime, when small-signal gain is very high, say > 1000. This situation is commonly encountered in regenerative and multi-pass amplifiers (Walker et al., 1994; Le Blanc et al., 1996). However, single-pass amplification in a couple of grazing-incidence Nd:YVO$_4$ slabs readily yields 60 dB overall small-signal gain, leading to severe pulse broadening if too short seed pulses are injected. In order to predict this effect quantitatively, we developed a numerical model based on Franz-Nodvik equations, adding gain filtering (of lorentzian form with fwhm $= \Delta\nu_g$) on each of the M slices into which the amplifier is sectioned. The basic assumption is that the gain depletion in each slice must be small enough (< 1%), so that we may apply the exact gainfilter model:

$$\hat{u}_{n+1}(\nu) = \hat{u}_{n+1}^{(u)}(\nu)\exp\left[\frac{\left(g_{n+1} + g_n\right)}{2}\frac{\gamma(\nu)}{2}\right]$$

(16)

$$\gamma(\nu) = \frac{1}{1 + \left(2\nu/\Delta\nu_s\right)^2}$$

(17)

$$u_n(t) = \int_{-\infty}^{+\infty} \hat{u}_n(\nu)e^{-i2\pi\nu t}\,d\nu$$

(18)

where $\hat{u}_{n+1}^{(u)}(v)$ is the Fourier transform of the amplified (unfiltered) field $u_{n+1}^{(u)}(t)$ at the end of the $(n+1)$-th section, according to the Franz-Nodvik model:

$$G_{n+1} = \frac{e^{E_n/E_s}}{e^{E_n/E_s} - 1 + 1/G_n^{(0)}}$$

(19)

$$E_n = \int_{-\infty}^{+\infty} |u_n^2(t)| dt$$

(20)

$$G_n^{(0)} = G_0^{1/M}$$

(21)

$$g_n = \ln(G_n)$$

(22)

Here G_0 is the total small-signal gain, and g_n is the result of the previous step. The condition for accurate computation is that

$$\frac{g_n - g_{n+1}}{g_n} \ll 1$$

(23)

Therefore M is chosen accordingly. We notice that computation time might be conveniently reduced provided one chose a logarithmic distribution of amplifier sections lengths, since most severe saturation occurs at the end of the amplifier. However, in order to model amplifier gains as high as 60 dB, we found that M = 500 yields sufficiently accurate results with only few seconds of computation time on a 2.4-GHz laptop

Let us consider typical data for a pulsed grazing-incidence amplifier: $\Delta v_g = 212$ GHz (a-cut Nd:YVO$_4$, fluorescence fwhm = 0.8 nm (Zayhowski & Harrison, 1997)), $E_S = 188$ µJ. We calculated the output energy and pulse width as a function of small-signal gain, input energy and seed pulse width. The results are summarised in Figures. 2-4.

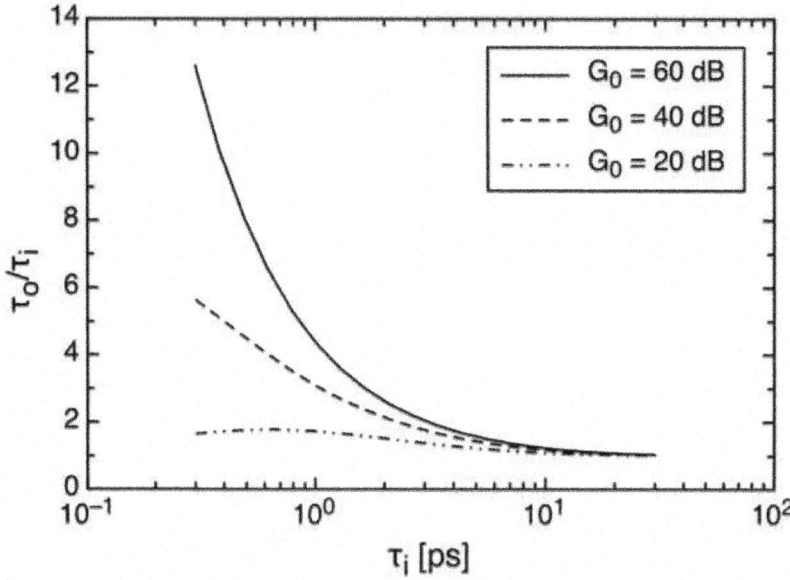

Figure. 2: Output pulse width broadening as a function of the seed pulse duration. Three smallsignal gain levels are considered. Injected seed energy of 1 nJ is typical of most picosecond seeders.

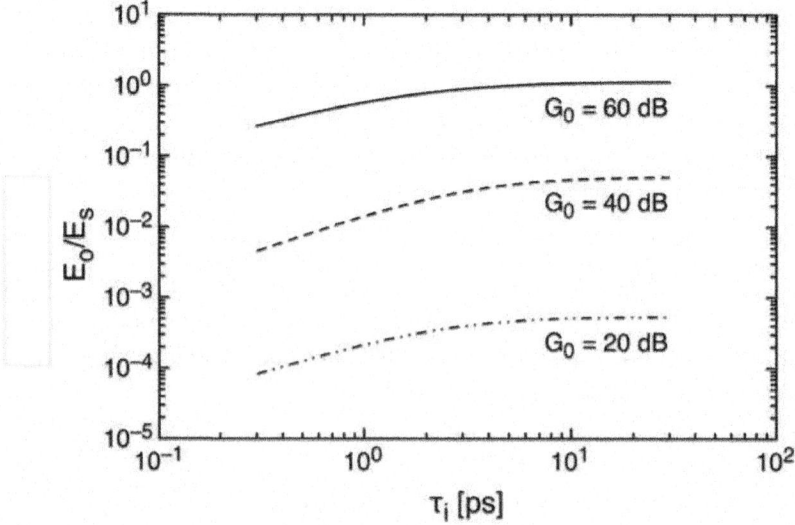

Figure. 3: Output energy as a function of the seed pulse width, for three small-signal gain levels. Injected seed energy is 1 nJ.

In conclusion, we note that seed pulse duration exceeding four times the minimum width set by the amplifier bandwidth (≈ 1.5 ps in this example, assuming $sech^2$ pulse shape) produces no significant broadening after amplification even at an unsaturated gain of 60 dB (Figure. 2). The gain narrowing effect is most clearly seen at shorter seed pulse widths, as expected. Notice that at relatively low gain values, $G_0 < 20$ dB, the output pulse width reaches a maximum broadening as the seed pulse shortens, then approaches the seed duration provided it is short enough. The physical explanation is that the gain modulation is not as strong as with higher gain, so that only a relatively small central region of the seed spectrum is amplified but the overall non-amplified spectrum energy dominates, leading to an output pulse width still approaching that of the seed pulse.

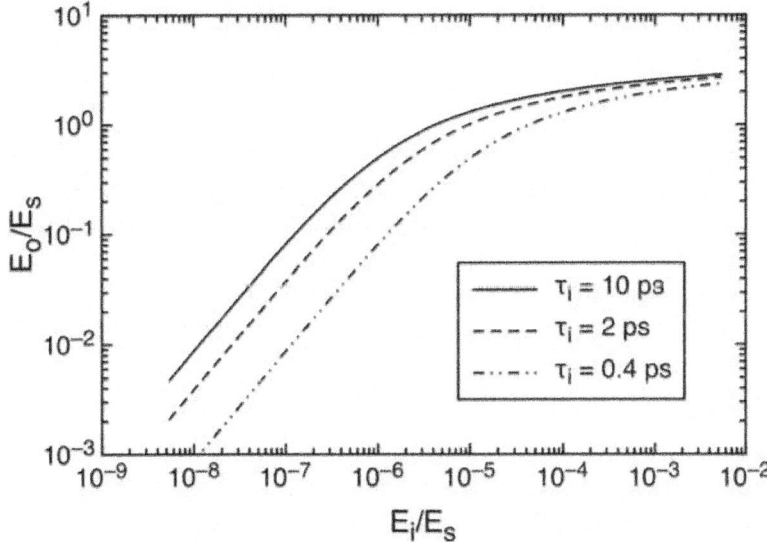

Figure. 4: Output energy as a function of the seed energy (both normalised to the saturation energy); three different seed pulse width have been considered.

As far as efficient energy extraction is of concern, Figure. 3 shows that with the given (typical) amplifier parameters a small-signal gain $G_0 = 50$-60 dB is required to reach saturation starting from the typical seed energy level ≈ 1 nJ. It is remarkable that, whilst this is usually achieved with either regenerative or extra-cavity multi-pass setups, typical gains of qcw grazing-incidence modules ≈ 30 dB/slab yield such gain levels with only two slabs or a single slab in two-passes, if only one is careful enough to suppress ASE. Figure. 4 shows that gain saturation in the $Nd:YVO_4$ slab is optimised with seed pulse width > 6 p_s: this means that, owing to gain narrowing, all the seed spectral energy is used for the effective amplification bandwidth.

The distortion can be readily calculated from the Franz-Nodvik model, given the power envelope P(t) of the pulse train, the small-signal gain G_0 and the saturation energy ES:

$$G(t) = \frac{e^{E_i(t)/E_s}}{e^{E_i(t)/E_s} - 1 + 1/G_0} \tag{24}$$

$$E_i(t) = \int_{-\infty}^{t} P_i(t')dt' \tag{25}$$

$$P_o(t) = G(t)P_i(t) \tag{26}$$

Therefore, the envelope modulation waveform required for pre-compensation of the saturation distortion can be easily computed, in order to yield a nearly flat-top output pulse from the amplifier. Owing to the strong amplifier saturation, the output energy reduction due to the pre-compensation is not very penalizing, notwithstanding the significant energy reduction of the modulated seed. This will be discussed in more detail in Section 5.

AMPLIFICATION OF A CW PICOSECOND LASER

Here we discuss the simplest setup that one can conceive, i.e. a single grazing incidence slab, side-pumped by a single diode array, for amplification of a low-power cw picosecond laser. In particular, the seeder is a Nd:YVO$_4$ oscillator (Figure. 5), longitudinally pumped by a 1-W laser diode at 808 nm. The laser is passively mode-locked with a SESAM, generating a 50- mW average power train of 6.8-ps, linearly polarized pulses at a 150-MHz repetition rate in each of the two diffraction-limited output beams.

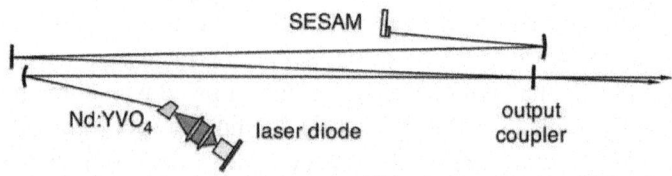

Figure. 5: Representative layout of a cw diode-pumped picosecond oscillator.

A 4 x 2 x 16 mm³, *a*-cut, 6°-wedged, 1%-doped Nd:YVO₄ slab is employed as amplification head, pumped by a TE-polarised 10-mm x $1\text{-}\mu m$ laser diode array, emitting at 808 nm and collimated along the fast-axis by a 0.9-mm focal length microlens. This laser diode is actively cooled and temperature controlled with a thermoelectric cooler and emits a beam of maximum power of 32 W. The amplifier crystal is antireflection (AR) coated at 1064 nm on the input and output faces and simply polished and uncoated on the pumped side face. The Fresnel loss of about 14% of the incident pump power limits the absorbed pump power at 28 W. The 4 x 16 mm² slab faces are placed in contact with a water-cooled heat exchanger by thin indium foils. A proper choice of the water temperature set point is crucial in optimising the amplifier performances. The reason is twofold:

- minimisation of the thermal stress inside the slab and
- reduction of the thermal red-shift of the fluorescence bandwidth of Nd:YVO₄, that may reduce the amplifier gain and performance. In this case the optimum operating temperature is 8°C. The collimated beam emitted by the pump diode is polarisation-rotated by a half-wave plate and focused into the Nd:YVO₄ slab with a 15-mm focal length cylindrical lens.

In order to estimate the vertical thickness of the gain sheet, the seed beam is collimated (0.9- mm half-width at 1/e ²) and sent to the amplifier without any focusing optics. In these conditions, because of the strong gain shaping of the pump beam in the high-gain amplifier and of the great mismatch between the seed diameter and the gain-sheet thickness, the beam divergence along the y-direction is nearly pump-insensitive. Considering a flat-top y-profile for the output beam leaving the amplifier, the thickness W of the gain-sheet can be readily estimated by measuring the full vertical divergence angle $\theta_y \approx 2\lambda_L/W$. . In this example it turns out $W \approx 70\ \mu m$. Notwithstanding the clearly non optimised beam size, the seed beam yields up to 3.5 W in a single pass through the amplifier for an optimum internal grazing incidence angle of $\approx 3°$.

The setup for the cw amplification experiments is shown in Figure. 6.

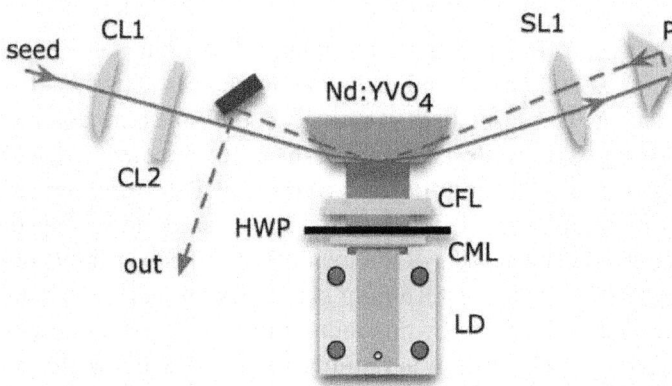

Figure. 6: Setup for the cw amplifier. CL1: 250-mm focal cylindrical focusing lens; CL2: 100-mm focal cylindrical focusing lens; SL1: 200-mm focal spherical lens; P: right-angle prism; CFL: 15-mm focal cylindrical pump focusing lens; HWP: half-wave plate; CML: collimation microlens (0.9-mm focal length).

Being the thickness W of the gain sheet set by the pump focusing lens and limited by the residual smile of the laser diode, the depth of the gain sheet in the horizontal plane is basically determined by the pump absorption depth, $1/\alpha P \approx 0.5$ mm, given the typical pump spectrum width ($\approx$ 2-3 nm) and the 1%-doping concentration of $Nd:YVO_4$. This generates an elliptical transverse gain profile that suggests the need for a different seed focusing in the horizontal and vertical directions. As shown in Figure. 6, the cylindrical lens CL1 focuses the oscillator beam down to wx $\approx$ 100 μm in the horizontal plane, whereas a cylindrical lens CL2 focuses the seed beam in the vertical plane to wy $\approx$ 36 μm. In order to choose the seed waist inside the amplifier, both the thickness of the pumped region and the Rayleigh range of the focused beam should be taken into account. In particular the pumped region length along the seed propagation direction should be shorter than two times the Rayleigh range of the focused beam in order to take advantage of the entire length of the active region. These considerations set a minimum acceptable dimension of $\approx$ 30 μm for the focused seed beam waist. Moreover, the grazing incidence angle has to be chosen as small as possible in order to maximise the gain while avoiding clipping effects.

With 34-mW seed injection the amplifier yields as much as 6.1 W after a single pass for the 28-W absorbed pump power, corresponding to a 22% optical-to-optical efficiency. The beam quality of the seed ($M^2 = 1.1$) is only slightly degraded to $M_x^2 = 1.3$ and $M_y^2 = 1.2$ at the maximum amplification level.

Figure. 7 shows the autocorrelation traces of both seed and amplified pulses and the correspondent optical spectra. The pulse duration after amplification is only slightly increased and the effect of spectral gain narrowing is quite evident. We notice that, owing to dishomogeneous broadening due to spatial hole burning (Flood et al., 1995), longitudinallypumped picosecond lasers with gain at the end usually show excess bandwidth with respect to the measured pulsewidth. Therefore, in this case gain narrowing improves the pulsewidth × bandwidth product from 0.68 for the seed to 0.42 for the amplified pulses, approaching the Fourier limit of 0.32 for a sech^2 pulse shape.

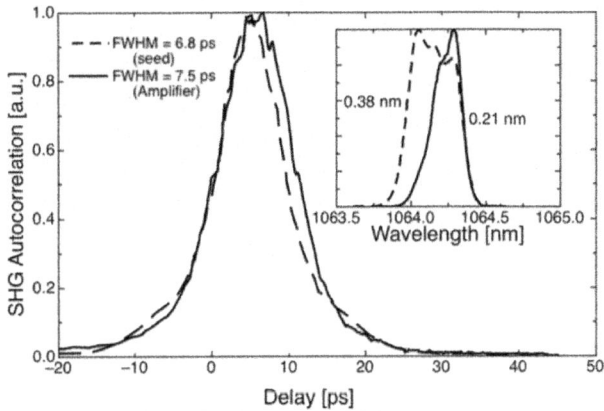

Figure. 7: Pulse autocorrelation of seed (dashed curve) and amplified (continuous curve) beam. Inset: corresponding optical spectra for seed and amplified pulses.

Double-pass amplification can be conveniently realised by re-imaging the single-pass output with a spherical lens SL1 (200-mm focal length) and a right-angle antireflection-coated prism P (see Figure. 5), thus maintaining the same spot size as in the first pass, but with a slightly increased grazing angle of $\approx 4°$. Beam extraction occurs before the lens CL2. The first consequence of realising a return path in the amplifier is the immediate growth of the amplified spontaneous emission (ASE) background. When the pump diode beam is focused for single-pass amplification, addition of the second pass boosts the ASE to > 1 W in this setup. In order to reduce the gain in the amplifier and hence the amount of ASE, the pump diode focusing must be relaxed, for example replacing the 15-mm cylindrical lens with a 20-mm lens.

This increases W to ≈ 95 μm. In these conditions ASE background reduces to < 200 mW and as much as 8.4 W are obtained at 28-W absorbed pump power after second pass amplification. The corresponding optical-to-optical efficiency increases to $\approx$ 30%. Owing to a relatively small horizontal beam width compared to the horizontal transverse gain profile dimension, off-axis thermal aberrations in the critical horizontal plane are reasonably low. Therefore beam quality is well conserved also after the second pass ($M_x^2 = 1.4$ and $M_y^2 = 1.3$). Results similar to those reported in Figure. 7 are obtained for both pulse autocorrelation and spectrum.

Several different conFigureurations were investigated, in single and double pass and also employing a single 300-mm focal spherical lens to focus the seed into the Nd:YVO$_4$ slab instead of the cylindrical optics shown in Figure. 6. Both the amplifier gain as a function of the injected seed power and the output power as a function of the absorbed pump power are summarised in Figures. 8 and 9.

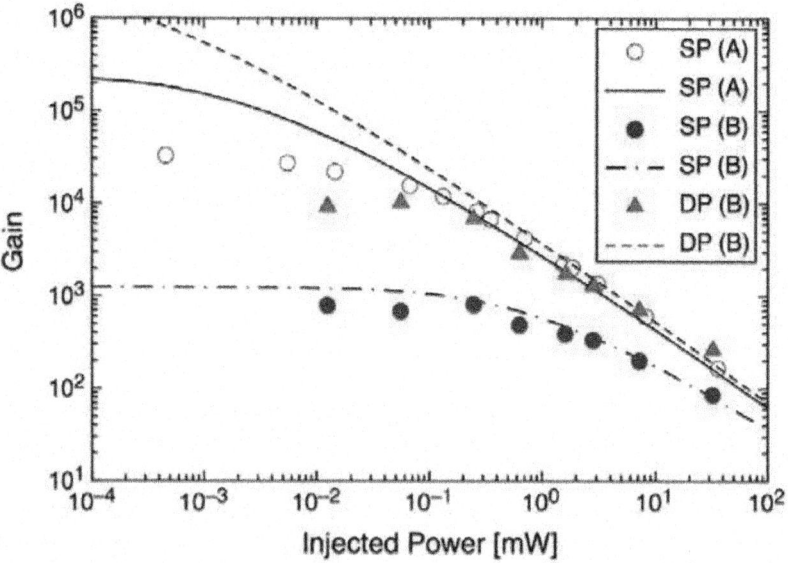

Figure. 8: Gain curves for the single- and double-pass amplifier. (A) Setup with pump focusing with 15-mm focal and seed focusing with cylindrical lenses of focal lengths f$_x$ = 250 mm and f$_y$ = 100 mm. (B) Setup with pump focusing with 20-mm focal and seed focusing with 300- mm focal spherical lens. SP single pass, DP double pass. Continuous curves: numerical results. Points: data.

Both in single or double pass setup, a small signal gain higher than 40 dB for the lowest input power is obtained. The model employed for fitting the experimental data is described in Section 2.1. The following physical parameters of the laser crystal can been used to fit the experimental data: $I_S = 2 \text{ kW/cm}^2$, $\alpha_{P0} = 24 \text{ cm}^{-1}$, $I_{SA} = 3.61 \text{ kW/cm}^2$. . The unsaturated absorption coefficient has been calculated assuming the absorption spectrum averaged over the laser diode emission spectrum. The best-fit saturation intensity value IS yields a product $\sigma\tau$ about 30% smaller than the one usually reported in literature. Such a behaviour, beside intrinsic model approximations, can be explained by a reduction of the fluorescence lifetime due to upconversion effects and by a reduction of the effective emission cross section due to the mismatch between the spectrum peaks of seed and amplified pulses (see the inset of Figure. 7).

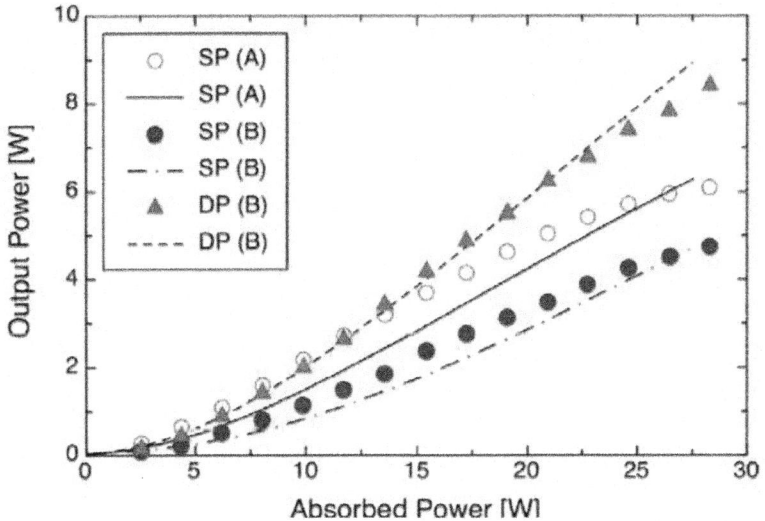

Figure. 9: Output power as a function of the absorbed pump power for the single- and double pass amplifier. (A) Setup with pump focusing with 15-mm focal and seed focusing with cylindrical lenses of focal lengths $f_x = 250$ mm and $f_y = 100$ mm. (B) Setup with pump focusing with 20-mm focal and seed focusing with cylindrical lenses $f_x = 250$ mm and $f_y = 100$ mm. SP single pass, DP double pass. Continuous curves: numerical results. Points: data.

As clearly shown by Figure. 8, the amplifier behaviour tends to deviate from the model for decreasing injected power levels, suggesting that the maximum available gain is lower than that predicted. For very small injected power, gain saturates around $\approx 10^4$ owing to ASE, while gain saturation with higher input seed power levels recovers a fair agreement between predictions and experiments, since stored energy is more effectively extracted from the amplified beam rather than being wasted into wide-angle ASE.

SINGLE-PULSE AMPLIFICATION UP TO 1 KHZ REPETITION RATE

A typical, representative layout of a laser system for single picosecond pulse amplification after pulse-picking is shown in Figure. 10.

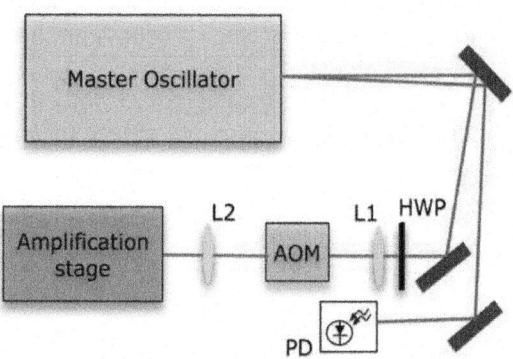

Figure. 10: Layout of a laser system for single pulse amplification. HWP: half-wave plate; L1: focusing spherical lens; L2: collimating spherical lens; AOM: acousto-optic modulator; PD: photodiode.

The master oscillator in this case is still similar to that used for cw experiments, providing two output beams each carrying out about 70-mW average power, 7.7-ps long pulses but with a reduced repetition rate ≈ 48-MHz. This last parameter is important since wellseparated pulses are easier to pick up with optoelectronic devices. The repetition frequency downscaling scheme for single pulse amplification employs an acousto-optic modulator (AOM) to pick up single pulses from the cw mode-locked oscillator. The major benefit of using AOM with respect to faster electro-optic modulators for pulse picking is the simpler and easier to customise driving electronics. Moreover AOMs are cheaper and do not require fast switching of high voltages that generates electromagnetic noise difficult to suppress. The pulse train sampled by the pulse picker has programmable repetition frequency, basically upper-limited by the duty cycle of the highpower diode arrays pumping the amplifier stage, that is typically 10%. In case of vanadate slab amplifiers, since the fluorescence time is ≈ 100 μs, the maximum operating frequency for qcw laser diodes bars is fixed at 1 kHz.

Given a temporal separation between two adjacent master oscillator pulses larger than the minimum deflection window allowed by the AOM, the key issues is to be able to drive the pulse picker synchronously with respect to the laser pulses. The correct timing can be realised conveniently by employing one of the two beam outputs from the master oscillator to generate a clock signal intrinsically synchronous to the pulse train for the pulse-picker driving electronics. In this example, a time separation between two adjacent pulses of about 20 ns allows a suppression ratio between selected and adjacent pulses better than 2% with a deflection efficiency exceeding 40%. If higher suppression ratio were an issue, a faster pulse picker, two acousto-optic pulse pickers in series or an oscillator running at a lower repetition rate might be employed.

The amplification stage setup is shown in detail in Figure. 11. The seed emerging from the pulse picker is injected into a grazing-incidence single-pass amplifier made by a couple of 4 x 2 x 16 mm³, a-cut, 5°-wedged, 1%-doped Nd:YVO$_4$ slabs. The slabs are AR-coated at 1064 nm at the input and output faces, while the pumped sides are AR-coated at 808 nm. Each slab is pumped by a 150-W peak power qcw laser diode array with an emitting size of 10 mm x 1 $^{\mu}$m. The radiation emitted by each diode is collimated by a microlens to ≈ 0.9-mm thickness x 10-mm width stripe and polarisation rotated by a half-wave plate to be aligned with the caxis of the vanadate slabs. Notice that, owing to the much higher peak power of the diode arrays, tight focusing into the slab is no more required as in the cw amplifier of Section 3.

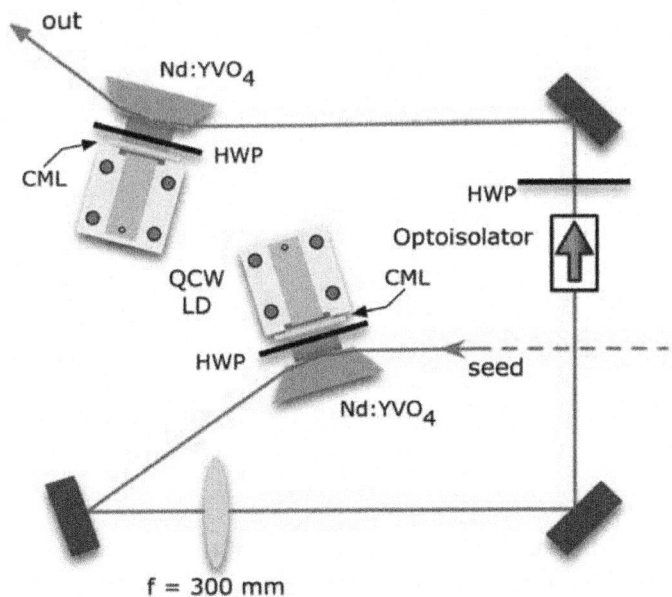

Figure. 11: Setup for the single-pulse amplification 2-slabs module. HWP: half-wave plate; CML: collimation microlens (0.9-mm focal).

In order to optimise the energy storage in the amplifiers the pump pulse duration is set to about 100 μs, slightly longer than the fluorescence time of the 1%-doped Nd:YVO$_4$ slab. The pulse picking occurs on the trailing edge of the pump current pulses, when the maximum gain is stored in the amplifiers medium. Pulse duration is slightly increased after amplification. As shown in Figure. 12, fwhm 7.7-ps long seed pulses are slightly stretched to 11 ps long. The 0.37-nm wide pulse spectrum is substantially preserved during amplification, improving the quality parameter pulse width × bandwidth.

In order to fully exploit the potential of high-gain amplifier modules, besides the optimisations concerning the grazing angle as well as the matching between the seed beam waist and the gain layer, the main issue is to be able to contrast successfully those effects, such as amplified spontaneous emission (ASE), that compete with the amplifying beam in depleting the population inversion in the gain media. Since the ASE generated in the one slab can seed the other, the two amplifiers must be kept sufficiently separated, about 30 cm in the final setup, in order to reduce the solid angle under which the first slab is seen by the second and consequently reduce the reciprocal ASE seeding.

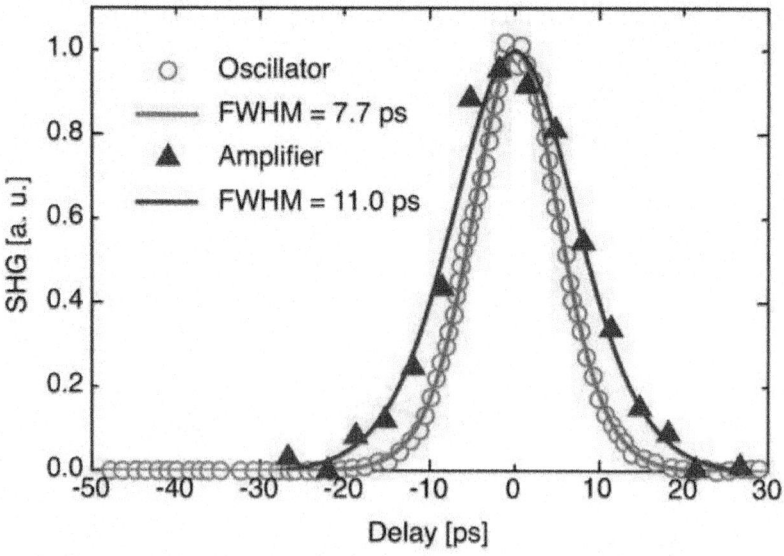

Figure. 12: Pulse autocorrelation of the oscillator (circles) and amplified (triangles) beam. Also shown are the best-fit traces corresponding to sech^2 intensity envelope.

This solution yields significant performance improvements with respect to more closely separated slabs only few centimetres apart (Agnesi et al., 2006b). Furthermore, an optical isolator is inserted between the two amplifiers to prevent self oscillations and back-injection in the direction of the master oscillator. A spherical lens (300-mm focal length) placed at about 250 mm from the second slab re-collimates the spatial mode before entering the second amplifier. A small-signal gain as high as $\approx 10^6$ (≈ 60 dB) is measured for this double-slab module. Injecting pulses with energy of about 0.4 nJ, yields amplified pulses with energy as high as 210 μJ with weak gain saturation. Figure. 13 shows the energy extraction curve of the amplifier. The model for experimental data fitting is described in Section 2.2. The following best-fit parameters values can be employed: saturation fluence FS = 133 mJ/cm^2 (correspondent effective emission cross-section $\mu \approx 15 \times 10^{-19}$ cm^2), absorption coefficient $\alpha_p = 22$ cm^{-1}, gain thickness W = 1.2 mm (flat-top approximation). For the seed beam it is assumed an equivalent flat-top profile diameter ws = 0.6 and 0.8 mm in the first and second slab respectively, while the grazing incidence angle θ is 1.6° in both slabs, according to a direct measurement of the deviation angle at the slabs output for the conFigureuration yielding the highest pulse energy. The calculated angular sensitivity of the second amplifier is shown in Figure. 14. The model confirms that the slab amplifiers are operated near the optimum predicted angle.

Moreover, the numerical model suggests some useful optimisation criteria for further improvements of the system performances. First of all, in order to fully exploit the gain available in the first slab it is necessary to reduce the transverse dimensions of the seed beam in order to match its confocal parameter with the length of the pumped region. Such a relatively small beam size allows to explore smaller grazing incidence angles without clipping the beam, hence maximising the small-signal gain in the first unsaturated amplifier.

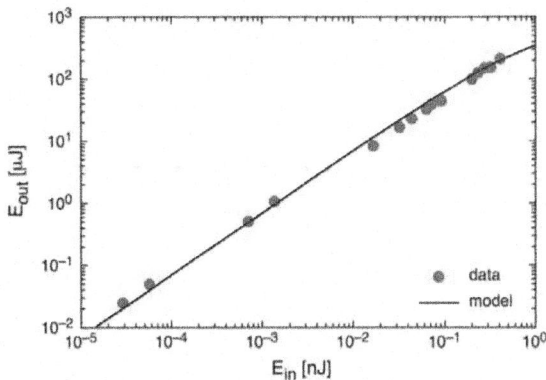

Figure. 13: Energy extraction curve of the amplifier (experiment and numerical model).

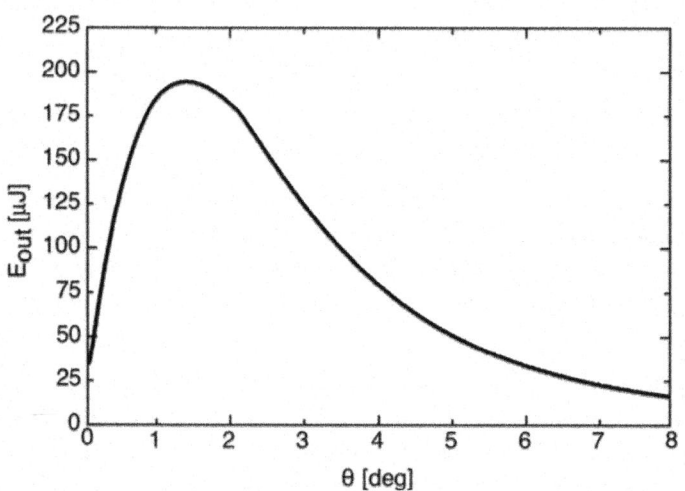

Figure. 14: Calculated energy extraction curve of the second amplifier as a function of the grazing incidence internal angle θ corresponding to a full pump power and maximum seeding injection.

Beam clipping should be carefully avoided, since it produces scattering and beam distortions that can substantially limit amplifiers performance. In the second slab the beam transverse dimensions and the grazing incidence angle should be increased in order to reduce non-linear effects and damage issues. The upper limit to input beam waist is given in this case by the gain region thickness W. According to these guidelines, output energy is predicted to increase up to 0.5 mJ with very small grazing incidence angles ($\approx 0.3°$) in the first slab and beam radius ≈ 0.5 mm in the second slab.

It is worth noticing that for an accurate measurement of the output pulse energy it is necessary to remove the contribution of both ASE and undesired background pulses. In fact, since the static extinction ratio of the AOM is $\approx$ 1:2000, this value can be considered the attenuation factor for pulses far away from the selected one, while pulses coming immediately before and after the selected one will surely experience an even lower extinction ratio. All of these contributions can become significant after amplification and affect a direct measurement carried out with a power meter. However, the post amplification pulse contrast will be significantly increased by second harmonic generation (SHG), since the integrated contribution of the small background pulses will be practically negligible and the only significant contribution will come from the main pulse. Following these considerations, the amplified pulse energy at the fundamental wavelength can be inferred from a measurement of both SHG conversion efficiency and second harmonic pulse energy. With this setup it was obtained up to 160 μJ at 532 nm with a conversion efficiency of 75% in an angularly phase-matched 3 x 3 x 8 mm³, type-I LiB_3O_5 (LBO) crystal without any focusing, owing to the high peak intensity level ($\approx$ 4.8 GW/cm²) of the nearly diffraction-limited amplified picosecond pulses. Fourth harmonic generation was also performed, employing the SHG pulse as the pump for a $_3$ x $_3$ x $_7$ mm³ β-$B_aB_2O_4$ (BBO) crystal. As much as 64 μJ at 266 nm were generated with 40% conversion efficiency. The high peak power makes such laser sources ideal for simple, travelling wave optical parametric generation (OPG). With only 10 μJ at 1064 nm, 55%-efficiency SHG in a 15-mm LBO crystal yields $\approx$ 8-ps pulses at 532 nm for pumping an OPG, consisting in a 15-mm type-II $KTiOPO_4$ (KTP) for phase-matching in the xz plane. Tight focusing to $\approx$ 40-μm spot radius yields signal generation in the range 770-1020 nm (idler range was $\approx$ 1110-1720 nm) with angle tuning, with small efficiency $\approx$ 1%. This OPG might be readily amplified by a second stage pumped by a more energetic pulse at 532 nm, with much higher conversion efficiency and energy

Amplification of high repetition rate pulse trains

Some applications requires amplification of bunches of pulses instead of single pulses or continuous trains. The repetition rate of the micro-pulses contained in the bunches is set by the master oscillator, while temporal shape, duration and repetition frequency of the micropulses envelope (macro-pulse) can be settled by a suitable pulse-picking device. High smallsignal single-pass gain diode side-pumped grazing incidence amplifier modules can be employed also for macro-pulse amplification, owing to their good performances in terms of gain, beam quality and micro-pulse duration preservation, their relatively simple optical arrangement and their cost effectiveness. The main issue, peculiar of this application, is related to the severe temporal distortions experienced by the envelope of the pulse burst during amplification, due to gain saturation and extremely high gain values. This effect is strongly dependent on the macro-pulse duration and in the end limits the possibility to obtain rectangular-shaped amplified envelopes for macro-pulses longer than few μs or even few hundreds of ns, depending on the amplifiers saturation level. In order to overcome amplification distortions it is necessary to inject in the amplifier chain a conveniently shaped macro-pulse with a temporal envelope designed to compensate saturation effects (Butterworth et al., 1996). This can be done by modulating the radio-frequency driving signal of the AOM pulse picker. An example of rectangular-shaped and custom-shaped macro-pulses is shown in Figure. 15.

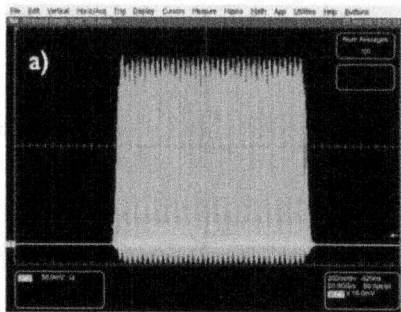 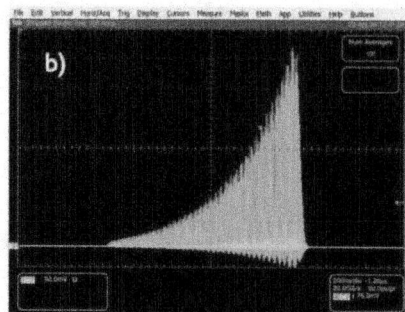

Figure. 15: Examples of two different deflected waveforms. a) Rectangular-shaped macro-pulse envelope; b) Macro-pulse envelope reshaped for amplification distortions compensation. Note that the energy of the modulated envelope is reduced to ~ 1/3.

In the example reported here, the master oscillator for pulse train amplification experiments is an high-frequency passively mode-locked Nd:YVO$_4$ oscillator (Agnesi et al., 2005). The ≈ 3-cm long V-folded resonator is pumped by a 1-W cw laser diode emitting at 808 nm, yielding ≈ 20-mW average power, ≈ 5-GHz repetition rate, 6-ps long mode-locking pulses. The AOM pulse-picker yields typically 500-ns long macro-pulses containing approximately 2500 picosecond micro-pulses from the continuous train (any macro-pulse length > 50 ns can be chosen). The macro-pulses are amplified employing three Nd:YVO$_4$ slabs, as shown in Figure. 16.

Each 14 x 4 x 2 mm^3, a-cut, 1% doped vanadate slab is side-pumped by a pulsed 150-W peak power laser diode array. A couple of optical isolators are employed to prevent self-lasing and dangerous, high-intensity back-injections in the direction of the master oscillator. The deflected laser beam emerging from the acousto-optic modulator needs to be carefully collimated to a vertical dimension optimally matched to the pumped layer, while the path between the three amplification heads is kept sufficiently long in order to reduce the amount of ASE generated in the first amplifier and injected into the others. All these optimisations contribute to achieve a single-pass gain of ≈ 70 dB. As shown in Figure. 17, pulse duration increases from 6 ps to 8.9 ps after amplification. A maximum output energy of about 2.5 mJ per macro-pulse is obtained after the three stage amplifier. The pulse spectrum after amplification is significantly narrowed: starting from a master oscillator 0.4-nm wide spectrum (fwhm), it shrinks to ≈ 0.19 nm at the output of the diode pumped amplifiers. Beam quality is well preserved ($M^2 ≈ 1.2$) as Figure. 18 clearly suggests.

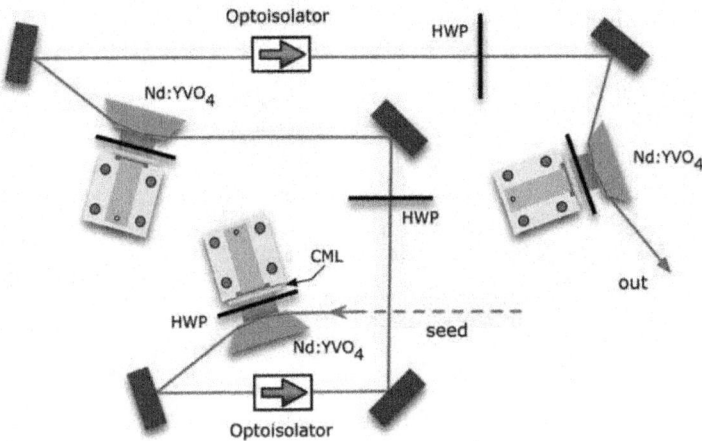

Figure. 16: Diode pumped amplifiers setup for pulse train amplification. Each laser diode beam is polarisation-rotated by a half-wave plate (HWP) and collimated by a 0.9-mm collimation microlens (CML).

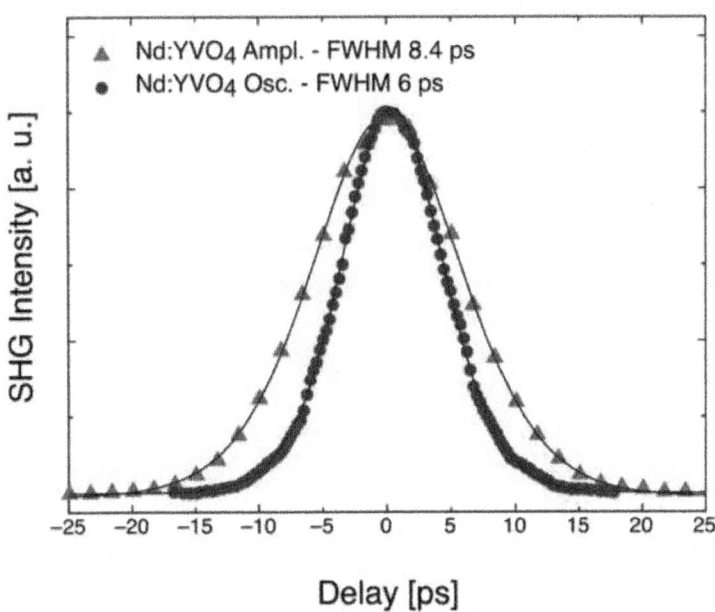

Figure. 17: Autocorrelation traces of both master oscillator (circles) and amplified pulses (triangles). Also shown the sech2 experimental data fitting.

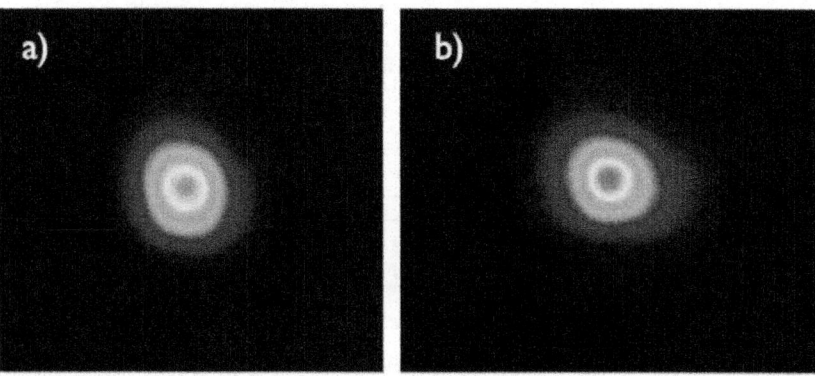

Figure. 18: Laser beam profile before amplification (a) and at the amplifier output (b).

Macro-pulse repetition rates up to 1 kHz can be chosen, owing to the duty cycle limitation of qcw diode arrays discussed previously. A particular high-energy application required that the macro-pulses emerging from the diode pumped amplification stage was further amplified in a couple of 12-cm long, 6-mm diameter, flash-lamp pumped Nd:YAG rods. At the maximum pump level, energy as high as 300 mJ was obtained for a 500-ns macropulse containing approximatively 2500, 12-ps long micro-pulses. The good beam quality ($M^2 = 1.5$) and peak power > 10 MW of the single pulses in the amplified burst allowed efficient nonlinear frequency conversion. SHG efficiency as high as 60% at 532 nm was achieved in a 16-mm long type-I LBO crystal. The second harmonic output beam was used to synchronously pump an optical parametric oscillator (OPO) plane-plane cavity. The nonlinear medium was a 12-mm long AR-coated KTP crystal cut for type II phase-matching. Both the OPO mirrors and the crystal coatings were designed for singly-resonant oscillation around 800 nm, even though much broader tuning behaviour might be obtained in principle, with a suitable choice of mirror coatings. At the maximum pump energy at 532 nm, with a 40% transmitting output coupler the OPO generated about 60-mJ at 800 nm, in 500-ns long, rectangular-shaped macro-pulses.

CONCLUSIONS

Effective amplification of diode-pumped picosecond laser oscillators employing grazingincidence side-pumped slab modules has been reviewed. Numerical models, useful for the design and the investigation of the amplifier characteristics, both in terms of energy extraction and pulse dynamics, have been illustrated. Some particularly representative examples of such amplification techniques, under various operating regimes, have been reported and discussed, comparing their results with the numerical model and putting them in perspective against other state of the art amplification schemes. The most distinctive features of grazing-incidence amplifiers can be summarised as follows:

- extremely high gain that allows considerable setup simplification;
- strong gain shaping allowing beam quality preservation along the fast axis of the diode array, notwithstanding the strong thermal lensing in that direction;
- compactness, modularity and easy scalability.

It has been pointed out that picosecond laser amplification might be now take advantage of simple, rugged cw mode-locked fibre lasers, adding DPSS amplifiers such as grazingincidence slabs. This is especially interesting for high-energy picosecond pulse amplification at multi-kHz up to MHz repetition rate. Higher power and energy levels can readily be obtained by adding pump diodes or even diode stacks, instead of using a single diode array to pump the bounce face of the slab (Minassian et al., 2005). Applications of this amplification technique can also be attractive for other laser materials such as Yb^{3+}-doped crystals (mostly Yb:YAG), Nd:glass, Nd:YAG and Nd:YLF. It is required that the absorption depth be small enough, and the absorbed pump power sufficiently high to produce a population inversion density yielding an integrated gain $\approx$ 20-30 dB per pass, in order to exploit all the useful features of the grazing-incidence slab. Particularly attractive are wide-band materials such as Nd:glasses that might allow high-energy femtosecond amplification with much simpler setups than usually reported in literature, at least using powerful qcw diode arrays at relatively low frequency (< 1 kHz). Quasi-three level materials such as Yb^{3+}-doped crystals also allow sub-picosecond and femtosecond amplification, however their re-absorption losses have to be taken into account properly and the pump geometry carefully designed to maximise efficiency. On the other hand, they are definitely more attractive than glasses for power upscaling, owing to their superior thermal and mechanical properties.

ACKNOWLEDGEMENT

We acknowledge the contribution of Bright Solutions, Srl, for supporting and discussion. This research received funding from the European Community's Seventh Framework Programme FP7/2007-2011 under grant agreement n° 224042.

REFERENCES

1. Agnesi, A.; Reali, G. C.; Kubecek, V.; Kumazaki, S.; Takagi, Y. & Yoshihara, K. (1993). β- barium borate and lithium triborate picosecond parametric oscillators pumped by a frequency-tripled passive negative-feedback mode-locked Nd:YAG laser. Journal of the Optical Society of America B, Vol. 10, No. 11, (November 1993) pp. 2211-2217, ISSN: 0740-3224

2. Agnesi, A.; Pirzio, F.; Tomaselli, A.; Reali, G. & Braggio, C. (2005). Multi-GHz tunablerepetition-rate mode-locked Nd:GdVO$_4$ laser. Optics Express, Vol. 13, No. 14, (July 2005) pp. 5302-5307, ISSN: 1094-4087

3. Agnesi, A.; Carrà, L.; Pirzio, F.; Reali, G.; Tomaselli, A.; Scarpa, D. & Vacchi, C. (2006a). Amplification of a low-power picosecond Nd:YVO$_4$ laser by a diode-laser sidepumped grazing-incidence slab amplifier. IEEE Journal of Quantum Electronics, Vol. 42, No. 8, (August 2006) pp. 772-776, ISSN: 0018-9197

4. Agnesi, A.; Carrà, L.; Pirzio, F.; Scarpa, D.; Tomaselli, A.; Reali, G.; Vacchi, C. & Braggio, C. (2006b). High-gain diode-pumped amplifier for generation of microjoule-level picosecond pulses. Optics Express, Vol. 14, No. 20, (October 2006) pp. 9244-9249, ISSN: 1094-4087

5. Agnesi, A.; Carrà, L.; Dallocchio, P.; Pirzio, F.; Reali, G.; Tomaselli, A.; Scarpa, D. & Vacchi, C. (2008a). 210- μJ picosecond pulses from a quasi-cw Nd:YVO$_4$ grazing-incidence two-stage slab amplifier package. IEEE Journal of Quantum Electronics, Vol. 44, No. 10, (October 2008) pp. 952-957, ISSN: 0018-9197 Agnesi, A.; Braggio, C.; Carrà, L.; Pirzio, F.; Lodo, S.; Messineo, G.; Scarpa, D.; Tomaselli, A.; Reali, G. & Vacchi, C. (2008b). Laser system generating 250-mJ bunches of 5-GHz repetition rate, 12-ps pulses. Optics Express, Vol. 16, No. 20, (September 2008) pp. 15811-15815, ISSN: 1094-4087

6. Bermudez, J. C. G.; Pinto-Robledo, V. J.; Kir'yanov, A.V. & Damzen M. J. (2002). The thermolensing effect in a grazing incidence, diode-side-pumped Nd:YVO$_4$ laser. Optics Communications, Vol. 210, No. 1-2, (September 2002) pp. 75-82, ISSN: 0030-4018

7. Bernard, J. E. & Alcock, A. J. (1993). High-efficiency diode-pumped Nd:YVO$_4$ slab laser. Optics Letters, Vol. 18, No. 12, (June 1993) pp. 968-970, ISSN: 0146-9592

8. Breitling, D.; Ruf, A. & Dausinger, F. (2004). Fundamental aspects in machining of metals with short and ultrashort laser pulses, Proceedings of SPIE Vol. 5339 Photon processing in microelectronics and photonics

III, pp. 49-63, ISBN 9780819452474, San Jose CA (USA), January 2004, SPIE, Bellingham WA (USA)

9. Burns, D.; Hetterich, M.; Ferguson, A. I.; Bente, E.; Dawson, M. D.; Davies, J. I. & Bland, S. W. (2000). High-average power (>20-W) Nd:YVO$_4$ lasers mode locked by straincompensated saturable Bragg reflectors. Journal of the Optical Society of America B, Vol. 17, No. 6, (June 2000) pp. 919-926, ISSN: 0740-3224

10. Butterworth, S. D.; Clarkson, W. A.; Moore, N.; Friel, G. J. & Hanna, D. C. (1996). Highpower quasi-cw laser pulses via high-gain diode-pumped bulk amplifiers. Optics Communications, Vol. 131, No. 1-3, (October 1996) pp. 84-88, ISSN: 0030-4018

11. Damzen, M. J.; Trew, M.; Rosas, E. & Crofts G.J. (2001). Continuous-wave Nd:YVO$_4$ grazingincidence laser with 22.5 W output power and 64% conversion efficiency. Optics Communications, Vol. 196, No. 1-6, (September 2001) pp. 237-241, ISSN: 0030-4018

12. Dausinger, F.; Hügel, H. & Konov, V. (2003). Micro-machining with ultrashort laser pulses: from basic understanding to technical applications, Proceedings of SPIE Vol. 5147 ALT'02 International Conference on Advanced Laser Technologies, pp. 106-115, ISBN 9780819450173, Adelboden (Switzerland), November 2003, SPIE, Bellingham WA (USA)

13. Edwards, G. S.; Hutson, M. S.; Hauger, S.; Kozub, J. A.; Shen, J.-H.; Shieh, C.; Topadze, K. & Joos, K. (2002). Comparison of OPO and Mark-III FEL for tissue ablation at 6.45 m, Proceedings of SPIE Vol. 4633 Commercial and biomedical applications of ultrafast and free-electron lasers, pp. 194-200, ISBN 9780819443724, San Jose CA (USA), April 2002, SPIE, Bellingham WA (USA)

14. Farrell, D. J. & Damzen, M. J. (2007). High power scaling of a passively modelocked oscillator in a bounce geometry. Optics Express, Vol. 15, No. 8, (April 2007) pp. 4781- 4786, ISSN: 1094-4087

15. Fermann, M. E. & Hartl, I. (2009). Ultrafast fiber laser technology. IEEE Journal of Selected Topics in Quantum Electronics, Vol. 15, No. 1, (January 2009) pp. 191-206, ISSN: 1077- 260X

16. Flood, C. J.; Walker, D. R. & Van Driel, H. M. (1995). Effect of spatial hole burning in a mode-locked diode end-pumped Nd:YAG laser. Optics Letters, Vol. 20, No. 1, (January 1995) pp. 58-60, ISSN: 0146-9592

17. Frantz L. M. & Nodvik, J. S. (1963). Theory of pulse propagation in a laser amplifier. Journal of Applied Physics, Vol. 34, No. 8, (August 1963) pp. 2346-2349, ISSN: 0021-8979

18. Gerhard, C.; Druon, F.; Blandin, P.; Hanna, M.; Balembois, F.; Georges, P. & Falcoz, F. (2008). Efficient versatile-repetition rate picosecond source for material processing applications. Applied Optics, Vol. 47, No. 7, (March 2008) pp. 967-974, ISSN: 0003- 6935

19. Keller U. (2003). Recent developments in compact ultrafast lasers. Nature, Vol. 424, (August 2003) pp. 831-838, ISSN: 0028-0836

20. Killi, A.; Dörring, J.; Morgner, U.; Lederer, M. J.; Frei, J. & Kopf, D. (2005). High speed electro-optical cavity dumping of mode-locked laser oscillators. Optics Express, Vol. 13, No. 6, (March 2005) pp. 1916-1922, ISSN: 1094-4087

21. Kleinbauer, J.; Knappe, R. & Wallenstein, R. (2005). 13-W picosecond Nd:GdVO4 regenerative amplifier with 200-kHz repetition rate. Applied Physics B, Vol. 81, No. 2-3, (July 2005) pp. 163-166, ISSN: 0946-2171

22. Koechner, W. (2006). Solid-State Laser Engineering, Springer, ISBN 9780387290942, Berlin Le Blanc, C.; Curley, P. & Salin, F. (1996). Gain-narrowing and gain-shifting of ultra-short pulses in Ti:sapphire amplifiers. Optics Communications, Vo. 131, No. 4-6, (November 1996) pp. 391-398, ISSN: 0030-4018

23. Mani, A. A.; Dreesel, L.; Hollander, P.; Humbert, C.; Caudano, Y.; Thiry, P. A. & Peremans, A. (2001). Pumping picosecond optical parametric oscillators by a pulsed Nd:YAG laser mode locked using a nonlinear mirror. Applied Physics Letters, Vol. 79, No. 13, (September 2001) pp. 1945-1947, ISSN: 0003-6951

24. McCarthy, M. J. & Hanna, D. C. (1993). All-solid-state synchronously pumped optical parametric oscillator. Journal of the Optical Society of America B, Vol. 10, No. 11, (November 1993) pp. 2180-2189, ISSN: 0740-3224

25. McDonagh, L.; Wallenstein, R. & Nebel, A. (2007). 111 W, 110 MHz repetition-rate, passively mode-locked TEM00 Nd:YVO4 master oscillator power amplifier pumped at 888 nm. Optics Letters, Vol. 32, No. 10, (May 2007) pp. 1259-1261, ISSN: 0146-9592

26. Minassian, A.; Thompson, A.; Smith, G. & Damzen, M. J. (2005). High-power scaling (>100 W) of a diode-pumped TEM00 Nd:GdVO4 laser system. IEEE Journal of Selected Topics in Quantum Electronics, Vol. 11, No. 3, (May/June 2005) pp. 621-625, ISSN:1077-260X

27. Nawata, K.; Ojima, Y.; Okida, M.; Ogawa, T.& Omatsu T. (2006). Power scaling of a picosecond Nd:YVO4 master-oscillator power-amplifier with a phase-conjugate mirror. Optics Express, Vol. 14, No. 22, (October 2006) pp. 10657-10662, ISSN: 1094- 4087

28. Nawata, K.; Okida, M.; Furuki, K. & Omatsu, T. (2007). MW ps pulse generation at sub-MHz repetition rate from a phase-conjugate Nd:YVO4 bounce amplifier. Optics Express, Vol. 15, No. 15, (July 2007) pp. 9123-9128, ISSN: 1094-4087

29. Neuhaus, J.; Kleinbauer, J.; Killi, A.; Weiler, S.; Sutter, D. & Dekorsy, T. (2008). Passively mode-locked Yb:YAG thin-disk laser with pulse energies exceeding 13 µJ by use of an active multipass geometry. Optics Letters, Vol. 33, No. 7, (April 2008) pp. 726- 728, ISSN: 0146-9592

30. Ojima, Y.; Nawata, K. & Omatsu T. (2005). Over 10-W picosecond diffraction-limited output from a Nd:YVO4 slab amplifier with a phase conjugate mirror. Optics Express, Vol. 13, No. 22, (October 2005) pp. 8993-8998,, ISSN: 1094-4087

31. Okhotnikov, O. G.; Gomes, L. A.; Xiang, N.; Jouhti, T.; Chin, A. K.; Singh, R. & Grudinin A. B.(2003). 980-nm Picosecond Fiber Laser. IEEE Photonics Technology Letters, Vol. 15, No. 11, (November 2003) pp. 1519-1521, ISSN: 1041-1135

32. Porta, J.; Grudinin, A. B.; Chen, Z. J.; Minelly, J. D. & Traynor, N. J. (1998). Environmentally stable picosecond ytterbium fiber laser with a broad tuning range. Optics Letters, Vol. 23, No. 8, (April 1998) pp. 615-617, ISSN: 0146-9592

33. Rufling, B.; Nebel, A. & Wallenstein, R. (2001). High-power picosecond LiB3O5 optical parametric oscillators tunable in the blue spectral range. Applied Physics B, Vol. 72, No. 2, (January 2001) pp. 137-149, ISSN: 0946-2171

34. Siebold, M.; Hornung, M.; Hein, J.; Paunescu, G.; Sauerbrey, R.; Bergmann, T. & Hollemann, G. (2004). A high-average power diode-pumped Nd:YVO$_4$ regenerative laser amplifier for picosecond-pulses. Applied Physics B, Vol. 78, No. 3-4, (February 2004) pp. 287-290, ISSN: 0946-2171

35. Snell, K. J.; Lee D.; Wall, K. F. & Moulton, P. F. (2000). Diode-pumped, high-power cw and modelocked Nd:YLF lasers, OSA Trends in Optics and Photonics Vol. 34, pp. 55-59, ISBN 1557526281, Davos (Switzerland), February 2000, Optical Society of America, Washington, D.C.

36. Sun, Z.; Ghotbi, M. & Ebrahimzadeh, M. (2007). Widely tunable picosecond optical parametric generation and amplification in BiB3O6. Optics Express, Vol. 15, No. 7, (April 2007) pp. 4139-4147, ISSN: 1094-4087

37. Walker, D. R.; Flood, C. J.; van Driel, H. M.; Greiner, U. J. & Klingerberg, H. H. (1994). High power diode-pumped Nd:YAG regenerative amplifier

for picosecond pulses. Applied Physics Letters, Vol. 65, No. 16, (October 1994) pp. 1992-1994

38.   Will, I.; Koss, G. & Templin, I. (2005). The upgraded photocatode laser of the TESLA Test Facility. Nuclear Instruments and Methods in Physics Research. Section A, Vol. 541, No. 3, (April 2005) 467-477, ISSN: 0168-9002

39.   Zayhowski, J. J. & Harrison, J. (1997). Miniature solid-state lasers, in: Handbook of Photonics, Gupta, M. C., (Ed.), pp. 326-392, CRC Press, ISBN 0849389097, Boca Raton FL (USA)

Chapter 2

DIODE-PUMPED HIGH ENERGY AND HIGH AVERAGE POWER ALL-SOLID-STATE PICOSECOND AMPLIFIER SYSTEMS

Jiaxing Liu[1], Wei Wang[2], Zhaohua Wang[1], Zhiguo Lv[3], Zhiyuan Zhang[4]and Zhiyi Wei[1,3]

[1]Beijing National Laboratory of Condensed Matter Physics, the Institute of Physics, Chinese Academy of Science, No. 8, 3rd South Street, Zhongguancun, Haidian District, Beijing 100190, China

[2]Department of Physics, Capital Normal University, 105 West Third Ring Road North, Haidian District, Beijing 100048, China

[3]School of Physics and Optoelectronic Engineering, Xidian University, 266 Xinglong Section of Xifeng Road, Xi'an 710126, Shanxi, China

[4]School of Science, China University of Mining & Technology, Ding No. 11 Xueyuan Road, Haidian District, Beijing 100083, China

ABSTRACT

We present our research on the high energy picosecond laser operating at a repetition rate of 1 kHz and the high average power picosecond laser running at 100 kHz based on bulk Nd-doped crystals. With diode-pumped solid state (DPSS) hybrid amplifiers consisting of a picosecond oscillator, a regenerative amplifier, end-pumped single-pass amplifiers, and a side-pumped amplifier, an output energy of 64.8 mJ at a repetition rate of 1 kHz was achieved. An average power of 37.5 W at a repetition rate of 100 kHz pumped by continuous wave laser diodes was obtained. Compact, stable and high power DPSS laser amplifier systems with good beam qualities are excellent picosecond sources for high power optical parametric chirped pulse amplification (OPCPA) and high-efficiency laser processing.

INTRODUCTION

High power picosecond lasers have attracted widespread attention due to their wide application in nonlinear frequency conversion, precision processing, and bio-medicine. A picosecond laser operating at a repetition rate of a few kHz, with both high peak power and high average power, is an excellent source for optical parametric chirped pulse amplification (OPCPA), an optical parametric oscillator (OPO), optical parametric generation (OPG), and UV light generation [1,2,3,4]. At ELI-Beamlines, two picosecond lasers with 1 kHz, >100 mJ, were used to pump an OPCPA system as a part of an X-ray pump laser [5]. A laser operating at a 193 nm wavelength could be also realized by mixing the fourth harmonic of 1064 nm (266 nm) and 708 nm, which was obtained by an OPO pumped with a picosecond 532 nm laser [6]. Recently, ultra-short laser pulses, especially picosecond pulses, have been widely used in the field of material processing. To achieve high processing speed and productivity, high average power picosecond lasers operating at repetition rates of hundreds of kHz are preferred. The demand for compact, stable, and low-cost picosecond lasers has increased rapidly. Laser diodes are preferred pumps for high power lasers. Diode-pumped solid-state lasers (DPSSLs) are smaller, more stable, and have a lower cost, thus making them more practical for more applications [7,8,9].

In order to achieve high output power, higher power has to be pumped into the gain medium. In this case, excessive heat will be generated in the gain medium due to a finite conversion efficiency. Thus, the key issue in a high power laser system is how to remove the heat from the gain medium effectively. In the past years, several designs of high power picosecond lasers have been studied. Using rod materials, pulses with 1.5 J energy and 110 ps duration at 10 Hz [10], with 130 mJ and 64 ps at 300 Hz [11], with 80 mJ and 50 ps at 1 kHz, and with 145 W average power and 200 ps duration at 3 kHz were demonstrated with Nd-doped YVO_4 and YAG [12,13]; pulses with 1 J and 5 ps at 100 Hz [4], with 115 mJ before compressing at 200 Hz [14], and with 58.5 mJ at 1 kHz were also recently reported with cryogenically cooled bulk Yb:YAG [9]. Using thin-disk technology, a 300 W picosecond laser operating at 10 kHz was demonstrated with 1.6 ps in 2013, and a 1.3 kW thin-disk multi-pass amplifier running at 300 kHz with sub-8 ps was reported in 2014 [15,16]. Recently, the team at ELI-Beamlines presented a 110 mJ picosecond thin-disk amplifier with good beam quality. A 1-kHz-repetition-rate thin-disk regenerative amplifier was reported with a pulse energy of 10 mJ and a pulse duration of 1.29 ps after compression [5,17]. With innoslab and fiber materials, an average power of 250 W at 12.5 kHz was obtained, and 97 W average output power at 5.47 MHz picosecond pulses was generated from a fiber master oscillator power amplifier (MOPA) system, respectively [18,19]. Combining some of the different technologies mentioned above, a hybrid CPA laser system was reported recently. The system consisting of fiber pre-amplifiers, a bulk material-based regenerative amplifier, and multi-pass amplifiers provided 70 mJ pulse energy at the repetition rate of 1 kHz and about 6 ps pulse duration after compression [20]. Although thin-disk and innoslab lasers have the advantage of easy heat dissipation, lasers based on bulk materials generally could have higher efficiencies and simpler structures, making them more attractive.

In this paper, the research on high energy and high repetition rate DSPP picosecond amplifier systems with bulk Nd-doped crystals is reported. Nd-doped laser media with large emission cross-sections and long upper level lifetimes are very suitable for laser amplifications. Due to its comparatively short upper state lifetime and broad pumping bandwidth, $Nd:YVO_4$ is appropriate for efficient oscillators and amplifiers. In addition, the 0.8 nm emission bandwidth of $Nd:YVO_4$ supports sub-10 or sub-100 ps pulses [21]. In 2010, an output power of 53 W with a peak power of 40 kW from a hybrid amplifier laser system using $Nd:YVO_4$ was reported, corresponding to an optical efficiency of 42.4% [22]. In 2012, using gratings to stretch and compress the pulses, a multi-pulse picosecond laser with 14 mJ and a pulse duration of 28 ps operating at a repetition rate of 1 kHz from a $Nd:YVO_4$ regenerative amplifier was reported [23]. In 2013, a laser beam with an average power of 4.7 W running at the repetition rate from 1 kHz to 10 kHz was delivered from a diode dual-end-pumped $Nd:YVO_4$ regenerative anplifier [24]. Most recently, 10.5 W, 14.2 ps pulses at 1064 nm with a repetition rate of 10 kHz from a $Nd:YVO_4$ amplifier system was obtained [25] and an average power of 44.5 W, a pulse duration of 8.8 ps at 100 MHz with an optical efficiency of 56% was achieved from a sapphire face-cooled $Nd:YVO_4$ slab amplifier [26]. However, $Nd:YVO_4$ has low mechanical fracture so the pump intensity must be kept under a threshold. Considering the characteristics of the $Nd:YVO_4$ crystal, we employed quasi-continuous wave laser diodes as end-pumping sources for $Nd:YVO_4$ crystals and a side-pumped Nd:YAG module as a power amplifier. At a repetition rate of 1 kHz, an output energy of 64.8 mJ centered at 1064.4 nm was achieved. An average power of 37.5 W at a repetition rate of 100 kHz pumped by continuous wave (CW) laser diodes was obtained. Compact, stable, and high power DPSS laser amplifier systems with good beam qualities are excellent picosecond sources for high power OPCPA and high-efficiency laser processing.

EXPERIMENTS AND RESULTS

Picosecond Amplifier System at 1 kHz

The picosecond laser system consisted of a picosecond oscillator, a regenerative amplifier, four stages of end-pumped single-pass amplifiers, and a side-pumped amplifier. The whole system was water-cooled and the timing was controlled by a homemade delay generator for synchronization. The generator had seven output channels and each channel provided a control signal at a repetition rate of 1 kHz with delay adjustable from 0.25 ns to 1 ms for the pump lasers and Pockels cells.

Picosecond Oscillator

The seed pulse was provided by a diode-pumped picosecond oscillator designed as shown in Figure 1. The laser crystal was an a-cut, 0.5 at. % Nd-doped YVO_4 crystal with a size of $3 \times 3 \times 5$ mm^3. The surfaces of the crystal were antireflection-coated at 808 nm and around 1 μm ($T > 99\%$). For heat dissipation, the crystal was wrapped with indium foil and mounted tightly in a water-cooled copper heat sink, where the water temperature was maintained at 18 °C.

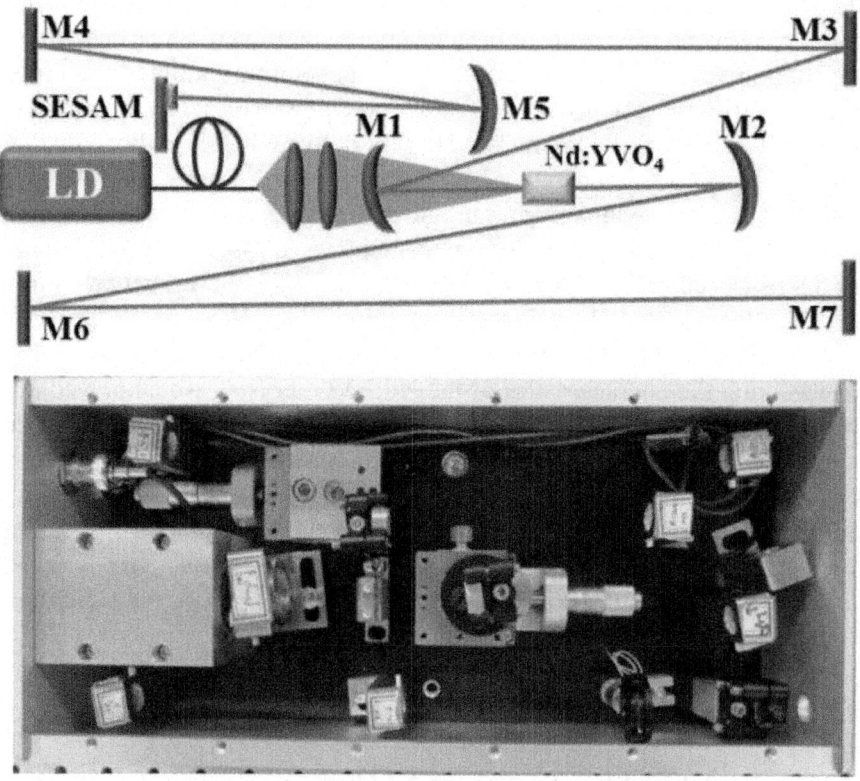

Figure 1: Set-up diagram and photograph of the picosecond oscillator. M1, M2: plane-concave pump mirrors, $R = 75$ mm; M3, M4, M6: high-reflection plane mirrors; M5: concave mirror, $R = 200$ mm; M7: output coupler, $T = 10\%$.

The pump was a fiber-coupled diode laser with a 50 μm core diameter (BWT, Beijing, China) and 2 W output power at 808 nm. With a semiconductor saturable absorption mirror (SESAM), the oscillator ran in a CW mode-locking state and provided about 400 mW average power at a repetition rate of 80 MHz as shown in Figure 2a. The stability of the output power was measured to be <0.7% rms as shown in Figure 2b. By controlling the dispersion of the cavity with Gires-Tournois interferometer (GTI) mirrors, the pulse duration could be changed from 10 ps to 15 ps. Figure 3 shows the intensity autocorrelation traces and spectra (AQ6315A, Ando Inc., Tokyo, Japan) of the mode-locked pulses with 10 ps and 15 ps, if sech^2-pulse shapes were assumed, measured by a commercial intensity autocorrelator (Femtochrome, FR-103MN). Considering the amplification efficiency and to avoid damaging optics, a 15 ps pulse was used as the seed for the regenerative amplifier. The oscillator worked at room temperature without cooling for its high efficiency and low pump power.

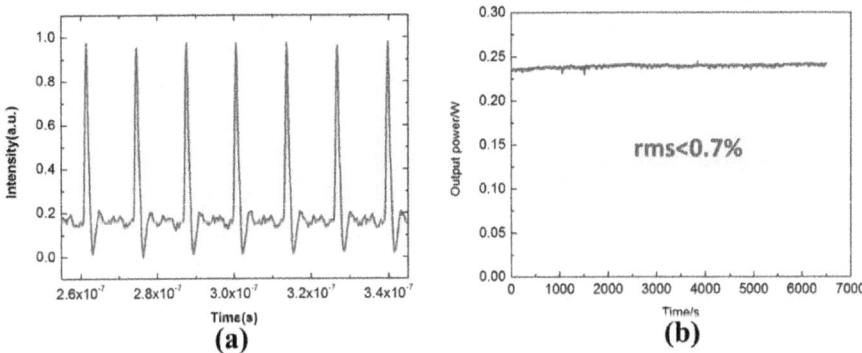

Figure 2: (a) Mode-locked pulse trains; and (b) power stability curve of the picosecond oscillator.

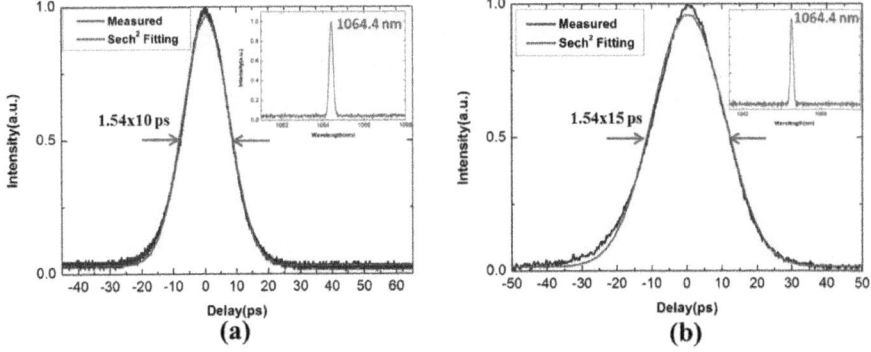

Figure 3: Intensity autocorrelation traces and spectra (insets) of the mode-locked pulses; (a) with dispersion compensation, pulse duration was 10 ps; (b) without dispersion compensation, pulse duration was 15 ps if sech^2-pulse shapes were assumed.

Regenerative Amplifier

The seed pulse was first amplified by the Nd:YVO$_4$ regenerative amplifier shown in Figure 4. The laser medium was a $4 \times 4 \times 10$ mm^3, a-cut, 0.3 at. % Nd:YVO$_4$ crystal, which was wrapped with indium foil and mounted tightly on a water-cooled copper heat sink at 14 °C. The surfaces of the crystal were antireflection-coated at 808 nm and around 1 μm ($T > 99\%$). Measures were taken in the regenerative amplifier to reduce the thermal effect. Firstly, the crystal is of low doping concentration and long length. It had high absorption for the pump laser (more than 85%) and good cooling capacity. Besides, the pump laser was a fiber-coupled quasi-CW laser diode with a 400 μm core diameter centered at 808 nm (DILAS, Germany), which produced less heat in the crystal than a CW laser diode. The pump laser was controlled by a trigger signal with a gate width of 120 μs at 1 kHz and focused into the crystal by an imaging system with a magnification of two. Finally, a thermal compensation cavity was constructed as shown in Figure 4. Since pulse energy in the cavity was very high, laser beams in the crystal and Pockels cell were relatively large to avoid damage. The beam diameters in the crystal and Pockels cell were about 1.2 mm and 1.4 mm. The intra-cavity transverse mode between the convex mirrors did not change when the focal length of the thermal lens changed from 1000 mm to infinity, which was caused by the increasing pump power, for ensuring the stability of the laser cavity. The stretcher and compressor were not used since the pulse duration of 15 ps was long enough in our laser system. In order to improve the contrast of the amplified pulses, the seed pulses passed through a Pockels cell to reduce the repetition rate from 80 MHz to 1 kHz before being injected into the amplifier. The seed pulse with about 1 nJ energy centered at 1064.4 nm was then amplified to 1.5 mJ at 1 kHz under the pump energy of 8 mJ in the regenerative amplifier. Figure 5 shows the pulse building process in the regenerative amplifier. Figure 6 shows the beam quality of the output pulses measured by a commercial instrument (M2-200S-FW, Ophir-Spiricon Inc., North Logan, UT, USA). The result ($M^2 < 1.15$) proved a good quality of the laser beam from the regenerative amplifier.

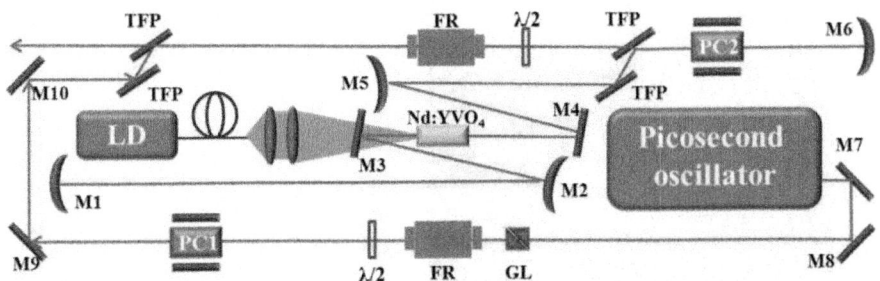

Figure 4: The scheme of the regenerative amplifier operating at a repetition rate of 1 kHz. M1, M6: Concave mirrors, $R = 900$ mm; M2, M5: Convex mirrors, $R = -1000$ mm; M3, M4, M7–M10: Plane high-reflection mirrors; PC: Pockels cell; TFP: Thin-film polarizer; FR: Faraday rotator; GL: Glan prism.

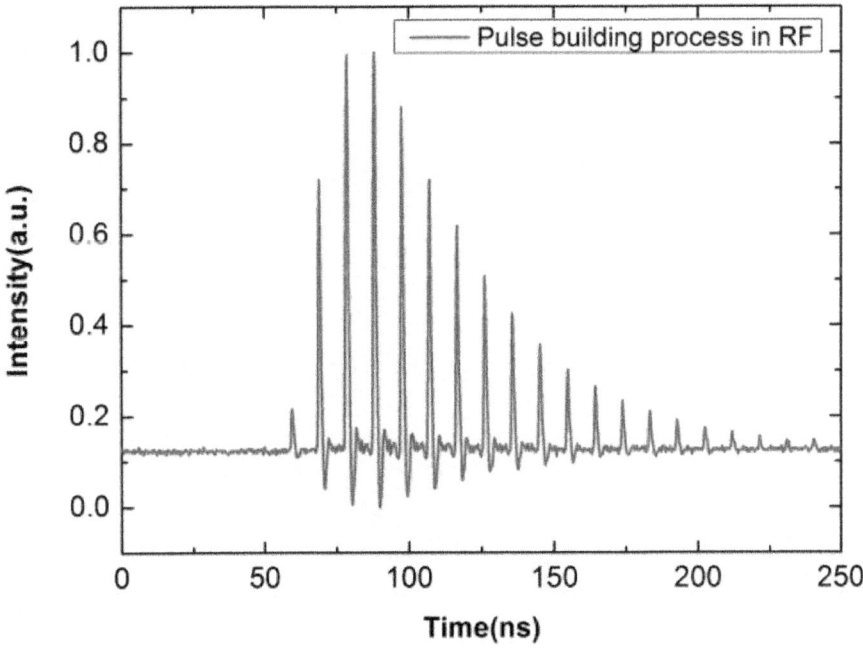

Figure 5: Pulse building process in the regenerative amplifier.

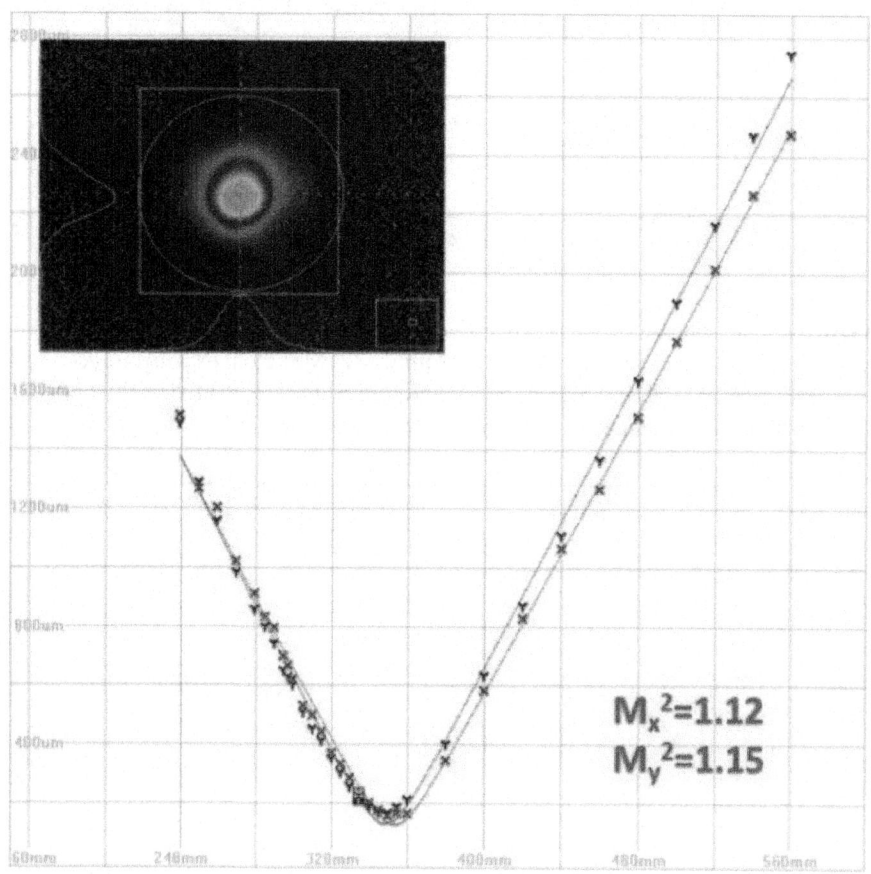

Figure 6: The beam quality of the output pulses from the regenerative amplifier.

End-Pumped Single-Pass Amplifiers

A 1.5 mJ, 1064.4 nm, 1 kHz pulse with high contrast and good beam quality was obtained by the regenerative amplifier as described in the previous section. Higher pulse energies are required for many applications [4,5,6]. As previously reported [12], multiple stages of side-pumped amplifiers were used to obtain 80 mJ picosecond laser pulses. However, due to the inhomogeneity in the directions of the side-pumped modules, the beam quality ($M^2 > 4$) was not good enough for some applications. To obtain a high pulse energy picosecond laser with good beam quality, we implemented a hybrid power amplifier consisting of four stages of end-pumped single-pass Nd:YVO$_4$ amplifiers and a side-pumped Nd:YAG amplifier as shown in Figure 7. In end-pumped amplifiers, 0.3 at. % Nd:YVO$_4$ crystals with a dimension of $4 \times 4 \times 10$ mm^3 were used as gain media and also wrapped with indium foil and mounted on water-cooled copper heat sinks at 14 °C. The fiber-coupled quasi-CW laser diodes with an 800 µm core diameter centered at 808 nm (DILAS, Mainz-Hechtsheim, Germany) were used as pump sources, and each of them produced a 120 µs pulse at 1 kHz and focused into the crystal by an imaging system with a magnification of two. For time synchronization of the whole system, a homemade delay generator provided trigger signals at the repetition rate of 1 kHz to drive the Pockels cells and quasi-CW pumps. The measured power curves and beam images are shown in Figure 8. The 1.5 mJ, 1064.4 nm, 1 kHz picosecond pulses were amplified to be more than 34 mJ after four stages of end-pumped single-pass amplifiers. The total pump energy was about 125 mJ, corresponding to an optical efficiency of 26%.

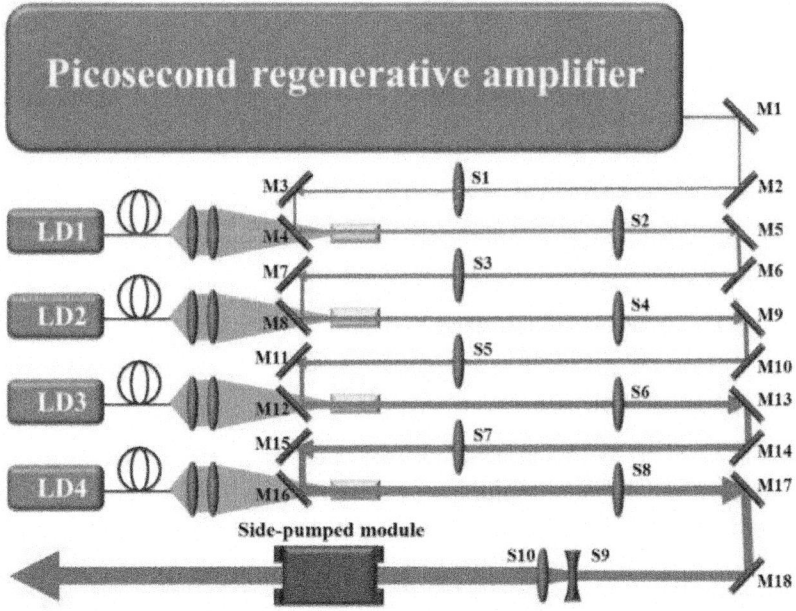

Figure 7: Diagram of the hybrid amplifier. M1–M3, M5–M7, M9–M11, M13–M15, M17–M18: Plane high-reflection mirrors; M4, M8, M12, M16: Plane high-reflection pump mirrors; S1–S8, S10: Plane-convex lens; S9: Plane-concave lens.

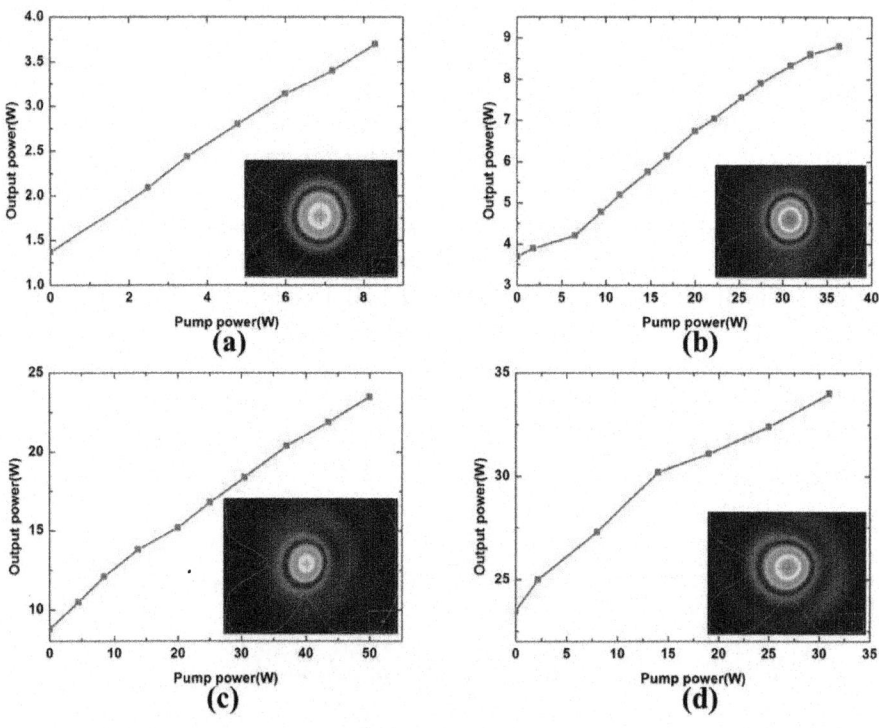

Figure 8: Power curves and beam images of each stage. (**a**) First stage; (**b**) second stage; (**c**) third stage; (**d**) fourth stage.

Side-Pumped Amplifier

Finally, the 34 mJ picosecond pulses from end-pumped amplifiers were injected into the last amplifier, a module consisting of a $\Phi 8 \times 185$ mm^3 Nd:YAG side-pumped by CW laser diodes. The maximum allowed pump power of 1000 W could be obtained at the wavelength of 808 nm. The cooling water system maintained at 14 °C was used for the heat dissipation of the crystal. In this experiment, a pump power of 750 W was used and a 64.8 mJ picosecond laser pulse at 1 kHz was obtained. Figure 9 shows the power curve and beam quality, where a good beam quality of $M^2 < 2$ is illustrated.

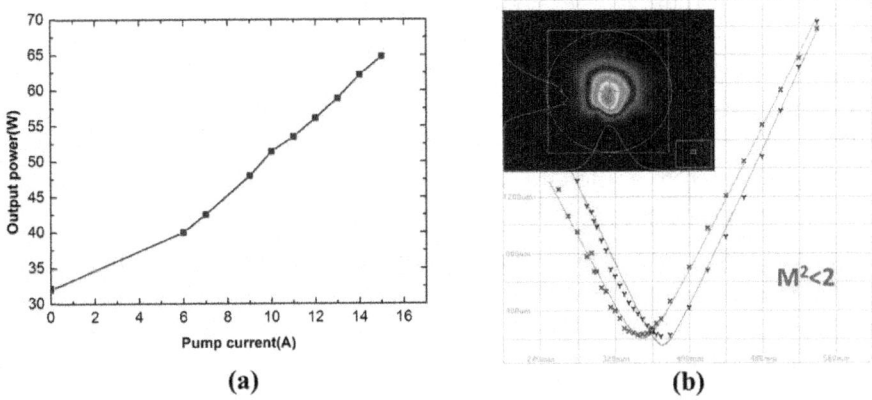

Figure 9: Power curve and beam quality of the side-pumped amplifier. **(a)** Power curve; **(b)** Beam quality.

High Average Power Picosecond Amplifier System at 100 kHz

As mentioned above, while the high energy picosecond laser operating at the repetition rate of 1 kHz is an excellent pump source for OPCPA, OPO, OPG, and UV light generation, a high power picosecond laser running at a higher repetition rate, such as 100 kHz, is preferred in the field of processing. Based on the experience introduced in Section 2.1, we demonstrated a 100 kHz, 37.5 W picosecond DPSS laser system, which was an excellent source for high efficiency laser processing.

At a higher repetition rate, the energy density is lower, but more heat is produced in the crystal. Reducing the diameter of the laser beam in the crystal is a good choice because of the smaller heat generation area and comparatively higher efficiency. Another reason for a smaller beam diameter was that a 3 mm diameter BBO Pockels cell was used as an electro-optic switch and worked at the repetition rate of 100 kHz, and the laser beam in the Pockels cell had to be smaller than that mentioned in Section 2.1.2. Thus, a thermal compensation cavity with a laser beam size smaller than that in Figure 4 was designed and constructed as a regenerative amplifier shown in Figure 10. The convex mirrors were replaced with concave mirrors (R = 600 mm). The laser beam was about 550 µm in the crystal and 730 µm in the Pockels cell as the focal length of the thermal lens changed from 300 mm to infinity. This cavity was more suitable for high-repetition-rate amplifier performance. A 0.3 at. %, a-cut, $3 \times 3 \times 10$ mm^3 Nd:YVO$_4$ crystal antireflection-coated at 808 nm and around 1 µm ($t >$ 99%) was used for this cavity. The crystal was wrapped with indium foil and mounted tightly on a water-cooled copper heat sink at 14 °C. The pumping source was a fiber-coupled CW laser diode with a 200 μm core diameter centered at 808 nm (Focuslight, China). Figure 11a shows the power curve of the regenerative amplifier. At the CW pump power of 15.5 W, 2.6 W average output power was obtained at the repetition rate of 100 kHz by injecting a 15 ps seed pulse from a 1064.4 nm, 80 MHz picosecond oscillator. Figure 11b shows the beam quality of the output pulses as good as $M^2 < 1.2$.

To further improve the average power, a hybrid power amplifier was also used. The hybrid power amplifier consisted of three stages of end-pumped single-pass amplifiers with CW laser diodes and a side-pumped module introduced in Section 2.1. The 2.6 W output pulses from the regenerative amplifier were then amplified by the single-pass amplifiers. The 0.3 at. %, a-cut, $3 \times 3 \times 10$ mm³ Nd:YVO$_4$ crystals were employed as gain media. At the total pump power of 88.5 W, an output average power of 21 W was obtained at 1064.4 nm, corresponding to a pulse energy of 210 μJ and an optical efficiency of 20.8%. At last, the picosecond pulses were amplified to 37.5 W by the side-pumped module at a pump power of 900 W. The pulse energy was as high as 375 μJ at a repetition rate of 100 kHz. Figure 12 shows the power curves and beam images of the power amplifiers.

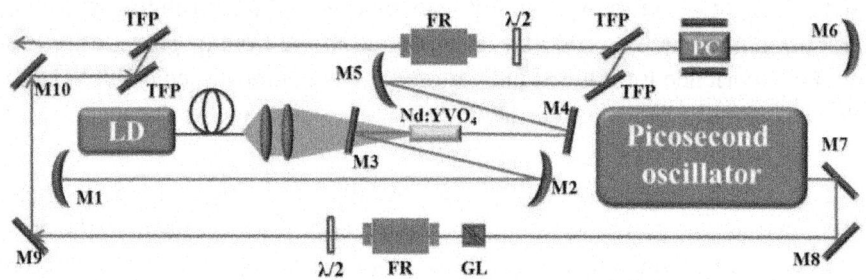

Figure 10: Diagram of the regenerative amplifier running at the repetition rate of 100 kHz. M1, M6: Concave mirrors, $R = 900$ mm; M2, M5: Concave mirrors, $R = 600$ mm; M3, M4, M7–M10: Plane high-reflection mirrors; PC: Pockels cell; TFP: Thin-film polarizer; FR: Faraday rotator; GL: Glan prism.

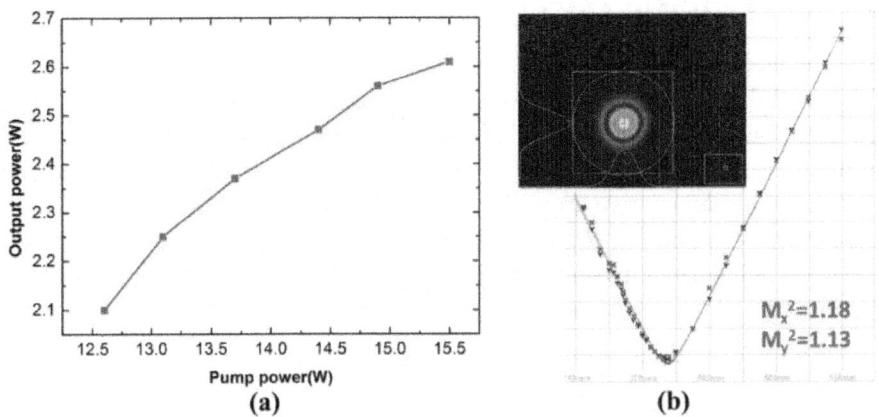

Figure 11; (**a**) Power curve of regenerative amplification; (**b**) Beam quality of the output pulses from the regenerative amplifier.

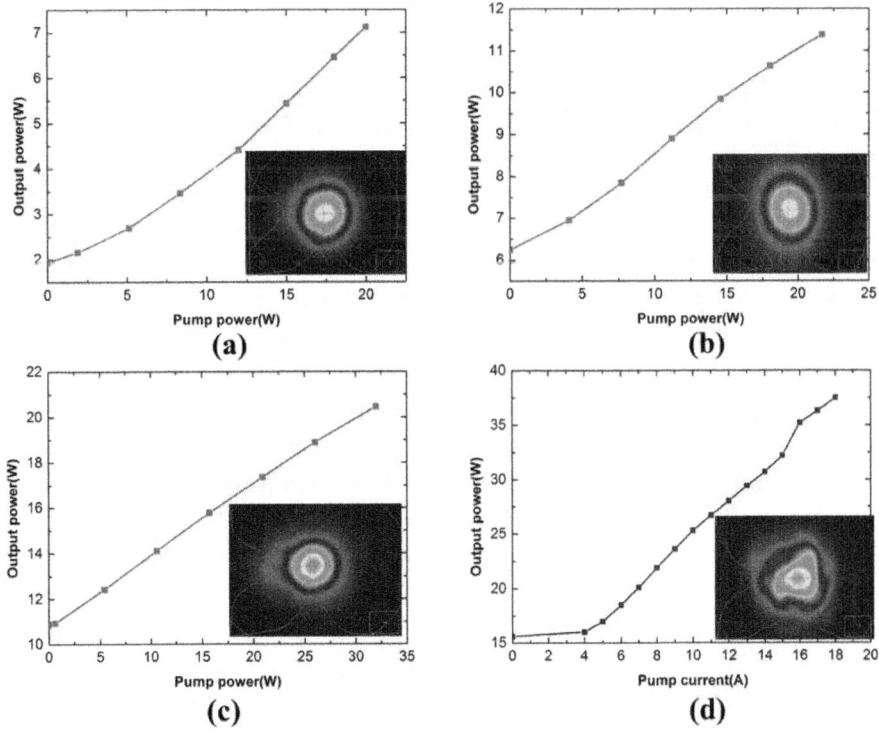

Figure 12: Power curves and beam images of each stage. (**a**) First stage; (**b**) second stage; (**c**) third stage; (**d**) side-pumped stage.

CONCLUSIONS AND DISCUSSIONS

In conclusion, we have introduced the latest developments in high power DPSS picosecond lasers. We presented our work on a high energy picosecond laser operating at a repetition rate of 1 kHz. and a high average power picosecond laser running at 100 kHz. With DPSS hybrid amplifiers consisting of a picosecond oscillator, a regenerative amplifier, end-pumped single-pass amplifiers, and a side-pumped amplifier, the output energy of 64.8 mJ at a repetition rate of 1 kHz was achieved. An average power of 37.5 W at a repetition rate of 100 kHz pumped by CW laser diodes was also obtained.

Although without the advantage of heat dissipation, laser systems based on bulk materials generally could have high conversion efficiency and simple structures. These characteristics make bulk materials attractive choices as laser media in high power DPSS picosecond lasers. In this paper, we demonstrated high energy and high average power picosecond laser systems with bulk Nd-doped crystals, proving them as promising candidates to achieve high power DPSS lasers.

To improve the system's capacity of heat dissipation, we took measures on the crystals, pumps, and cavity design. We used long crystals with low doping concentration and quasi-CW laser diodes as pumps. A thermal compensation cavity was designed and used as a regenerative amplifier. The crystals were water-cooled at 14 °C. In the experiments, we found that the efficiency dropped significantly when the pump power increased to a certain value beyond the damage threshold of the crystals. The main influence factor, we believed, was the heat accumulated in the crystals. In the next step, we will replace the water-coolers with thermal electronic coolers (TEC) or cryogenic coolers to improve the capacity of heat dissipation to achieve higher power or energy.

By the figures showing the beam qualities of the output pulses in this paper, we noticed that the beam quality deteriorated significantly after including the side-pumped amplifier despite providing a huge pump power. Its inhomogeneity in space had a great influence on the beam quality. However, the beam quality of the picosecond laser is especially important for certain applications. The side-pumped module will have to be replaced by an end-pumped amplifier for better beam quality.

ACKNOWLEDGMENTS

This work was partially supported by the National Major Equipment Development Project of the Ministry of Science and Technology of China (2012YQ120047) and the National Natural Science Foundation of China (11174361, 61575217).

AUTHOR CONTRIBUTIONS

Zhiyi Wei and Zhaohua Wang designed and supervised the research; Jiaxing Liu, Zhiguo Lv and Zhiyuan Zhang performed the research; Jiaxing Liu and Wei Wang analyzed the data; and Jiaxing Liu wrote the paper with discussions from all authors.

CONFLICTS OF INTEREST

The authors declare no conflict of interest.

REFERENCES

1. Ursula, K. Recent developments in compact ultrafast lasers. *Nature* **2003**, *424*, 831–838.

2. Kleinbauer, J.; Knappe, R.; Wallenstein, R. A powerful diode-pumped laser source for micro-machining with ps pulses in the infrared, the visible and the ultraviolet. *Appl. Phys. B* **2005**, *80*, 315–320.

3. Kienle, F.; Chen, K.K.; Alam, S.U.; Gawith, C.B.; Mackenzie, J.I.; Hanna, D.C.; Richardson, D.J.; Shepherd, D.P. High-power, variable repetition rate, picosecond optical parametric oscillator pumped by an amplified gain-switched diode. *Opt. Express* **2010**, *18*, 7602–7610. [PubMed]

4. Brendan, A.R.; Cory, B.; Keith, W.; Herman, B.; Mark, W.; Alden, C.; Federico, J.F.; Bradley, M.L.; Dinesh, P.; Carmen, S.M.; *et al.* 1 Joule, 100 Hz repetition rate, picosecond CPA laser for driving high average power soft X-ray lasers. In Proceedings of the 2014 Conference on Lasers and Electro-Optics (CLEO), San Jose, CA, USA, 8–13 June 2014.

5. Jakub, N.; Pavel, B.; Jonathan, T.G.; Zbynek, H.; Bedrich, R. 100 mJ thin disk regenerative amplifier at 1 kHz as a pump for picosecond OPCPA. In Proceedings of the 2015 Conference on Lasers and Electro-Optics (CLEO), San Jose, CA, USA, 10–15 May 2015.

6. Sadovskii, S.P.; Chizhov, P.A.; Bukin, V.V.; Brendel, V.M.; Dolmatov, T.V.; Polivanov, Y.N.; Orlov, S.N.; Garnov, S.V.; Vartapetov, S.K. Picosecond laser system with a wavelength of 193 nm based on a solid-state Nd:YAG laser, parametric oscillator, and ArF amplifier. *Phys. Wave Phenom.* **2014**, *22*, 223–226.

7. Rudiger, M.; Michael, K.; Christian, G.; Klaus, K.; Wilfried, P.; Ulrich, T.S.; Joachim, W. Laser processing of gallium nitride-based light-emitting diodes with ultraviolet picosecond laser pulses. *Opt. Eng.* **2012**, *51*.

8. Bai, Z.X.; Ai, Q.K.; Duan, J.P.; Chen, M.; Li, G. The research on the micro

processing used all-solid-state picosecond laser. *Proc. SPIE* **2012**, *8334*.

9.   Chang, C.L.; Krogen, P.; Liang, H.K.; Stein, G.J.; Moses, J.; Lai, C.J.; Siqueira, J.P.; Zapata, L.E.; Hong, K.H. Multi-mJ, kHz, ps deep-ultraviolet source. *Opt. Lett.* **2015**, *40*, 665–668. [PubMed]

10. Daniel, H.; Laszlo, V.; Raphael, T.; Franz, T.; Karl, S.; Vladimir, P.; Ferenc, K. Generation of sub-three-cycle, 16 TW light pulses by using noncollinear optical parametric chirped-pulse amplification. *Opt. Lett.* **2009**, *34*, 2459–2461.

11. Daniel, W.E.N.; Stefan, W.; Jonas, M.; Robert, K.A.; Kjeld, S.E.E. High-energy, high-repetition-rate picosecond pulses from a quasi-CW diode-pumped Nd:YAG system. *Opt. Lett.* **2013**, *38*, 3021–3023.

12. Michailovas, K.; Smilgevicius, V.; Michailovas, A. High average power effective pump source at 1 kHz repetition rate for OPCPA system. *Lith. J. Phys.* **2014**, *54*, 150–154.

13. Nathan, B.; Benjamin, W.; Michael, C.; Lawrence, S.; Martin, R. 145 W, 3 kHz picosecond amplifier for OPCPA pumping. In Proceedings of the 2015 Conference on Lasers and Electro-Optics (CLEO): STu4O.5, San Jose, CA, USA, 10–15 May 2015.

14. Hemmer, M.; Reichert, F.; Zapata, K.; Smrz, M.; Calendron, A.R.; Cankaya, H.; Hong, K.H.; Kartner, F.; Zapata, L. Picosecond, 115 mJ energy, 200 Hz repetition rate cryogenic Yb:YAG bulk-amplifier. In Proceedings of the 2015 Conference on Lasers and Electro-Optics (CLEO): STu4O.3, San Jose, CA, USA, 10–15 May 2015.

15. Catherine, Y.T.; Marcel, S.; Robert, B.; Matthias, H.; Stephan, P.; Dirk, S.; Thomas, M. 300 W picosecond thin-disk regenerative amplifier at 10 kHz repetition rate. In Proceedings of the 2013 Advanced Solid-State Lasers Congress Postdeadline: JTh5A.1, Paris, France, 27 October–1 November 2013.

16. Jan-Philipp, N.; Andreas, V.; Marwan, A.A.; Dominik, B.; Dirk, S.; Alexander, K.; Thomas, G. 1.3 kW average output power Yb:YAG thin-disk multipass amplifier for multi-mJ picosecond laser pulses. In Proceedings of the 2014 Conference on Lasers and Electro-Optics (CLEO): STu1O.2, San Jose, CA, USA, 8–13 June 2015.

17. Yoshihiro, O.; Keisuke, N.; Momoko, M.; Masaaki, T.; Fumiko, Y.; Nanase, K.; Michiaki, M.; Akira, S. Yb:YAG thin-disk chirped pulse amplification laser system for intense terahertz pulse generation. *Opt. Express* **2015**, *23*, 15057–15064.

18. Schulz, M.; Riedel, R.; Willner, A.; Mans, T.; Schnitzler, C.; Russbueldt, P.; Dolkemeyer, J.; Seise, E.; Gottschall, T.; Hadrich, S.; *et al.* Yb:YAG Innoslab amplifier: Efficient high repetition rate subpicosecond pumping system for optical parametric chirped pulse amplification. *Opt. Lett.* **2011**, *36*, 2456–2458. [PubMed]

19. Chan, H.Y.; Alam, S.U.; Xu, L.; Bateman, J.; Richardson, D.J.; Shepherd, D.P. Compact, high-pulse-energy, high-power, picosecond master oscillator power amplifier. *Opt. Express* **2014**, *22*, 21938–21943. [PubMed]

20. Chang, C.L.; Krogen, P.; Hong, K.H.; Zapata, L.E.; Moses, J.; Calendron, A.L.; Liang, H.; Lai, C.J.; Stein, G.J.; Keathley, P.D.; *et al.* High-energy, kHz, picosecond hybrid Yb-doped chirped-pulse amplifier. *Opt. Express* **2015**, *23*, 10132–10144. [PubMed]

21. Zayhowski, J.J.; Harrison, J. *Handbook of Photonics*; Gupta, M.C., Ed.; CRC Press: Boca Raton, FL, USA, 1997; pp. 326–392.

22. Wushouer, X.; Yan, P.; Yu, H.; Liu, Q.; Fu, X.; Yan, X.; Gong, M. High peak power picosecond hybrid fiber and solid-state amplifier system. *Laser Phys. Lett.* **2010**, *7*, 644–649.

23. Bai, Z.N.; Bai, Z.X.; Yang, C.; Chen, L.Y.; Chen, M.; Li, G. High pulse energy, high repetition picosecond chirped-multi-pulse regenerative amplifier laser. *Opt. Laser Technol.* **2013**, *46*, 25–28.

24. Zhang, Z.L.; Liu, Q.; Yan, P.; Xia, P.; Gong, M.L. Laser diode end-pumped Nd:YVO4 regenerative amplifier for picosecond pulses. *Chin. Phys. B* **2013**, *22*.

25. Peng, Z.G.; Chen, M.; Yang, C.; Chang, L.; Li, G. A cavity-dumped and regenerative amplifier system for generating high-energy, high-repetition-rate picosecond pulses. *Jpn. J. Appl. Phys.* **2015**, *54*.

26. Masashi, A.; Hiroki, S.; Maya, K.; Yuta, S.; Katsuhiko, M.; Takashige, O. High average power, diffraction-limited picosecond output from a sapphire face-cooled Nd:YVO4 slab amplifier. *J. Opt. Soc. Am. B* **2015**, *32*, 714–718.

Chapter 3

DESIGN OF A NOVEL HIGH POWER V-BAND HELIX-FOLDED WAVEGUIDE CASCADED TRAVELING WAVE TUBE AMPLIFIER

Tianxiang Zhuge and Yulu Hu

School of Physical Electronics, University of Electronic Science and Technology of China, Chengdu 610054, China

ABSTRACT

A design of a V-band Helix-Folded Waveguide (H-FWG) cascaded traveling wave tube (TWT) is presented. In this cascaded structure, a digitized nonlinear theory model is put forward first to simulate these two types of the tubes by common process. Then, an initial design principle is proposed, which can design these two different kinds of tubes universally. Using this principle, a high-gain helix TWT is carefully designed as a first stage amplifier followed by a FWG TWT to obtain high power. Simulations predict that a peak power of 800 W with saturated gain of 60 dB from 55 GHz to 60 GHz can be achieved.

INTRODUCTION

The V-band frequency range (50–75 GHz) is a region of the millimeter wave spectrum with great potential application for intersatellite communications and high-performance radar applications, including atmospheric studies, space debris detection, precise tracking, and high resolution imaging [1–3]. However, it has to face to the size and power limitation at this frequency range, when using traditional helical traveling wave tube [1, 4] and some novel structure TWTs [5–7] on such high-frequency band. However, some metal structure TWTs (Coupled Cavity (CC) TWT [8], FWG TWT [9, 10], and so on) can obtain high power but low gains in single tube for strong backward wave oscillation (BWO) instability, also for some FWG cascaded TWTs [11, 12] and CC-FWG cascaded TWTs [13].

In order to obtain a high power and high-gain V-band traveling wave tube in feasible manufacture way, we designed a novel H-FWG cascaded traveling wave tube amplifier. In this design, we use helical traveling wave tube as a first stage amplifier, which is famous for its wide bandwidth and high gain and then exported the amplified signal into the FWG traveling wave tube to get high power. The FWG traveling wave tube has large power capability for its full metal structure and is chosen to get high power in the V-band frequency range.

A digitized beam-wave interaction theory model [14, 15] is used to analyze the performance of the designed helix-FWG cascaded traveling wave tube, which has been developed and included in MTSS [16]. In this model the digitized fields, getting the data from electromagnetic (EM) simulation software for arbitrary SWSs, are used to interact with the beam keeping the energy translation and conservation between the beam and the EM wave. Due to its general way to deal with the beam-wave interaction process, this digitized theory model can be used to simulate and analyze the nonlinear performance of traveling wave tube with different SWS.

A helix-FWG cascaded traveling wave tube testified the design method and is optimized. Finally, the simulation of the saturated output power is above 34 W with the saturated gain above 33 dB in 5 GHz bandwidth in the first stage helical TWT. And, for the cascaded FWG traveling wave tube in same bandwidth, the saturated gain is above 18 dB achieving the output power to be 800 W. After matching the input-output power in the connection, the total cascaded tube achieves 60 dB gain and 800 W in 5 GHz band. The illustration of the helix-FWG cascaded TWT model is presented in Figure 1.

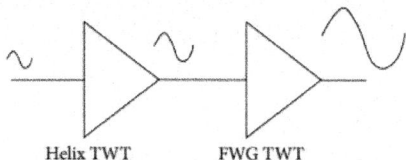

Helix TWT   FWG TWT

Figure 1: (Black and white) H-FWG cascaded TWT model.

The rest of the paper is organized as follows. Section 2 presents the sketch of the digitized nonlinear beam-wave interaction theory model, and Section 3 introduces the principle of the TWT design. Based on the design principle and the digitized theory model, the design and analysis of the H-FWG cascaded TWT are detailed in Section 4. A conclusion is followed at the end of the paper.

THE DIGITIZED NONLINEAR THEORY MODEL [14]

The digitized nonlinear theory model is a quasi-3D nonlinear theory model developed and included in MTSS [16]. It gets electromagnetic field distribution of arbitrary SWSs from the numerical simulation software such as HFSS, CST, and HFCS and then uses the energy conservation principle to get the energy exchange between the beam and the EM wave. Consequently, this digitized theory model is allowed to analyze the nonlinear performance of many kinds of traveling wave tube using different periodic high-frequency structures.

In the digitized nonlinear theory, RF fields acting on the electron beam propagating in a SWS are represented as

$$\mathbf{E}_{rf}(\mathbf{x}, t) = \sum_n \mathbf{E}_{rf,n}(\mathbf{x}, t) = \sum_n \left[f_n(z) \, \mathbf{e}_n(\mathbf{x}) \, e^{-i\omega_n t} \right],$$

$$\mathbf{H}_{rf}(\mathbf{x}, t) = \sum_n \mathbf{H}_{rf,n}(\mathbf{x}, t) = \sum_n \left[f_n(z) \, \mathbf{h}_n(\mathbf{x}) \, e^{-i\omega_n t} \right]. \tag{1}$$

The RF electric field $\mathbf{E}_{rf}$ and magnetic field $\mathbf{H}_{rf}$ are composed of many different time harmonics, whose angular frequency is ω_n and complex amplitude varying along the z axial is $f(z)$. Subscript n indicates the harmonic number and $e_n(\mathbf{x})$ and $h_n(\mathbf{x})$ are the RF field profiles without electron beam loaded, which will be obtained by EM simulation of HFSS [17], HFCS [16], CST [18], and so forth. Once obtaining the RF field profiles, the evolution of the RF field amplitude can be developed based on the law of energy conservation, and finally we get

$$\left[\frac{\partial}{\partial z} + \alpha_n(z) \right] f_n(z) = -\frac{1}{2} \int_z^{z+\lambda_h} \frac{dz}{\lambda_h} \int_t^{t+T} \frac{dt}{T}$$

$$\cdot \iint_A \mathbf{j} \cdot \mathbf{e}_n^*(\mathbf{x}_\perp, z_0) \, e^{-im\varphi_n} e^{i\omega_n t} ds, \quad z_0 \in [0, \lambda_h). \tag{2}$$

Here α_n means the attenuation constant, λ_h is the axial period length of the structure, and T is the period time. $\mathbf{j} = \rho_k$ is the beam current density. m indicates the number of the period among which z is solving. And φ_n is the phase shift for the nth RF signal in one period.

For different type of slow-wave structure (SWS), $e_n(\mathbf{x})$ and $h_n(\mathbf{x})$ are quite different. Without care for the analytic model of the slow-wave structure, we can use (1) to get the EM field and then to push the particles in the beam.

The beam current then affects the field amplitude $f(z)$, which will inevitably change the status of the beam and thus the current. Following this circulation, we can trace the process of the beam-wave interaction and get the nonlinear performance of the TWT.

DESIGN PRINCIPLE OF TWT

As a first step, to get the desired output power and gain, the beam voltage and the beam density in the tunnel (FWG or helix) in beam-wave interaction region have to be carefully selected, which are of the most important factors limiting the output power and gain. As it is well known, the output power P_o can be estimated by beam voltage V, beam current I_0, and electrical efficiency η; that is,

$$P_o = \eta V I_0.$$

(3)

From (3), we get beam current:

$$I_0 = \frac{P_o}{\eta V}.$$

(4)

And the beam density in the tunnel (FWG or helix) in beamwave interaction region is

$$J = \frac{I_0}{\pi b^2} = \frac{I_0}{\pi (\rho r_a)^2} = \frac{P_o}{\pi (\rho r_a)^2 \eta V}.$$

(5)

Here b is the beam radius, ρ is the beam filling ratio, and r_a is the radius of the tunnel (FWG or helix). On the other hand, the relativity factor $\gamma (= 1/\sqrt{1 - v_0^2/c^2})$ can be estimated by

$$\gamma - 1 = \frac{e}{m} \frac{V}{c^2},$$

(6)

where e/m is the electron charge to mass ratio and c is speed of light, from which we get the initial beam speed V_0:

$$v_0 = \sqrt{\frac{e}{m} \left(\frac{\gamma + 1}{\gamma^2} \right)} V.$$

(7)

Define a slow-varying and dimensionless parameter Γ as

$$\Gamma = \frac{\omega}{v_p} r_a \approx \frac{\omega}{v_0} r_a.$$

(8)

And combining (7) and (8), the tunnel radius r_a can be presented as

$$r_a = \frac{\Gamma}{\omega} \sqrt{\frac{e}{m} \left(\frac{\gamma + 1}{\gamma^2} \right)} V.$$

(9)

Inserting (9) into (5), the beam density can finally be written as filling ratio, and parameter Γ, a proper voltage and the tunnel radius can be selected to meet our design.

$$J = \frac{m P_o}{e \pi \eta (\gamma + 1)} \left(\frac{\gamma \omega}{\rho \Gamma V} \right)^2.$$

(10)

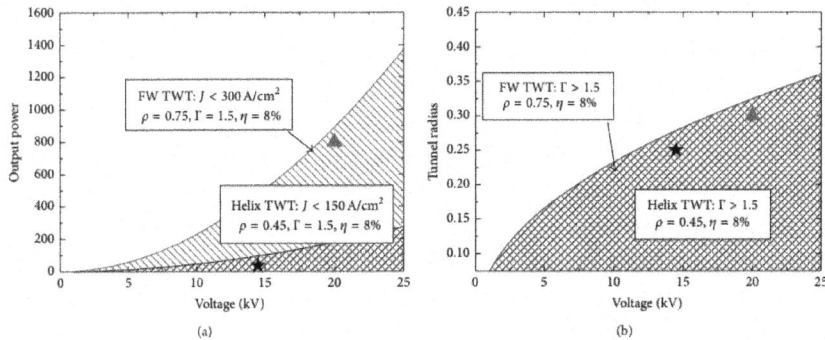

Figure 2: (a) Relationship between output power and voltage for helix TWT (black slashed region) and FWG TWT (red backslash region), which are partially overlapped. (b) Relationship between tunnel radius and voltage for helix TWT (black slash region) and FWG TWT (red backslash region), which are completely overlapped.

DESIGN OF THE H-FWG CASCADED

Traveling Wave Tube In the design of H-FWG cascaded traveling wave tube, we first select a proper voltage and tunnel radius and then optimize the SWS and beam-wave interaction profile of the helix TWT and the FWG TWT separately. Cascading these two sections, an output power of 800 W and a gain of 55 dB are expected in the V-band of 55 GHz to 60 GHz.

Choice of Voltage

In this V-band cascaded TWT design, we first set a proper parameter Γ and electronic efficiency η (here we set η=8% and $\Gamma = 1.5$ as initial value). The beam density of helix TWT is chosen less than 150 A/cm² and the filling ratio is set to be 0.45 for helix TWT. For FWG TWT with greater power capacity, the beam density is chosen as 300 A/cm² and the filling ratio is 0.75. Using (9) and (10), we get the relationship between output power and voltage, as well as the tunnel radius and voltage, as shown in Figure 2

After careful optimization, a voltage of 14.5 KV is selected for the first stage helix TWT, whose corresponding output power is 34 W and helix radius is 0.25 mm, shown as the black star in Figure 2. For the FWG TWT, the voltage is set to 20 KV, corresponding to an output power of 800 W and tunnel radius of 0.3 mm, demonstrated as the red triangle in Figure 2. And the current is to be 32.5 mA for helix TWT and 450 mA for FWG TWT.

Design of Helix TWT

A simple helix SWS with three APBN sector support rods is selected (shown in Figure 3), which can be manufactured easily. The design principle of the other SWS parameters, except fixed helix radius to 0.25 mm, is seeking the suitable dimensions to maximize the coupling impedance and to obtain flat dispersion between 55 GHz and 60 GHz. After careful optimization, the helix SWS structure parameters are shown in Table 1. The helix pitch ranges from 0.45 mm to 0.55 mm. For later beam-wave interaction optimization, the helix SWS with different helix pitch will be simulated with a step of 0.02 mm.

Table 1: Parameters of the helix slow-wave structure

Parameter	Value
Helix radius (R_a)	0.25 mm
Helix tape thickness	0.08 mm
Helix tape width	0.20 mm
Helix pitch	0.45–0.55 mm
Rod material	APBN
Rode wedge angle (θ)	30 deg
Shield radius (R_c)	1.2 mm

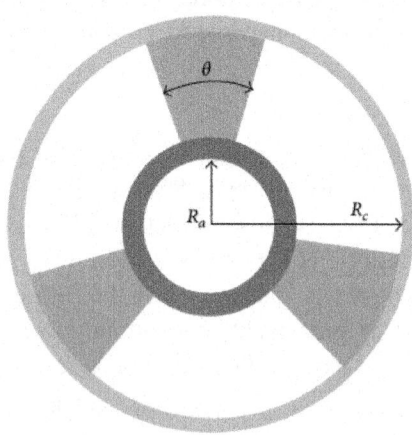

Figure 3: Illustration of the cross section of the helix SWS.

The normalized velocity, coupling impedance, and attenuation constant for the helix SWS with helix pitch ranging from 0.45 mm to 0.55 mm are shown in Figure 4.

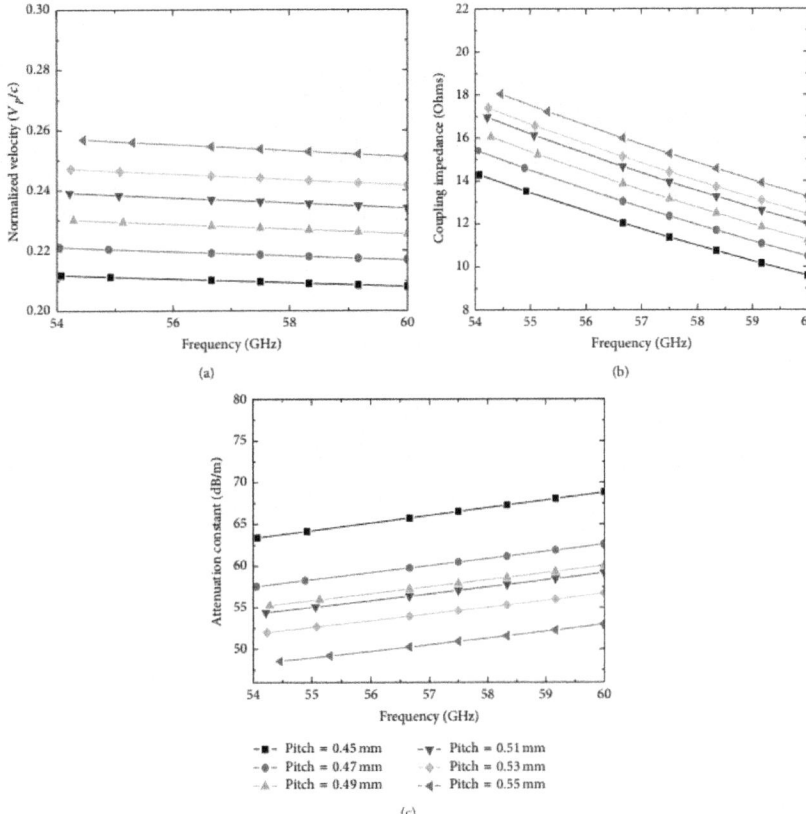

Figure 4: (a) Dispersion, (b) coupling impedance, and (c) attenuation constant for the helix SWS with pitch ranging from 0.45 mm to 0.55 mm.

It is simulated and optimized of the beam-wave interaction with pitch and attenuation profile (shown in Figure 5), using the simulation code developed on the model discussed in Section 2. Also BWO is considered by using an additional attenuator in input circuit and pitch step in output circuit. The pitch of the circuit is finally set to be 0.505 mm and then step to 0.49 mm, respectively. The profile of output power and gain along -axis at 57 GHz is shown in Figures 6(a) and 6(b). And the saturated output power and saturated gain in band are shown in Figure 6(c). Obviously, the output power of the first stage helix TWT can achieve 34 W and the saturated gain of 35 dB from 55 GHz to 60 GHz.

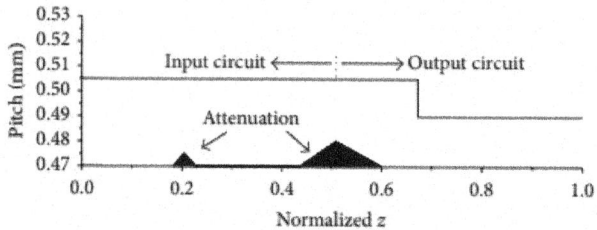

Figure 5: Pitch and attenuation profiles in helix TWT (black and white).

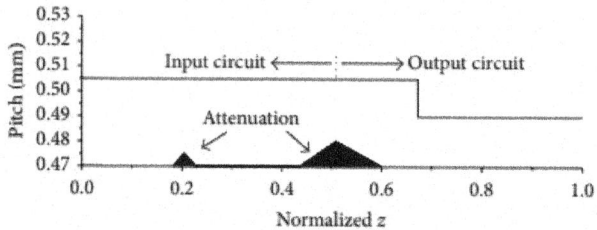

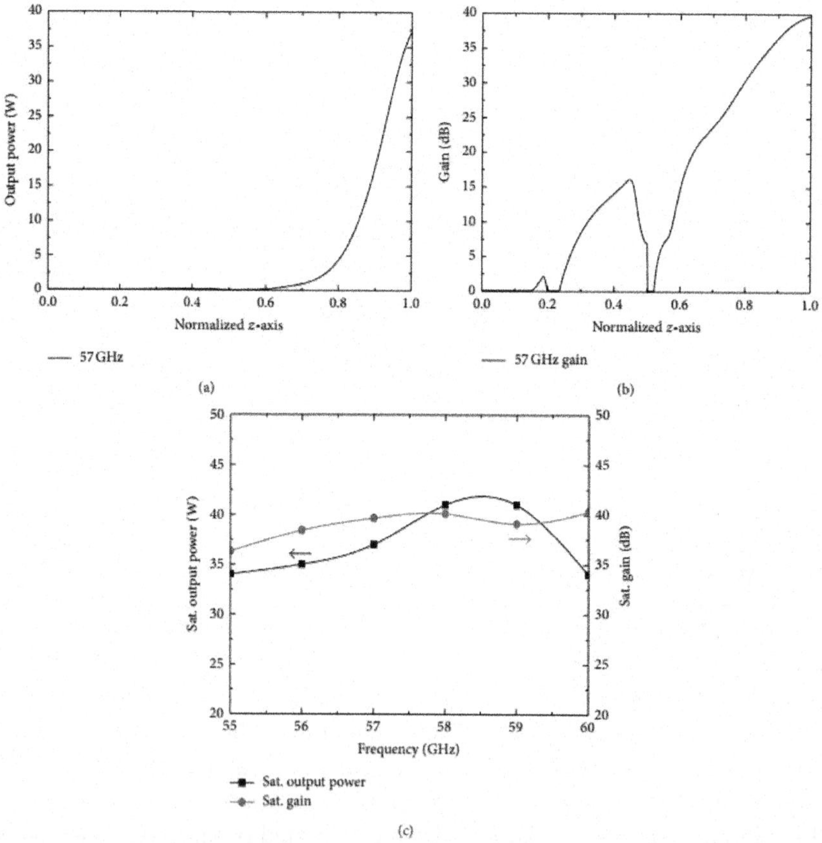

Figure 6: (a) Output power versus z at 57 GHz, (b) gain versus z at 57 GHz, and (c) saturated output power and saturated gain ranging from 55 GHz to 60 GHz.

Design of FWG TWT

Similar optimization is done for the FWG TWT (structure is shown in Figure 7). To get large coupling impedance and suitable synchronized velocity, we scan the structure parameters and finally get the FWG SWS with parameters listed in Table 2.

Table 2: Parameters of the folded slow-wave structure

Parameter	Value
r	0.3 mm
b	3.15 mm
a	0.85 mm
H	0.8 mm
L	2.0 mm

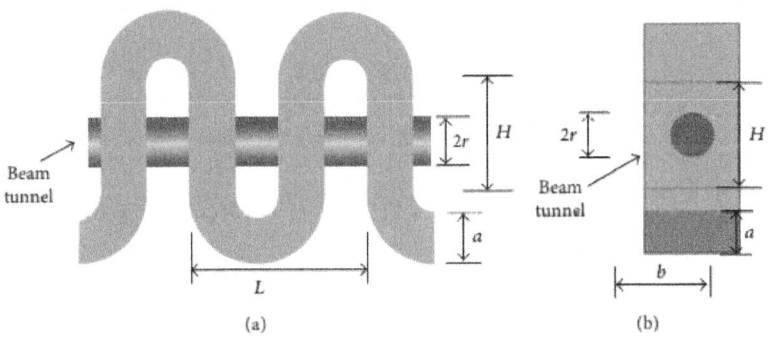

(a)   (b)

Figure 7: Illustration of (a) side view and (b) front view of the FWG SWS.

The normalized velocity, coupling impedance, and attenuation constant of the optimized FWG SWS are plotted in Figure 8. We can see that the relative discrepancy of normalized velocity is less than 1% from 55 GHz to 60 GHz in Figure 8(a), which indicates well dispersion flatness and can satisfy the synchronization condition. As shown in Figure 8(b), the coupling impedance is greater than 2.5 Ohms within the designed V-band 55 GHz to 60 GHz and a strong beam-wave interaction can be expected. It can be found that the FWG SWS has much lower loss than helix one comparing Figures 8(c) and 4(c). Because it is the surface wave traveling along the helix with strong field in helix TWT, it is light field on the boundary of the waveguide in FWG SWS.

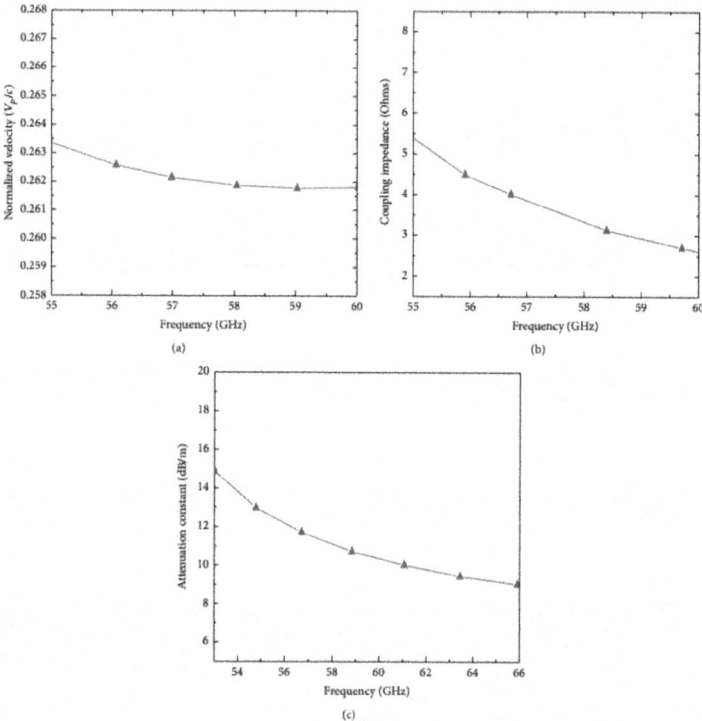

Figure 8: (a) Dispersion curve, (b) coupling impedance, and (c) attenuation constant for FWG SWS.

A large amount of optimization processes has been done to determine the profile of the FWG TWT and a high power pitch profile is proposed in Figure 9. A sever is select to cut the input circut short. Note that the pitch here refers to half-period of the FWG SWS.

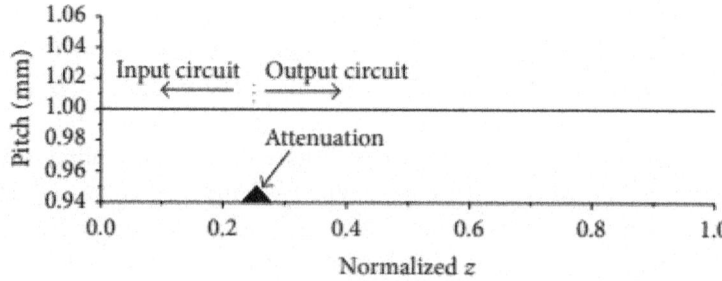

Figure 9: Pitch and attenuation profile of the FWG TWT (black and white).

The performance of the output power and gain at 57 GHz are provided in Figure 10. And the saturated output power and the corresponding input power, as well as the saturated gain, are shown in Figure 11. Obviously, the saturated output power is greater than 800 W and the saturated gain is larger than 18 dB from 55 GHz to 60 GHz.

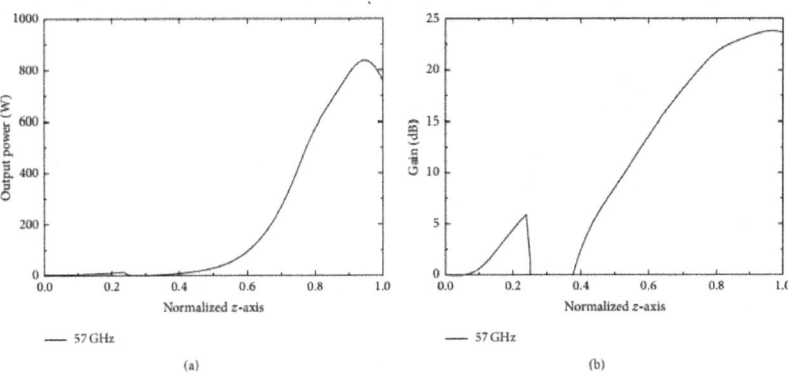

Figure 10: (a) Output power versus z and (b) gain versus z at 57 GHz of the FWG TWT (black and white).

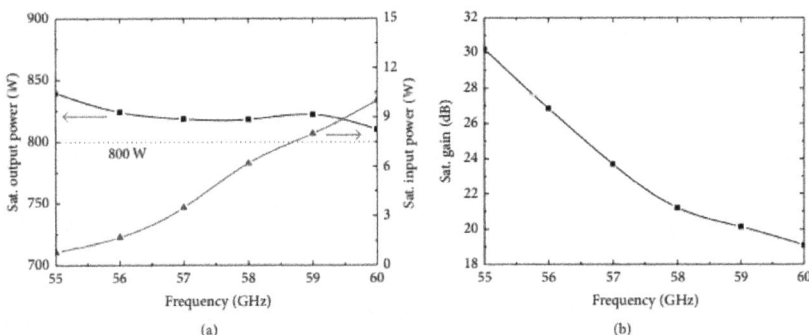

Figure 11: (a) Saturated output power and input power and (b) saturated gain of the FWG TWT ranging from 55 GHz to 60 GHz.

After design of the first stage helix TWT and FWG TWT separately, we match the output power of helix TWT to be the same of the input power FWG TWT by modifying the helix input power, keeping the output power of FWG saturated (see Figure 12). Then, the saturated cascaded gain can be obtained by saturated FWG gain plus gain of helix TWT. As shown in Figure 13, the cascaded gain above 60 dB can be expected within the V-band of 55 GHz to 60 GHz.

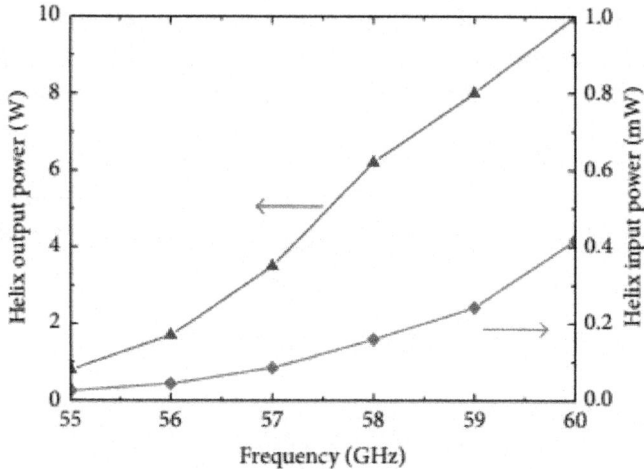

Figure 12: Helix output and input power curves to meet FW saturated input power.

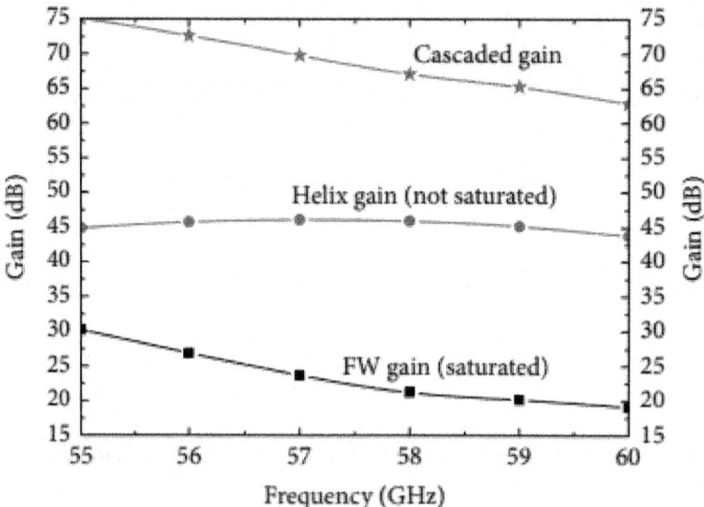

Figure 13: Cascaded gain, helix gain, and FW gain curves in band.

CONCLUSIONS

Based on the digitized nonlinear theory model developed in MTSS, a novel high power V-band cascaded helix-FWG TWT is designed, which is expected to generate an output power of 800 W and a gain of 60 dB from 55 GHz to 60 GHz. The digitized nonlinear theory model, which can process the beam-wave interaction universally, is briefly introduced. And then the principle of the TWT design is derivate, by which a proper voltage and tunnel radius can be selected. The first stage helix traveling wave tube is designed to have an output power above 34 W and gain above 35 dB. For the cascaded FWG traveling wave tube, the saturated output power is designed to be greater than 800 W and the saturated gain is larger than 18 dB in the 5 GHz bandwidth. As a result, we get the expected output power of 800 W and gain of 60 dB by cascading these two traveling wave tubes.

CONFLICT OF INTERESTS

The authors declare that there is no conflict of interests regarding the publication of this paper.

ACKNOWLEDGMENT

This work was supported by the National Natural Science Foundation of China (Grant no. 61201003).

REFERENCES

1.   G. K. Kornfeld, E. Bosch, W. Gerum, and G. Fleury, "60-GHz space TWT to address future market,"IEEE Transactions on Electron Devices, vol. 48, no. 1, pp. 68–71, 2001.

2.   C. K. Chong and W. L. Menninger, "Latest advancements in high-power millimeter-wave helix TWTs,"IEEE Transactions on Plasma Science, vol. 38, no. 6, pp. 1227–1238, 2010.

3.   Staprans, E. W. McCune, and J. A. Ruetz, "High power linear-beam tubes," Proceedings of the IEEE, vol. 61, no. 3, pp. 299–330, 1973.

4.   L. Li, J. J. Feng, B. Qu, and Y. H. Shang, "Design and experiment of a V-band helix TWT," inProceedings of the IEEE 14th International Vacuum Electronics Conference (IVEC' 13), pp. 1–2, Paris, France, 2013.

5.   L. Liu, Y. Wei, F. Shen et al., "A novel winding microstrip meander-line slow-wave structure for V-band TWT," IEEE Electron Device Letters, vol. 34, no. 10, pp. 1325–1327, 2013.

6.   F. Shen, Y. Y. Wei, H. R. Yin et al., "A novel V-shaped microstrip meander-line slow-wave structure for W-band MMPM," IEEE Transactions on Plasma Science, vol. 40, no. 2, pp. 463–469, 2012.

7.   Y. Gong, W. Wang, Y. Wei, and S. Liu, "Theoretical analysis of ridge-loaded ring-plane slow wave structure by variational methods," IEE Proceedings—Microwaves, Antennas and Propagation, vol. 145, no. 5, pp. 397–405, 1998.

8.   J. R. Legarra, J. Cusick, R. Begum, P. Kolda, and M. Cascone, "A 500-W coupled-cavity TWT for Ka-band communication," IEEE Transactions on Electron Devices, vol. 52, no. 5, pp. 665–668, 2005.

9.   S. Liu, "Folded waveguide circuit for broadband MM wave TWTs," International Journal of Infrared and Millimeter Waves, vol. 16, no. 4, pp. 809–815, 1995.

10. H. Gong, J. Xu, T. Tang et al., "A 1-kW 32-34-GHz folded waveguide traveling wave tube," IEEE Transactions on Plasma Science, vol. 42, no. 1, pp. 8–12, 2014.

11. K. Nguyen, D. K. Abe, L. Ludeking et al., "High-power broadband cascaded-TWT development," inProceedings of the IEEE 13th International Vacuum Electronics Conference (IVEC '12), pp. 121–122, Monterey, Calif, USA, April 2012.

12. K. Nguyen, D. Pershing, J. Pasour et al., "Development of high-power broadband Ka-band cascaded-TWT," in Proceedings of the 14th IEEE International Vacuum Electronics Conference (IVEC '13), Paris, France, May 2013.

13. B. Levush, D. Abe, I. A. Chernyavskiy et al., "A 1.8 kW broad band Ka-band TWT power booster," inProceedings of the 14th IEEE International Vacuum Electronics Conference (IVEC '13), pp. 1–2, IEEE, Paris, France, May 2013.

14. W. F. Peng, Y. L. Hu, Z. Cao, and Z. H. Yang, "Digitized nonlinear beam and wave interaction theory of traveling wave tube amplifiers," Progress In Electromagnetics Research M, vol. 28, pp. 73–88, 2013.

15. W. F. Peng, Z. H. Yang, Y. L. Hu et al., "Time-dependent nonlinear theory and numerical simulation of folded waveguide traveling wave tubes," Journal of the Korean Physical Society, vol. 62, no. 6, pp. 949–953, 2013.

16. B. Li, J.-Q. Li, Q. Hu et al., "Recent developments to the microwave tube simulator suite," IEEE Transactions on Electron Devices, vol. 61, no. 6, pp. 1735–1741, 2014.

17. Ansoft Corporation, Ansoft HFSS User's Reference, Ansoft Corporation, 2015,http://www.ansoft.com.cn/.

18. CST Microwave Studio, 3-D Electromagnetic Simulation Software, CST Corporation, Darmstadt, Germany, 2015.

Chapter 4

DYNAMICS OF CONTINUOUSLY PUMPED SOLID-STATE REGENERATIVE AMPLIFIERS

Mikhail Grishin[1] and Andrejus Michailovas[1]

[1]Institute of Physics & EKSPLA uab, Lithuania

INTRODUCTION

Regenerative amplifiers are extensively used for amplifying pulses generated by mode-locked oscillators (Koechner, 2006). This is a powerful technique providing several orders of magnitude gain, virtually unlimited by amplified spontaneous emission, (the well known nemesis for multi-pass amplifiers) (Forget et al., 2002). Such uniquely high gain is achieved due to multiple passes of optical pulse through the gain medium. Multiple passes are organized by placing the gain media in an optical resonator. The number of round trips is typically controlled by an electro-optic switch (Nickel et al., 2005) [occasionally with acousto-optic modulator (Norris, 1992)]. The optical switch also provides quality control of the optical cavity, suppressing lasing and reducing the time period when parasitic amplification of spontaneous emission takes place. In addition, the optical cavity provides "filtering" of spatial mode (there is no spatial imperfection accumulation during multiple passes of amplifying pulse). Consequently, the possibility exists to obtain perfect beam quality. At the same time the stable resonator does not allow expanding mode diameter too much (Magni, 1986), restricting capabilities for amplifying optical pulses to very high intensity - when it is required this job is placed to subsequent high aperture amplifiers (Siebold et al., 2008). An auxiliary technique, vitally important to amplify femtosecond pulses to high energies, is chirped-pulse amplification (the stretcher-compressor scheme) (Strickland & Mourou, 1985, Mourou & Umstadter, 1992).

In respect to the gain, the regenerative amplifiers probably have only one competitor – fiber amplifiers (Fermann et al., 2002, Liu H. et al., 2008). These amplifiers as well as all lasers based on optical fibers exhibit impressive progress over the last decade (Jeong et al., 2004). However, the extremely small mode diameter and large medium length intrinsic for optical fibers lead to significant influence of detrimental nonlinear effects. These distort the amplifying signal's optical spectrum eventually resulting in serious limitation of peak power so that 0.7 MW before compression is one of the best achievements (Röser et al., 2007). Regenerative amplifiers are commonly designed with bulk materials allowing larger mode diameter that moves away the peak power limit well above tens of megawatts (Kleinbauer et al., 2008).

Most frequently used material for amplification of femtosecond pulses (and thoroughly dominating below 100 fs) is titanium doped sapphire. Broad gain bandwidth, exceptionally good lasing properties and opto-mechanical characteristics place this medium in such a special position. High pulse energy and high average output power have been achieved by using Ti:saphire amplifiers (Walker et al., 1999, Matsushima et al., 2006). The disadvantages of this gain medium are related to possible means of pumping. First, corresponding absorption lines are located in the green spectrum range, where suitable laser diodes with reasonable power are not available. Second, the short gain relaxation time (3.2 μs) requires pulsed pumping with high pulse energy (usually with Q-switched frequency-doubled neodymium lasers) in order to store substantial population inversion and consequently to obtain high output energy. We should remark here that the issues described in present chapter are not valid for Ti:saphire regenerative amplifiers since we use approximations which suitable only for long lifetime media. From the other hand due to very short lifetime the Ti:saphire amplifiers do not exhibit extraordinary dynamic properties on which we are mainly focusing in the chapter.

Another family of popular laser materials is ytterbium doped crystals and glasses. Their wide spectrum supports amplification of ultrashort pulses (however not as short as Ti:sapphire supports). Moreover ytterbium materials allow direct laser diode pumping with intrinsically small quantum defect (typically 10% or even less). The latter enhances overall power efficiency and reduce heat generation in active elements, that in turn alleviates thermal effects which inhibit average power increase. Long upper-level lifetime of ytterbium ions virtually in all the crystals and glasses allows good capacity of stored energy under convenient continuous laser diode pumping. What somewhat challenges operation with ytterbium materials is relatively low gain (typically less than 10% per pass) which originated from peculiar to this materials small stimulated emission cross section. From the other hand regenerative amplification is indeed a way of efficient energy extraction at low gain; just special attention should be given to reduce inrtracavity losses (Biswal et. al., 1998). An alternate way to improve stimulated emission features is cryogenically cooled active elements (Kawanaka et al., 2003) although this method is bulky and, as a rule, it narrows gain bandwidth. Then special attention should be placed to host crystal selection not to limit development towards shorter pulses (Pugžlys et al., 2009).

Ytterbium doped media are able to withstand very intense optical pumping without detriment to exited-state population which in other materials can be limited by quenching effects (e.g. exited state absorption or up-conversion). This favorable property permits use of active elements in thin disc geometry. In particular, Yb:YAG thin disk lasers are scalable to very high average power and to high pulse energies (Speiser & Giesen, 2008). Extremely short optical pass within the thin disc reduces nonlinear effects (in essence the Kerr effect) allowing high peak power pulses even without using stretcher-compressor technique (Kleinbauer et al., 2008).

The amplifiers based on neodymium gain media have their specific advantages. High stimulated emission cross section simplifies system design and reduces requirements for optical components. Good energy storage capabilities allow high output energy and power efficient operation when pumping with laser diodes. Neodymium laser materials are well suited for picosecond pulse durations and are competitive for moderate average power. Systems based on $Nd:YVO_4$ and $Nd:GdVO_4$ crystals routinely produce more than 10 W of output power (Kleinbauer et al., 2005, Clubley et al., 2008).

The regenerative amplifiers of solid state lasers designed for scientific applications usually operate at low or moderate repetition rates (not exceeding several kHz). Presently, there is rising demand for high repetition rate ultrashort-pulse solid state lasers for material micro-processing (Meijera et al., 2002). On the other hand, a new generation of fast electro-optic switches became available such as Pockels cells based on -barium borate along with improved high-voltage electronics (Nickel et al., 2005, Siebold et al., 2004). As a result, picosecond and femtosecond lasers with repetition rate of the order of 100's kHz have come onto the market (Raciukaitis et al., 2006). Regenerative amplifiers are an important part of most ultrafast industrial solid state laser systems. Both high system efficiency and stable output parameters over a wide range of pulse repetition rates are essential for this actively developing field. For creation of power-efficient systems, neodymium and ytterbium laser gain media which may be directly pumped by laser diodes is advantageous. Long lifetime of the upper laser level typical of both these ions supports accumulation of substantial population inversion under continuous laser diode pumping. However, this long inversion lifetime may also cause stability problems at high repetition rates. Continuously pumped regenerative amplifiers demonstrate peculiar pulse amplification dynamics when the pulse repetition period becomes comparable or shorter than the gain relaxation time (Müller et al., 2003). Period doubling bifurcations develop generating periodically alternating energy pulses or even sequences of pulses having chaotic energy distribution.

Complex dynamic behavior is well known phenomenon in laser physics (Haken, 1975). Generally, nonlinear differential equations describing laser dynamics tend to have unstable solutions containing multi-stabilities and bifurcations when a number of independent variables representing system states are equal or more than three (Lorenz, 1963). By no means complete list of laser systems exhibiting complicated dynamics includes Q-switched gas lasers (Arecchi et al., 1982) passively Q-switched solid state lasers (Tang et al., 2003), optically injected solid state lasers (Valling et al., 2005). The specifics of high repetition rate regenerative amplifiers is such that their operation can be described with two differential equations, but periodic disturbance caused by release of the amplified pulse complicates the system behavior. Unlike many lasers which have been created specially to study dynamic phenomena and chaotic behavior, dynamics of regenerative amplifiers needs to be understood from a more utilitarian position in order to comprehensively optimize real systems. To date, only a few articles have been dedicated to this phenomenon despite its critical influence on the performance of regenerative amplifiers. Period doubling regime passing to chaotic operation has been observed for a system based on ytterbium doped glass and the role of bifurcations has been investigated both theoretically and experimentally (Dörring et al., 2004). However, one of the important parameters, the seed pulse energy, was left beyond the scope, and so applicability of the obtained results was restricted. The experiments were confined to studying cavity dumping of the Q-switched laser, an approximately equivalent system to the regenerative amplifier seeded by extremely low pulse energy. Our recent theoretical work has presented a generalized picture of stability features of a continuously pumped high repetition rate regenerative amplifier based on laser media with long relaxation time. The regions exhibiting different system behavior have been mapped in the space of non-dimensional control parameters: repetition rate and round trip number (Grishin et al., 2007). Additionally this analysis revealed the importance of the seed pulse energy and demonstrated that increase in the seed energy helps in eliminating the instabilities. Comprehensive utilization of these theoretical results has promoted top performance obtained from multi-kilohertz Yb:YAG disk amplifier (Metzger et al., 2009). Experimental study, performed with Nd:YVO$_4$ system, has thoroughly confirmed theoretical conclusions and on its basis the concept of regenerative amplifiers optimization has been formulated (Grishin et al., 2008, Grishin et al., 2009).

In the present chapter we summarize information from our already published papers and expand this description on the basis of our recent results. In the first theoretical part of the chapter we present a concept for modeling of continuously pumped solid state regenerative amplifiers. Simplified rate equations in normalized form allow reducing the system analysis to a task which has been thoroughly accomplished in theory of discrete-time dynamical systems (Alligood et al., 1996). Then we focus more attention to the influence of the main governing parameters on dynamics of regenerative amplification and limitations of the practical system performances occurred due to instabilities. We show why seed pulse energy plays such an important role at repetitive regime in contrast to low repetition rates. Influence of parasitic intracavity losses is shown to be a factor which not only decreases energy extraction efficiency (as in all laser systems) but also enhances instability range. We analyze numerically obtained stability diagrams which allow determination of optimal operation regime when maximum output energy can be extracted from the amplifier as stable pulse train. Important laser parameters which influence the system performance indirectly will be considered too. Heating of intracavity optical components depends not only on the range of output average power and the optical components quality but also on the amplification regime. The Kerr nonlinearity which limits system performance for short optical pulses will be evaluated taking into account multiple passes.

The verification of the created model is presented in the experimental part of the chapter. We investigate operation peculiarities of the system consisting of mode-locked master oscillator, the preamplifier and the regenerative amplifier based on Nd:YVO$_4$ crystals and analyze criteria of stable operation limitation. The experimental dependences of the output energy versus round trip number and repetition rate well agree with theoretical data. We demonstrate that increase of the seed pulse energy up to the value predicted by numerical model ensures stable operation within full range of repetition rates. The chapter is concluded by summarizing of presented theoretical and experimental results and suggesting of further investigations.

THEORY OF REGENERATIVE AMPLIFICATION IN REPETITIVE REGIME

Principle Of Operation

Regenerative amplifier can be regarded as a system in which an optical resonator provides multiple passes, a gain medium is responsible for amplification, and an electro-optic switch serves as a valve in-turn admitting weak input pulse and releasing amplified pulse. At that, spatial properties of the amplified radiation are primarily determined by the optical resonator; the output energy mostly depends on the population inversion stored in the gain medium whereas the amplified pulse duration is imposed by the input seed pulse. A schematic diagram of a conventional solid-state diode-pumped regenerative amplifier is depicted in Fig. 1. This optical layout by no means differs from that for a common Q-switched laser and consequently there are certain similarities in operation (Murray & Lowdermilk, 1980). The resonator quality (Q-factor) is controlled by an electro-optic switch. The electro-optic switch usually consists of a Pockels cell, a quarter-wave plate and a polarizer. An operational cycle of the regenerative amplifier consists of two successive stages: low-Q and high-Q. When voltage is not applied to the Pockels cell the wave plate along with the polarizer provides high intracavity losses (low-Q state of the resonator). Laser action is suppressed by high losses, and the gain medium, being under continuous pumping, accumulates population inversion. The amplification takes place during high-Q stage. It starts when quarter-wave voltage is applied to the Pockels cell and the seed pulse is injected into the resonator. The intracavity losses become minimal and are kept low for some pre-set time while the optical pulse circulating in the resonator is amplifying, simultaneously consuming a certain part of the stored energy. The intracavity pulse energy grows until the gain becomes equal to the resonator losses and then the pulse energy decays. As soon as the intracavity energy reaches a desired level the Pockels cell voltage is switched off. This dumps the amplified pulse out of the cavity as the output pulse. The system returns into the initial state. Then during the next low-Q stage the depleted part of the inversion population is restored by uninterrupted pumping and the cycles iterate.

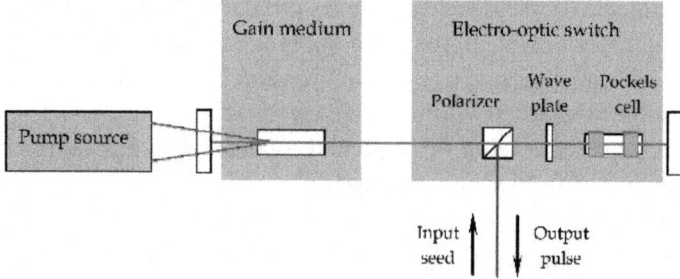

Figure 1: Optical layout of typical solid-state diode pumped regenerative amplifier.

At higher repetition rates, when the pump stage is comparable or shorter than the gain relaxation time, the operation cycles become interdependent. The equilibrium between population inversion depletion caused by amplification and inversion restoration caused by pumping may become unstable. This often leads to breakdown of the single-energy regime and to generation of periodically alternating high/low energy pulses, or more complicated instability patterns. We will consider this and related phenomena theoretically in the first part of the present chapter.

Rate Equations And Basic Terms

The process of regenerative amplification is essentially determined by the interaction of the intracavity radiation with the laser medium excited by pumping. The rules of this interaction can be established by using a simple phenomenological notion of a dynamic balance between basic processes: pumping and relaxation, amplification and extinction. Thus the system evolution can be described by coupled differential equations for the population inversion density N and photon number. We use the space independent rate equations which have been formulated for idealized four-level laser medium with a homogenously broadened line (Svelto, 1998):

$$\frac{dN}{dt} = R_p - \frac{\sigma c}{V}\phi N - \frac{N}{T_1}$$

(1)

$$\frac{d\phi}{dt} = \left(\frac{\sigma c L_a A_a}{V} N - \frac{1}{T_c}\right)\phi \; .$$

(2)

The density of population inversion grows due to pumping (proportionally to the pumping rate Rp) and decays with the constant rate inversely proportional to the upper laser level lifetime T_1. This balance is significantly influenced by stimulated emission, the contribution of which depends on the gain medium characteristics and particularly on the stimulated emission cross section . The second equation establishes a rule of stimulated photons multiplication. The photon number increase proportionally to the population inversion and at the same time it decreases because of intracavity losses. The losses here are defined in terms of the photon lifetime $T c$ which determines the rate of decay for the light field in the optical resonator. Geometry of the amplifier is accounted for in terms of the mode volume within the optical resonator (V), the active medium length (La) and the beam area in the active medium (Aa). The velocity of light c is a constant coupling temporal and spatial terms of the equations.

Since we use the rate equations approximation for modeling, the range of proposed model applicability is limited by the range of the particular rate equation's validity. Concerning general validity of the rate equation approach, we refer to (Svelto, 1998) for details. Here we just mention that this is a conventional method of laser dynamics study. It gives sufficiently accurate results for most practical purposes and in particular for analysis of multi-pass and regenerative amplification (Murray & Lowdermilk, 1980, Lowdermilk & Murray, 1980)

We shall re-arrange the equations to a form containing macroscopic non-dimensional terms. This is a common way to reduce the number of control parameters in order to provide simplified picture of system behavior without losing information. A primary normalization coefficient is the term proportional to the pump level, $G_0 = R_p T_1 \sigma L_a$ The physical meaning of G_0 is steady state gain coefficient per pass. Such gain is achieved at the equilibrium of population inversion density $(dN/dt=0)$ obtained under constant pumping $(dR_p/dt=0)$, constant rate of spontaneous decay (T_1) and in absence of lasing $(\phi=0)$ Thus we can proceed to equations for the normalized gain $g = N\sigma L_a/G_0$ and normalized energy $\varepsilon = \phi\sigma/(A_a G_0)$:

$$\frac{dg(t^\sim)}{dt^\sim} = -\varepsilon(t^\sim)g(t^\sim) + \frac{1-g(t^\sim)}{\tau_1} \tag{3}$$

$$\frac{d\varepsilon(t^\sim)}{dt^\sim} = \varepsilon(t^\sim)\left[g(t^\sim) - g_t\right]. \tag{4}$$

Additionally, we have modified the time scale by introducing normalized time $t^\sim = t G_0/T_0$, a product of the current number of cavity round trips and the steady state gain coefficient. Here the denominator T_0 represents the round trip time. As we will see later, the term $t^\sim$ assigns a natural time scale to the high-Q phase of operation. The governing parameter, which defines the amplification period (effective round trip number), will be introduced on this basis. The normalized relaxation time of the gain, $\tau_1 = T_1 G_0/T_0$ is essentially determined by the upper laser level lifetime T_1. A term accounting for parasitic losses of the optical resonator is represented by normalized threshold gain, $g_t = T_0/(T_c G_0)$. This parameter can be also expressed in terms of loss coefficient l as l as $g_t = -\ln(1-l)/G_0$, by using common threshold criterion of gain and losses equality.

Note that equations formulated with those new variables do not contain pumping characteristics in explicit form. Parameter G_0, proportional to the pumping rate, is hidden in the composition of basic variables. The pump effect (as well as other control parameters effects) is easy to restore when applying modeling results by performing reciprocal transformation from normalized to real parameters. Moreover, further in the theoretical part of this chapter we often omit the words "effective" and "normalized" just for shortening. Although Eqs. (3) and (4) have been obtained in the approximation of a four-level system, the identical equations can be formulated for the more general case of quasi-three-level systems [the initial rate equations and conditions of their applicability can be found in (Svelto, 1998)]. In the latter case the explicit expressions of normalized and effective terms look slightly more complicated but their physical meaning remains unchanged.

Consideration Of A Single Operation Cycle

Below in this section we define relations between basic system parameters deriving the rate equations for separate stages of regenerative amplifier operation. First we consider the low-Q phase. Since during this phase of operation amplification is suppressed while pumping takes place, it is also called pump stage. Inasmuch as there is no lasing during this stage (i.e. $\varepsilon = 0$), a set of Eqs. (3) and (4) transforms into a single equation:

$$\frac{dg(t^\sim)}{dt^\sim} = \frac{1 - g(t^\sim)}{\tau_1}.$$

(5)

Initial conditions specify the gain at the beginning of the pump phase g_{pi}. Taking into account that $t^\sim / \tau_1 = t / T_1$, we can find the relation between the initial gain g_{pi} and the final gain g_{pi} for a certain pump phase duration T:

$$g_{pf} = 1 - (1 - g_{pi})\exp\left(-\frac{T}{T_1}\right)$$

(6)

In the next section we will often use diminished form of the equation (6):

$g_{pf} = \hat{g}_p(g_{pi}, T / T_1)$, where the function $\hat{g}_p$ just establishes a rule of the gain transformation $g_{pi} \to g_{pf}$ for the pump stage.

Then let us proceed to the high-Q phase. The equations for this operation stage can be simplified as well. We remind here that we explore laser media having long relaxation time, the case functionally important for diode-pumped systems. Since the buildup time of the optical pulse is usually short in comparison with the pump phase duration and the upper laser level lifetime, the population inversion change due to pumping and relaxation processes is much smaller than the inversion depletion caused by amplification. Hence, the terms containing spontaneous decay and optical pumping can be neglected in Eq. (3). Also, as we have assumed negligible pump contribution during high-Q phase, this stage of operation can be called in a more common and more informative manner as the amplification stage.

The next assumption presumes low intra-cavity losses $(g_r \to 0)$. This approximation substantially simplifies the basics of the presented method. The simplification appears not only due to existence of the analytical solution for the rate equations but also, more importantly, because of the reduced number of parameters governing the system. An influence of the parasitic losses on amplification dynamics will be accounted in section 2.7 after the essence of the theoretical approach has been presented. For the present, we come to the situation at which Eqs. (3) and (4), when describing amplification stage, reduce to the following:

$$\frac{dg(t^\sim)}{dt^\sim} = -\varepsilon(t^\sim)g(t^\sim)$$

(7)

$$\frac{d\varepsilon(t^\sim)}{dt^\sim} = \varepsilon(t^\sim)g(t^\sim).$$

(8)

Initial conditions specify the system state at the beginning of the amplification phase: the initial gain g and the energy of the input pulse from which the amplification starts – the seed energy, ε_s. Also it is natural to constrain our consideration to the case of low seed energy with respect to the stored energy $(\varepsilon_s \ll g_{ai}$ in terms of normalized parameters). As a consequence of these assumptions, the solutions of coupled Eqs. (7) and (8) can be found in analytic form. Such solutions obviously describe temporal evolution of the gain and intracavity energy. One can also find other physical sense, more convenient for further consideration. In case of fixed amplification phase duration, temporal evolution is terminated at the moment $t^\sim = \tau$. Then the solutions also express how the system parameters on the amplification phase completion (final gain g_{af} and output pulse energy ε_f) depend on the initial conditions and governing parameters:

$$g(t^\sim) = \frac{g_{ai}}{1 + \frac{\varepsilon_s}{g_{ai}}\exp(g_{ai}t^\sim)} \Rightarrow g_{af} = \frac{g_{ai}}{1 + \frac{\varepsilon_s}{g_{ai}}\exp(g_{ai}\tau)}$$

(9)

$$\varepsilon(t^\sim) = \frac{g_{ai}}{1 + \frac{g_{ai}}{\varepsilon_s}\exp(-g_{ai}t^\sim)} \Rightarrow \varepsilon_f = \frac{g_{ai}}{1 + \frac{g_{ai}}{\varepsilon_s}\exp(-g_{ai}\tau)}.$$

(10)

Here the terms and ε_s are actual parameters controlling the amplification, whereas the initial gain is a variable coupling the equations. The normalized amplification stage duration is a product of cavity round trip number and steady state gain G_0. Further we will call this term in a more comprehensible manner an "effective round trip number". Essentially, the term also represents total multi-pass small signal gain factor for the amplifier. The Equations (9) and (10) can be presented in diminished form:

$$g_{af} = \hat{g}_a(g_{ai}, \varepsilon_s, \tau) \text{ and } \varepsilon_f = \hat{\varepsilon}(g_{ai}, \varepsilon_s, \tau)$$

simplifying further operation with these formulas. Now, as the basic terms have been introduced, we can summarize a rule for subscript notations. The pump and amplification operation phases are designated with p and a subscripts respectively; the initial and final states are notated with i and f subscripts correspondingly.

COUPLING OF SUCCESSIVE CYCLES. DISCRETE-TIME DYNAMICAL SYSTEM APPROACH

Evaluation of the output energy is a trivial task for low repetition rates, i.e. when the pump phase duration significantly exceeds the inversion relaxation time, $T >> T_1$. The gain in this case reaches saturation before the amplification phase starts, that is the initial gain g_{ai} tends to unity. Consequently, the output energy versus round trip number is strictly defined by Eq. (10) alone. In general case, and particularly at high pulse repetition rate the initial conditions for the current operation stage depend on the previous system state. Hence, the initial gain for each cycle depends not only on operation parameters but also on the system pre-history. In order to determine the gain at the beginning of amplification we shall relate final and initial states of successive operation cycles.

Now we proceed from evaluation of the gain and pulse energy within single operation stages to a description of those stages as a whole by analyzing the temporal evolution of the boundary values. We assign initial and final gains and pulse energy as variables defining the system state. The term $g_{ai}(1)$ is introduced as the initial gain of the amplification phase for the first cycle of operation. This stage finishes with the final gain denoted as $g_{af}(1)$. The subsequent pump phase of the current cycle obviously begins from the same gain value, $g_{pi}(1)=g_{af}(1)$. Similarly, the gain evolution continuity should be taken into account for coupling of all operation cycles. There is a boundary relation $g_{af}(k)=g_{pi}(k)$ within the cycle number k and for subsequent cycles: $g_{pf}(k)=g_{ai}(k+1)$. The legend of the gain evolution in a discrete time scale can be presented as follows:

$$...g_{ai}(k) \rightarrow g_{af}(k) = g_{pi}(k) \rightarrow g_{pf}(k) = g_{ai}(k+1) \rightarrow g_{af}(k+1) = g_{pi}(k+1) \rightarrow g_{pf}(k+1)...$$

$$(11)$$

The corresponding time boundary points of neighboring operation stages can be described as $t_{ai}(k)=(k-1)(\tau T_0/G_0+T)$ and $t_{af}(k)=k\tau T_0/G_0+(k-1)T$. Note that in an assumption of short amplification phase, the duration of complete cycle (dumping period) is approximately equal to the pump phase duration T. Hence the term $(T/T_1)^{-1}$ represents normalized pulse repetition rate of the regenerative amplifier (also called dumping frequency).

Analogous transition to the discrete time scale can be applied to energy evolution. However, unlike continuity of the gain evolution, the build-up of intra-cavity energy $\varepsilon_s \to \varepsilon_f(k)$ interrupts at the end of the amplification phase of each cycle at the moment of pulse dumping and then it begins again from the constant level which corresponds to the seed pulse energy ε_s. Hence the term $\varepsilon_f(k)$, determining the output energy, does not depend of its own pre-history. It is dependent on the gain and can be found from Eq. (10) for any operation cycle provided that the gain is known. Therefore, the gain becomes the only independent variable that needs to be analyzed.

We can consider the expressions obtained earlier [$\hat{g}_p$ for the pump phase and $\hat{g}_a$ for the amplification phase explicitly represented by Eq. (6) and Eq. (9) as the rules of the system state updating. These solutions of the rate equations serve as transformation rules that take the current state as input and update it by producing a new state. This new output state becomes the input for the next stage of operation. Then it is possible to combine amplification and pump phases within certain operation cycle and to form a joint gain transformation rule. We introduce $\hat{g}_\Sigma$ as the composition of functions $\hat{g}_a$ and $\hat{g}_p$ exhibiting the gain transformation rule for the complete cycle,

$$\hat{g}_\Sigma = \hat{g}_p(\hat{g}_a).$$

Then using expressions of the inner functions we can present an explicit form of the function $\hat{g}_\Sigma$

$$\hat{g}_\Sigma\left(g_{ai}\right) = 1 - \left[1 - \frac{g_{ai}}{1 + \dfrac{\varepsilon_s}{g_{ai}}\exp\left(g_{ai}\tau\right)}\right]\exp\left(-\frac{T}{T_1}\right)$$

(12)

Thus, we have reduced the regenerative amplification to the evolution of single variable (system state, g_{ai}) in a discrete time scale; and also we have found a rule of this variable updating. The basic properties of this updating function fit to the mathematical definition of so called maps [functions whose domain (input) space and range (output) space are the same]. Then the regenerative amplification can be described by using the theory of one-dimensional discrete-time dynamical systems (one-dimensional maps) (Alligood et al., 1996). The sequence of the system states $g_{ai}(1), g_{ai}(2), \dots g_{ai}(k) \cdots$ ... is called an orbit in terms of this theory. The orbits can be calculated by using a recurrent formula determining the subsequent state of the system in terms of the present state:
$g_{ai}(k+1) = \hat{g}_\Sigma[g_{ai}(k)]$.

It is obvious that in a regular single-energy regime the gain depletion during the amplification phase should be compensated by restoring the population inversion during the pump phase. In terms of states evolution, the initial gain of the amplification stage eventually should iterate, i.e. there is a certain gain value (designated as g_1) such that the subsequent gains stabilize upon reaching that value $g_{ai}(k+1)=g_{ai}(k)=g_1$. Consequently, the system eigenstate satisfying the condition $g_1 = \hat{g}_\Sigma(g_1)$ should exist. The solution of this equation is known as a fixed point in the discrete-time dynamical system theory. Being exactly equal to the fixed point, the system state reproduces itself after each cycle that leads to operating in a regular manner. However, requirement of technical feasibility of such a regime establishes a more strict condition to be fulfilled. The system should return to the fixed point after some perturbation has occurred, in other terms, more common for theory of dynamical systems, the fixed point should be attracting. Thus, study of regenerative amplification is reduced to the analysis of conditions of the fixed point existence and its stability characterization.

Since the equation determining the fixed points is transcendental we start analysis of the system state evolution with the graphical illustration. For more intuitive presentation, the fixed point existence condition [$g_1 = \hat{g}_\Sigma(g_1)$ with the explicit form of $\hat{g}_\Sigma$ given by Eq. (12) is rearranged into the following form:

$$1-(1-g_1)\exp\left(\frac{T}{T_1}\right)=\frac{g_1}{1+\dfrac{\varepsilon_s}{g_1}\exp(g_1\tau)}.$$

$$(13)$$

The right-hand part of this expression represents the gain transformation function during amplification, $\hat{g}_a$ [see Eq. (9)]. The left-hand part may be regarded as an inverse function of the gain recovery during the pump phase transformed Eq. (6) gives $g_{pi} = \hat{g}_p^{-1}(g_{pf})$]. Since gain continuity implies equality of the boundary states $(g_{ai}=g_{pf}$ and $g_{af}=g_{pi})$ we can combine both curves g_{af} versus g_{ai} and g_{pi} versus g_{pf} on a common diagram (Fig. 2). A space of system states, defined by this means, can give an intuitively simple but strict and fruitful picture of the system state evolution. The intersection of those curves, having clear physical meaning, gives solution of Eq. (13), i.e. it determines the fixed point of the system. It is important that the intersection of these curves always exists and it is always single for any set of control parameters. Really, the amplification stage curve is single-peaked, it begins from zero and always lies under the state space diagonal $(g_{af}=g_{ai})$ The letter is natural because during the amplification stage (provided, as we assumed, negligible pumping contribution) the population inversion is depleting by transforming to the intracavity pulse energy and respectively the gain can only decrease, $g_{af} < g_{ai}$. The pump stage curve is a straight line whose slope depends on the repetition rate. This curve begins in the right upper corner of the state space [(1, 1) point] and also always locates under the state space diagonal. Note, the state space, due to proper normalizing, has dimensions of (0-1). Thus, these curves cannot help intersecting and they intersect only once. Moreover, since the basic properties of the curves are universal, a fixed point existence and uniqueness is not only the result of mathematical speculations obtained under certain approximations but also the consequence of inherent physical properties of regenerative amplification.

One of two necessary requirements for existence of a stable single-energy regime, namely the fixed point uniqueness, is fulfilled and then the main concern is the fixed point stability study. Figures 2(a) and 2(b) represent diagrams of system state evolution for two typical cases. Figure 2(a)) presents the orbit converging into an attracting fixed point. Such a convergence means that the regenerative amplifier eventually (after sufficient number of reiterations, when initial value of the orbit is "forgotten") starts producing regular pulsing. It is intuitively seen that the behavior of the resulting orbit (convergent or non-convergent) depends on the slope of the "amplification" curve in the fixed point with respect to the slope of the "pump" curve. Strictly speaking, the fixed point becomes attracting if the derivative of $\hat{g}_\Sigma$ function in this point satisfies the requirement $|\hat{g}'_\Sigma(g_1)| = 1$ (Alligood et al., 1996). The condition $|\hat{g}'_\Sigma(g_1)| > 1$ represents the transition point between stable and unstable operation. In case of $|\hat{g}'_\Sigma(g_1)| > 1$ the fixed point is repelling, and consequently stable operation becomes unfeasible.

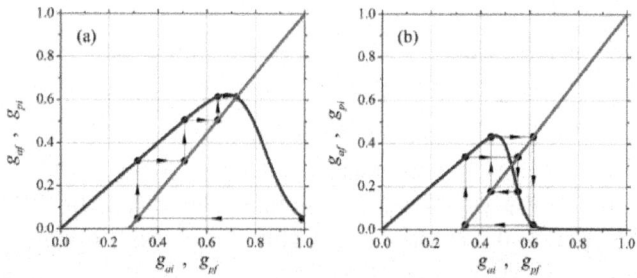

Figure 2: Graphical presentation of the orbits in state space. The fixed point is the intersection of the "amplification" and "pump" (blue and red) curves. Transition to the stable (attracting) fixed point (a) at $\varepsilon_s = 3 \times 10^{-7}$; $\tau = 18.0$; $T_1/T = 3.0$. Period-4T orbit (b) at $\varepsilon_s = 10^{-10}$;

It becomes apparent that in the latter case the system is unable to reproduce its own state after one cycle of operation. However, such a iteration may occur after two or several cycles. The corresponding set of system states is called periodic orbit. An example of periodic orbit is depicted inFig. 2(b). The condition for existence of the orbit with double period, 2T can be written by introducing an appropriate composite function. Define $\hat{g}_{\Sigma}^{2} = \hat{g}_{\Sigma}(\hat{g}_{\Sigma})$ to be the result of applying the map-function $\hat{g}_{\Sigma}$ to the system state two times. The system state g_{2}, such that $g_{2} = \hat{g}_{\Sigma}^{2}(g_{2})$, is the fixed point analogue but suited for two successive operation cycles. Generally the orbit with the period of mT exists if there is a system eigenstate gm satisfying the equation:

$$g_{m} = \hat{g}_{\Sigma}^{m}(g_{m}).$$

Here the term m is an integer number exhibiting a factor of output pulse repeatability for the corresponding multi-energy regime. If such a regime is realized, the system produces quasi-periodic sequence of the output pulses. The pulses of identical magnitude in this sequence appear each time in a multiplied period equaled to mT. In much the same way as the existence of a fixed point does not ensure stable operation, the existence of a periodic orbit does not in itself mean that the corresponding regime is realizable. Additional analysis of the orbit stability is required, that, similar to the fixed point case, reduces to evaluation of the corresponding map-function derivative. The orbit $g_{ai}(k)$ of period-m is stable provided that

$$|(\hat{g}_{\Sigma}^{m})'(g_{m})| < 1$$

Computation of the derivative for this composite function is feasible since its value eventually (at $k \rightarrow \infty$) tends to the product of its inner function derivatives at points along the orbit: $(\hat{g}_{\Sigma}^{m})'(g_{m}) = \hat{g}_{\Sigma}'[g_{ai}(1)]\hat{g}_{\Sigma}'[g_{ai}(2)]\cdots\hat{g}_{\Sigma}'[g_{ai}(k)].$

If the absolute value of the product of the derivatives is larger than one, then periodicity of the orbit becomes unfeasible meaning that the system exhibits chaotic behavior. This is the Lyapunov number criterion of deterministic chaos (Alligood et al., 1996).

We can remind here that the map-function $\hat{g}_{\Sigma}$ is itself a function of system parameters $(\tau, \varepsilon_{s}, T/T_{1})$. As one of the governing parameters is varied the corresponding fixed point passes through different states of stability. A pass through the position $|(\hat{g}_{\Sigma}^{m})'[g_{m}(k)]| = 1$ causes qualitative change of the system operation. Such transitions (e.g. transition from stable to unstable regime at m=1) is usually called a bifurcation. A set of control parameters (operation point) at which the bifurcation occurs is referred to as the bifurcation point.

Correspondingly the diagram of the output parameter versus one of the control parameters for the system exhibiting bifurcations is often called a bifurcation diagram. Among many possible types of bifurcations, known for dynamical systems, we have met here the bifurcation of period doubling. This relatively simple type of dynamic behavior is one of the consequences of the fixed point uniqueness. This attribute gives also primary unambiguity of the system behavior. The dynamics of regenerative amplification and output characteristics of the system are determined by the set of control parameters alone in contrast to e.g. bi-stability effects where initial value of the orbit, $g_{ai}(1)$ may also influence the operation. One can imagine the latter as qualitative change of system behavior caused by a way of switching it on (e.g. in practice either one has the pumping source enabled first and then seeding or other way around). The unambiguous relation between control parameters and operation regimes is quite an important property of regenerative amplifiers and our further analysis always implies this property without necessarily mentioning it.

DIAGRAMS OF DYNAMIC REGIMES IN THE PARAMETER SPACE

Simplifying assumptions and non-dimensional effective parameters, introduced for the basic rate equation model, reduce the number of independent control parameters of the system to the set of three. These are the normalized repetition rate, T_1/T, amplification phase duration expressed in terms of the effective round trip number and normalized seed energy ε_{s}. Analysis of stability for orbits of the initial gain of amplification phase [$g_{ai}(k)$] at each given control parameter provides thorough picture of regenerative amplifier behavior. The orbits were calculated by iterating of Eq. (12) in the range of control parameters wide enough to comprehend all the relevant dynamics features: $0.2 < (T_1/T) < 20;\ 10 < \tau < 110;\ 10^{-11} < \varepsilon_s < 10^{-3}$. We used as much as 3000 iterations, the sufficient number to be confident that the results are independent of the system state at the beginning of iterating. The orbits were analyzed in two stages. At first, the minimal number of cycles between repeating system states was revealed for each orbit in parameter space. It was performed by direct comparison of the system state sequences with themselves but shifted by a certain cycle number, $g_{ai}(k)$ versus $g_{ai}(k+m)$. In that way the periodic orbits up to m=32, including regular ones (m=1), were identified. Then the Lyapunov number criterion was applied to the residual unidentified orbits. They were separated into two fundamentally different bunches: chaotic and eventually periodic.

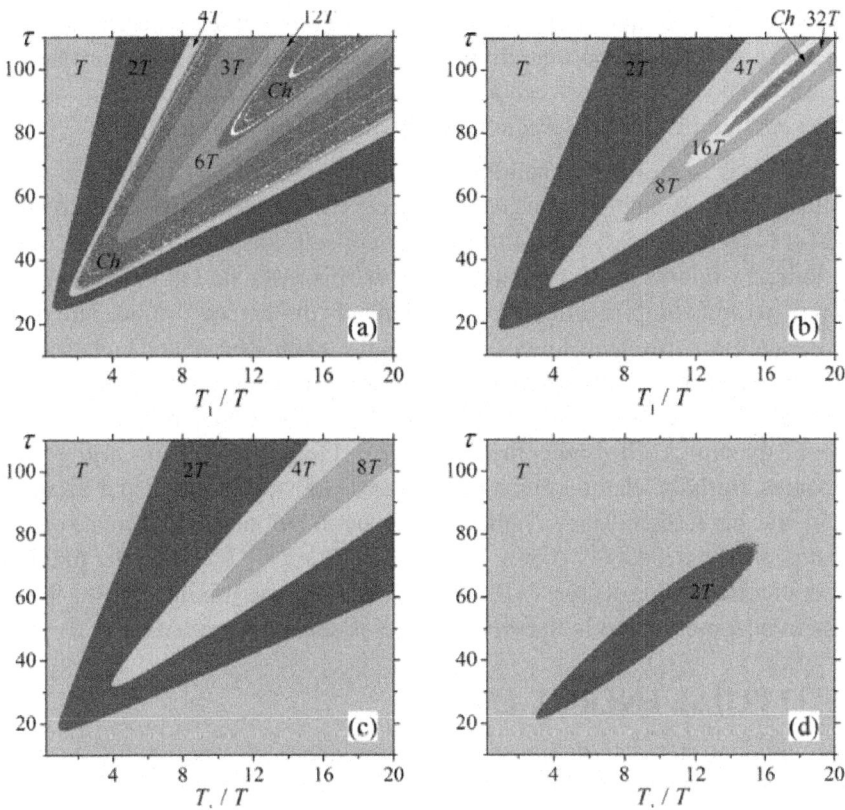

Figure 3: Diagrams of amplification dynamics in parameter space for different seed energies: $\varepsilon_s=10^{-10}$ (a); $\varepsilon_s=2.5\times10^{-7}$ (b); $\varepsilon_s=3\times10^{-7}$ (c); $\varepsilon_s=1.3\times10^{-4}$ (d).

Thus, the following dynamic regimes were distinguished in three-dimensional space of control parameters in accordance with the orbits properties: (1) The orbits evolving into stable fixed points (m=1) corresponding to the regular system behavior (single pulse energy output, i.e. 1T-regime). (2) Periodic orbits corresponding to multi-energy regimes with repeatability coefficients in the range of 2≤m≤32. (3) Eventually periodic orbits having larger repeatability factor (m>32), for which the m-number itself is not identified. (4) Regime of deterministic chaos in accordance with the Lyapunov number criterion.

The regions of different dynamics are mapped in space of the repetition rate round trip number (Fig. 3). The major part of the parameter space is occupied by the regions corresponding to the following regimes: single-energy (1T); quasi-periodic with fundamental period of two (2T, 4T, 8T, 16T, and 32T); quasi-periodic with fundamental period of three (3T, 6T, and 12T); and chaotic behavior. These domains are marked with different colors, whereas the rest of the space containing the remaining zones of eventually periodic orbits is left white. The boundaries between adjacent colors (i.e. between different regimes) represent manifolds of bifurcation points in parameter space.

As it is seen, the dynamics turned out multifarious. Chaotic regime ordinarily comes out from the chain of successive period doubling bifurcations: T-2T-4T-8T-16T-32T... The chaotic zone itself has fine structure. Quasi-periodic "windows" with various periods are disseminated in it. The dynamics of regenerative amplification strongly depends on the seed value. The pattern is complex for low seed level ($\varepsilon_{s\cdot} < 10^9$), the parameter space contains more than one clearly distinguishable chaotic regions [Fig. 3(a)]. Quasi-periodic regimes with fundamental period of three are observed between zones of chaotic dynamics. The higher the seed energy, the simpler the instability pattern becomes. Initially, chaotic domain shrinks to ellipse [Fig. 3(b)] and disappears from the parameter space. Furthermore, period doubling bifurcations with fundamental period of two only remain for $\varepsilon_{s\cdot} > 2.5210^7$. Then the maximum order of bifurcations decreases [Figs. 3(c) and 3(d)] and finally, at $\varepsilon_{s\cdot} > 1.910^4$ the system becomes stable in the whole range of control parameters.

SEED PULSE ENERGY EFFECT

The obtained results, demonstrating dependence of the operation on the seed pulse energy, are in essence not quite trivial. This phenomenon is in controversy with intuitive comprehension of regenerative amplification. The following speculations seem to demonstrate the negligible extent of the seed influence or at least to evidence much simpler looking relations. Imagine, initially low seed pulse energy s_1 after certain number of round trips ($^\Delta\tau$) is amplified to energy s_2 of several orders of magnitude larger but still much less than energy stored in the gain media, $g_{ai} \gg \varepsilon_{s2} \gg \varepsilon_{s1}$. Then further amplification should give the same output as if the amplification has initially started with seed energy ε_{s2} because the previous stage $(\varepsilon_{s1} \to \varepsilon_{s2})$ virtually has not changed the stored energy and, as a consequence, the system gain. This logic leads to an inference that a lack of seed energy can be compensated with additional round trips. Consequently, the operation diagrams have to look identical but shifted in the coordinate of round trip number for different seed values. Accurate computations give absolutely different results, Fig. 3.

Let us consider in details some subtleties which lead to this difference. In time-domain the amplification process looks exactly as simple logic predicts, that can be confirmed with straightforward calculations. Equation 10 gives equal but shifted in time intracavity energy evolutions,

$\varepsilon(g_{ai},\ \varepsilon_{s1},\ \tilde{t})=\varepsilon(g_{ai},\ \varepsilon_{s2},\ \tilde{t}-\Delta\tau)$ and this shift, $\Delta\tau$ can be determined

as $\qquad g_{ai}\Delta\tau=\ln\dfrac{g_{ai}}{\varepsilon_{s1}}-\ln\dfrac{g_{ai}}{\varepsilon_{s2}}$.

. This logic is accurate when we are considering amplification phase as "isolated" with given initial gain; this is absolutely true for low repetition rates.

Previously it was commonly accepted that regenerative amplification virtually is independent of the seed energy because only low repetition rates were under consideration. In reality, even for low repetition rate systems the seed energy value should not be too low. However, the reason for that is rather different of what we are describing. Simply, competing parasitic processes of amplification of spontaneous emission always exists in regenerative amplifiers. Thus, the seed energy should be well above the spontaneous emission level in order to get the amplified seed at the output instead of amplified spontaneous emission. However, at low repetition rates the sufficient pulse seed energy is extremely small, e.g. down to 10^{15} J in accordance to the experiments presented in the classical paper (Murray & Lowdermilk, 1980). Although, some exceptions may take place; they involve special applications which require very clean, high contrast optical pulse, e.g. parametric chirped pulse amplification (Ross et al., 2007).

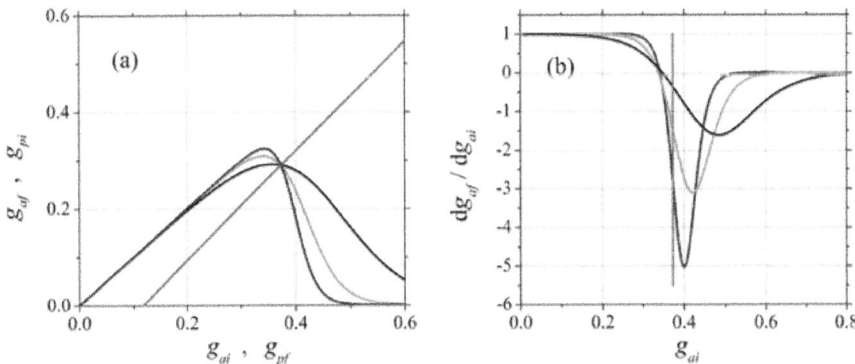

Figure 4: Typical state space diagrams of regimes having equal fixed points at different seed energies (a) and derivatives of the „amplification" curves (b).

It becomes apparent that at high repetition rates the initial gain depends on preceding operation cycles and can be determined indeed by taking them into account. This procedure is in essence nothing else than the fixed point determination. Let us return to geometric presentation of fixed points in state space. Figure 4(a) represents diagrams of the final gain against initial gain for three seed energy values. Corresponding numbers of round trips have been selected so that the fixed points at certain repetition rate are identical. All the curves intersect in a single point which corresponds to the fixed point for the repetition rate $T_1/T=8.0$ This means that decrease in seed energy is compensated by increasing of round trips but only in a sense of equal position of the fixed points. However the shapes of curves g_{af} versus g_{ai} are different and their derivatives in the intersection point are dependent on the seed energy [Fig. 4(b)]. Such derivatives indeed determine the regenerative amplification stability as we have described in section 2.4 by referring to the theory of discrete-time dynamical systems.

The identity of the fixed points can be realized by compensation of the seed energy difference by appropriate selection of the round trip number. Output energies corresponding to those fixed points are equal too. However the peculiarity of operation at high repetition rates is such that dynamical system behavior is absolutely different. Thus at high repetition rates the seed pulse energy becomes one of those critical parameters which determine the operation regime of the regenerative amplifier. The specific operation points, which have been analyzed here, can be also found in the diagrams of dynamic regimes (Fig. 3). With regard to stability, they were classified as stable, 2T-periodic and chaotic for seed values of $1.3 10^4$, $2.5 10^7$ and 10^{10} respectively. The dynamic regimes which are in general possible to obtain (by changing the round trip number) within a certain range of seed values are summarized in Table 1.

Table 1: Possible regimes versus seed pulse energy range

Existing regimes	Seed value range	
	$g_1=0$	$g_1=0.028$
Chaos and "all" periods	$<2.52\ 10^{(7}$	$<1.4\ 10 5$
T , 2 T , 4 T , 8 T , 16 T , 32 T...	$2.52\ 10^{(7} - 2.56\ 10^{(7}$	$1.4\ 10^{(5} - 1.5\ 10^{(5}$
T , 2 T , 4 T , 8 T , 16 T	$2.56\ 10^{(7} - 2.72\ 10^{(7}$	$1.5\ 10^{(5} - 1.74\ 10^{(5}$
T , 2 T , 4 T , 8 T	$2.72\ 10^{(7} - 3.56\ 10^{(7}$	$1.74\ 10^{(5} - 2.5\ 10^{(5}$
T , 2 T , 4 T	$3.56\ 10^{(7} - 1.39\ 10^{(6}$	$2.5\ 10^{(5} - 4.1\ 10^{(5}$
T , 2 T	$1.39\ 10^{(6} - 1.90\ 10^{(4}$	$4.1\ 10^{(5} - 3.5\ 10^{(3}$
T (table)	$"/1.90\ 10^{(4}$	$"/3.5\ 10^{(3}$

Influence Of Parasitic Intracavity Losses On Dynamic Pattern

The approximation of negligible losses is a good way to present the main ideas for application of the discrete-time dynamics method for regenerative amplification and to understand the dynamic patterns most relevant at high repetition rates. However, this approximation has limited application in practice. The output pulse energy grows in the lossless system monotonically together with amplification phase duration and reaches saturation at the level of $\varepsilon_{max}=1-\exp(-T/T_1)$ that corresponds to full conversion of stored energy (population inversion) to the output pulse energy. Consequently the number of round trips can be increased, without detriment to output energy, to the values high enough for operation behind the bifurcation zone that in turn assures stable operation. Actually, the system is always (i.e. irrespective of losses) stable provided that the population inversion is well depleted during the amplification phase. In this case the initial gain tends to the constant, determined only by the repetition rate, $[g_{af} \to 0 \Rightarrow g_{pi} \to 0;$; then from Eq. 6 follows that $g_{ai}=g_{pf} \to 1-\exp(-T/T_1)]$. Consequently, the interdependence of operation cycles vanishes that results in eliminating of immediate cause of unstable behavior. In reality, parasitic losses prevent utilization of this property since because of losses the mode of complete gain depletion becomes inefficient.

Well known efficiency criterion, to dump optical pulse off the resonator at the moment when the current gain has dropped down to the threshold gain $(g_{af}=g_t)$, is not applicable to repetitive operation as relating to only "isolated" operation cycles. Power efficiency enhancement takes place at high repetition rates when stored energy is left partially under-depleted $(g_{af}>>g_t)$ forming a substantial gain background after several operation cycles. The proportion of gain to losses, which eventually determines extraction efficiency of the stored energy, can be well improved by this means. However incomplete depletion is an origin of operation cycles interdependence therefore in presence of losses the system efficiency in some sense collides with the system stability.

Parasitic losses in laser systems are given by optical components imperfection and diffraction losses of the optical resonator. The latter are objects of resonator geometry optimization. In case of solid-state lasers pumped longitudinally (virtual absence of hard apertures) high order optical aberrations (spherical e.g.) may become the main origin of diffraction losses, especially at high pumping intensities (Liu C. et al. 2008). Among optical components, the electro-optic switch is usually the most critical part; contributions of the Pockels cell and the polarizer to the loss factor surpass the remaining components (Müller et al., 2003). Practically, the level of total parasitic losses can vary in quite a wide range, but the typical value should not exceed a few percent per roundtrip for high-quality, well optimized systems. Here we should remark that the losses, inherent for quasi three-level gain media and related to partially populated ground state, are not dissipative and they do not belong to the parasitic losses which we are considering.

At the account of intracavity losses, general, qualitative pattern of amplification dynamics (fixed points uniqueness, variety of orbits for the repulsive fixed points, significance of the seed pulse energy) remains the same, but naturally the quantitative difference factors in. The intracavity losses of regenerative amplifier not only reduce efficiency (that is natural for lasers), but also substantially interfere in total system stability.

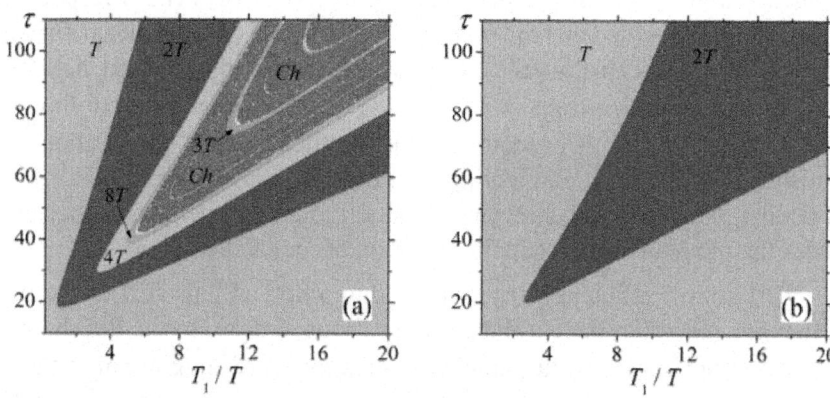

Figure 5: Diagrams of amplification dynamics for threshold gain g_t=0.028 and seed pulse energies ε_s=3×10^{-7} (a); ε_s=1.90×10^{-4} (b).

The diagrams of regenerative amplification dynamics in parameter space (round trips – repetition rate) for intracavity losses corresponding to the threshold gain gt=0.028 are presented in Fig. 5. Fixed point calculation and their stability evaluation were performed analogously as described in sections 2.4 and 2.5. The only difference is function $\hat{g}_a$, relating final and initial gains [solutions of Eqs. (4) and (7)], was calculated numerically since analytical solution is unknown in case of nonzero losses. The diagram presented in Fig 5(a) can be compared with that given for the same seed pulse energy s=310^{7}but for zero losses [Fig 4(c)]. The influence of losses results in a more complicated dynamic pattern; the zone of chaotic dynamics evolves and the high order bifurcations shift closer to the tip of the instability zone (towards lower repetition rate and lower round trip number). The second diagram [Fig 5(b)] was calculated for the seed energy s=1.9010^7, which at zero losses provided stable operation in the whole range of control parameters. Now a period doubling zone (2T) has occupied a certain part of the parameter space and noticeably narrowed the range of stable operation.

A more cumulative picture of dynamical regimes is presented in Table 1. One can see some general change for the worse for system stability with respect to the zero-loss case. The decrease in stability caused by losses is not an obvious phenomenon (why not increase?). The reason for that bears similarity to the seed energy effect. The losses decrease pulse energy addition per round trip that can be compensated by increasing of round trip number, but only in a sense of fixed point identity. The derivative magnitude of the gain transformation function, $|\hat{g}'_\Sigma(g_1)|$, in this point has changed so that system stability becomes worse as the losses are increasing. This phenomenon is not obvious but the conclusion is straightforward – the parasitic losses should be minimized as much as technically possible not only for efficiency but also for better stability.

System Performance Representation In Parameter Space

For understanding regenerative amplifier performance, not only dynamical regimes but also output pulse energy should be represented in relation to the parameters governing the system. In previous sections we have described in detail the method of fixed points and related orbits determination. Actually, during calculation of the initial gain orbit, $g_{ai}(k)$, the orbit of output energy $\varepsilon_f(k)$ comes out automatically as the second solution of coupled Eqs. (4) and (7), $\varepsilon_f(k) = \hat{\varepsilon}[g_{ai}(k), \varepsilon_s, \tau, g_t]$. If governing parameters are such that operation is unstable then the initial gain varies from cycle to cycle by specific means and consequently the output energy becomes a multi-valued function of governing parameters. Representation of such a function on 2D diagram is unfeasible, it looks like a mess. Therefore we present typical dependencies of output energy versus round trip number (bifurcation diagrams) at several repetition rates for the fixed loss factor (corresponding to the threshold gain $\varepsilon_s = 7.7 \times 10^{-7}$ (red line) and $\varepsilon_s = 7.7 \times 10^{-9}$ (Fig. 6).

The selected seed pulse energy belongs to the same range that typical for functionally important class of seed lasers operating in CW mode-locking regime with moderate average power (around hundreds milliwatt). The repetition rates are chosen according to the diagram of dynamic regimes [Fig. 6(a)] so that typical bifurcation diagrams up to 16T-regime are demonstrated.

It is seen that the output energy variation in the presence of period doubling is so high that it virtually leaves no opportunity to use such a regime in practical applications. For example at repetition rate $T_1/T = 4.6$ and in 30-60 round trips range the output energy alternates between high and low value so badly that the output pulse train looks almost as at twice less repetition rate [Fig. 6(c)]. It is even more pity that such bad stability often appears in regimes which potentially capable of providing high output energies.

The maximum capability of the system can be determined by calculating the output energy exactly in the fixed point: $g_{ai}{=}g_1 \Rightarrow \varepsilon_1 = \hat{\varepsilon}(g_1, \varepsilon_s, \tau, g_t)$. This energy is always a single-valued function of the governing parameters regardless of whether the fixed point is attractive or repulsive.

In case of a repulsive fixed point (i.e. unstable operation), the corresponding "fixed point" output energy becomes an artificial parameter but it can serve as a convenient reference for evaluating the power efficiency reduction caused by instability effects.

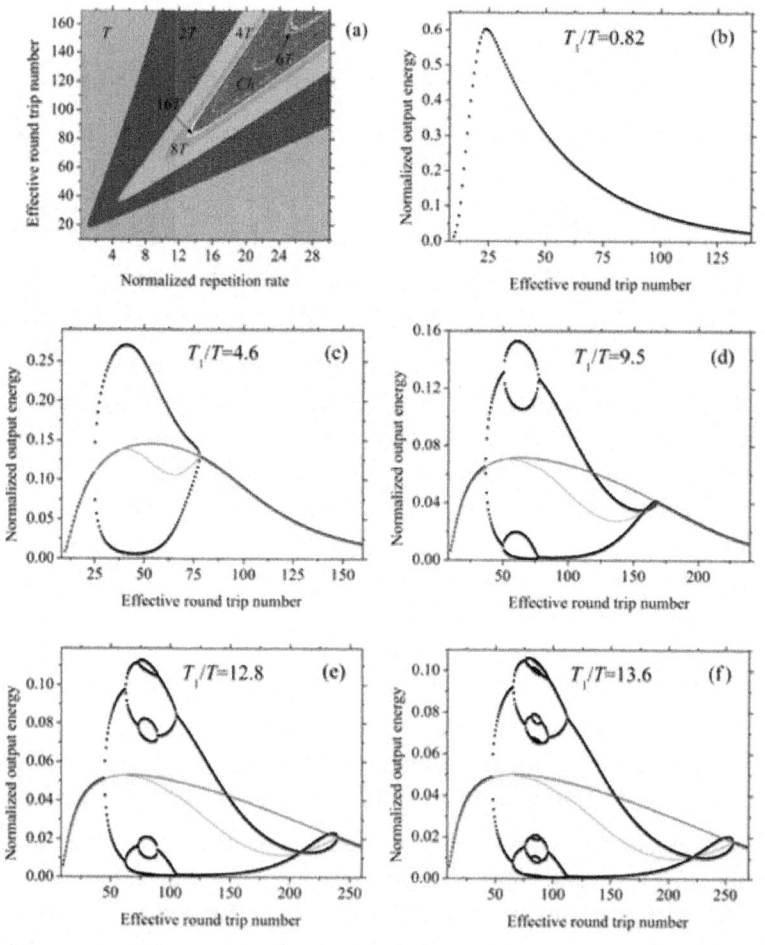

Figure 6: Diagram of dynamic regimes (a) and corresponding bifurcation diagrams for selected repetition rates (b–f) at $\varepsilon_s{=}7.7{\times}10^{-7}$ and $g_t{=}0.028$. Pulse energy, averaged pulse energy and "fixed point" energy correspond to black, green and red lines respectively.

We also determined real (accounting multi-energy nature) output energy averaged over large number of operation cycles $\langle \varepsilon_f(k) \rangle$ (Fig. 6). Interestingly, the real averaged energy is considerably lower than the reference "fixed point" energy in regimes exhibiting pronounced period doubling. The question is "what can the origin of this energy defect be?" One can suppose that in a multi-energy regime a relatively larger part of the resonator energy circulating is redistributed to the channel of parasitic losses. Some evidence for this explanation is that the same curves calculated in case of zero losses coincide in spite of bifurcations. The only alternate reason is stored energy depletion through spontaneous emission during the pump phase. However, the latter effect becomes dominant when the population inversion is on average high (i.e. at low round trips), that is not the case now.

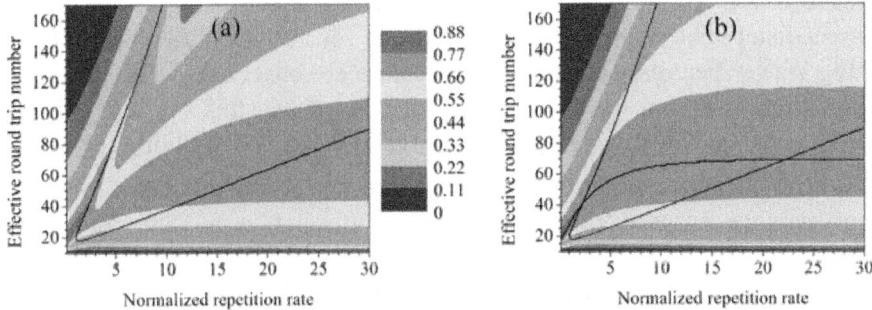

Figure 7: Efficiency of stored energy extraction with respect to the separatrix (a); the reference ("fixed point") efficiency with respect to the separatrix and τ_{max} curve (b); both diagrams are for $\varepsilon_s = 7.7 \times 10^{-7}$ and $g_t = 0.028$.

Single-valued functions characterizing output energy allow presentation of an informative picture of system performance in 2D diagrams. The extraction efficiency of stored population inversion may serve as a more convenient parameter for this purpose. The extraction efficiency can be defined as averaged output energy divided by maximum stored energy available at a given repetition rate, $\langle \varepsilon_f(k) \rangle / \varepsilon_{max}$, where $\varepsilon_{max} = 1 - \exp(-T_1/T)$. Then the value calculated in the fixed point $(\varepsilon_1 / \varepsilon_{max})$ can be regarded as the maximum attainable efficiency at a given round trip number. Typical diagrams of these parameters in the space of repetition rate versus round trip number are presented in Fig. 7. Additional normalization to ε max introduced for already normalized terms is not meshing. Quite the contrary, it simplifies the practical use of theoretical data. The efficiency is the ratio which has the same value for real "dimensional" and normalized parameters.

Apparently, stable single-energy operation is the only suitable regime for routine use of regenerative amplifiers. For completeness sake we can note that one may successfully use certain unstable regimes for specific applications provided that there is comprehensive understanding of the essence of period doubling (Metzger et al., 2009). However this is rather the exception than the common rule. Thus, we can omit a detailed picture of dynamical regimes in order to move towards more pragmatic considerations. It is sufficient to leave only one curve in the parameter space defining the range of operating points in which operation is stable. This curve (further referred to as a separatrix) represents a manifold of the first order bifurcation points (1T-2T boundary) and actually separates zones of stable and unstable operation in the parameter space. Both real and "fixed point" energies are equal within a range of control parameters providing stable operation. Some drop of real efficiency with respect to that obtained in the assumption of stable operation is observed immediately below the upper branch of the separatrix, in the instability zone [Fig 7(a)]. The cross-sections of this feature are observed in Fig. 6 at several repetition rates and we have already concluded that this "valley" originates from enhanced influence of parasitic loses in the period doubling regime.

The distribution of the "fixed point" extraction efficiency $(\varepsilon_1/\varepsilon_{max})$ in the parameter space contains sufficient information to determine system performance when accompanied with the separatrix curve. This curve confines the space of stable operation that is the range where the "fixed point" data correspond to reality. For system optimization it is important to find operation points potentially providing maximum performance at each given repetition rate. The manifold of such points represents a curve of round trip number versus repetition rate which is mapping the peak value of the "fixed point" extraction efficiency (2D distribution in parameter space) and further referred to as τ_{max} curve [Fig 7(b)].

Thepotential performance becomes real at repetition rates in which the potential performance becomes real at repetition rates in which the τ_{max} curve is outside of the instability region and when the round trip number is set equal to τ_{max}. If the τ_{max} curve is inside the instability zone then the system capabilities are underexploited. The corresponding range of repetition rates can be called the "critical range". Within the critical range there are two possible positions of the operating point which may provide maximum output in the stable regime – the points along the lower and upper branch of the separatrix [see Fig 7(b)]. According to the diagram, at lower repetition rates in the critical range the upper branch has an advantage from an efficiency point of view. Operation at lesser round trip number (lower branch of the separatrix) becomes preferable at higher repetition rates. Thus optimization of the regenerative amplifier is actually reduced to selection of the round trip number which provides maximum output pulse energy and at the same time allows stable operation for the required repetition rate range(imposed by the system specifications). The corresponding round trip number is logical to call optimum (τ_{opt}). Obviously, τ_{opt} is equal to τ_{max} outside the critical range. curve is outside of the instability region and when the round trip number is set equal to τ_{max}.

If the τ_{max} curve is inside the instability zone then the system capabilities are underexploited. The corresponding range of repetition rates can be called the "critical range". Within the critical range there are two possible positions of the operating point which may provide maximum output in the stable regime – the points along the lower and upper branch of the separatrix [see Fig 7(b)]. According to the diagram, at lower repetition rates in the critical range the upper branch has an advantage from an efficiency point of view. Operation at lesser round trip number (lower branch of the separatrix) becomes preferable at higher repetition rates. Thus optimization of the regenerative amplifier is actually reduced to selection of the round trip number which provides maximum output pulse energy and at the same time allows stable operation for the required repetition rate range(imposed by the system specifications). The corresponding round trip number is logical to call optimum (τ_{opt}). Obviously, τ_{opt} is equal to τ_{max} outside the critical range.

Instructive inference of considered above regenerative amplification properties is that within critical repetition rates the round trip number takes on optimum value either near the lower or near the upper separatrix branch but always at the margin of unstable operation. In general, operation at the margin of stability incurs challenges for robust operation in real systems which undergo technical noises. Even slight changes to control parameters may result in system instability. Therefore reliably stable operation generally requires setting the operating parameters well away from the instability border, but this in turn leads to a reduction of the system performance.

There is an important parameter related to intracavity losses which may influence performance of regenerative amplifiers indirectly: the amount of intracavity energy dissipated during the amplification stage. Accumulated over round trips, the fraction of intracavity energy, dissipated through parasitic losses is subject to the specific operation regime. In particular, multiple passes of the already amplified optical pulse lead to substantial enhancement of energy dissipation. This, in turn, may give unacceptably high heating of intracavity components caused by the absorbed part of the dissipated power. One of the critical components in this respect can be the Pockels cell crystal. It may lose contrast under excessive heating possibly resulting in failure of regenerative amplifier operation. This effect is especially pronounced for systems intended for high power applications. The energy defect, arisen due to parasitic losses ((ε_l),), can be determined as a product of intracavity energy, integrated over the amplification stage, and loss factor expressed by the threshold gain g_t. Using this definition and Eqs. (4) and (7) we get:

$$\varepsilon_l = g_t \int_0^t \varepsilon(t^\sim)dt^\sim = \int_0^t \varepsilon(t^\sim)g(t^\sim)dt^\sim - \varepsilon_f = g_{ai} - g_{af} - \varepsilon_f,$$

(14)

where the integration variable $t^\sim$ is the current normalized time. The same result can be obtained from the energy conservation law since during short (as we have initially assumed) amplification stage there are only two energy consumption channels – useful signal and parasitic losses. In case of multi-energy regime the effective lost part of the energy can be found by averaging:

$$\langle \varepsilon_l \rangle = \frac{1}{k} \sum_k \left(g_{ai}(k) - g_{af}(k) - \varepsilon_f(k) \right)$$

(15)

The diagram of dissipated energy (also normalized to maximum stored energy ε_{max}), for the set of governing parameters used before, is presented in Fig. 8(a). We can see that the large number of passes, typical for the upper separatrix branch, substantially contributes to this parameter. So the dissipated energy is about 7.5 times higher at the repetition rate when theoretical efficiencies for both branches are equal (at $T_1/T \approx_{6.5}$). This increase with respect to the lower branch often makes operation at high round trip number, well above τ_{max} point inefficient, despite theoretical preference.

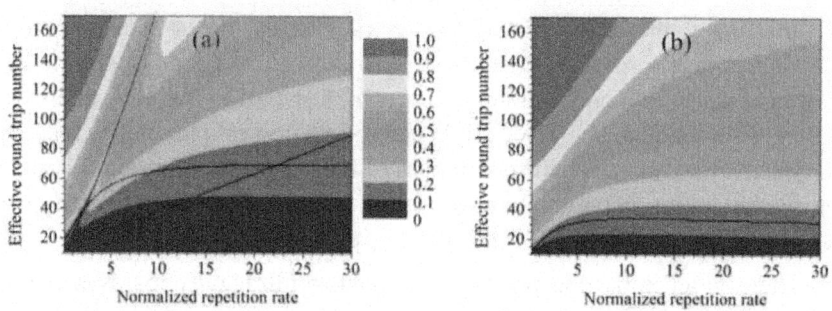

Figure. 8: Dissipated intracavity energy, separatrix and τ_{max} curves for $g_t=0.028$ and ε s$=7.7\times10^{-7}$ (a); dissipated energy, and corresponding τ_{max} curve for the seed energy $\varepsilon_s=2.46\times10^{-4}$ (b).

All the issues described in this section are regarded to one certain seed energy $\varepsilon_s=7.7\times10^{-7}$. The corresponding diagrams give a typical but single section of multidimensional space of control parameters. However, as we can already conclude from amplification dynamics data presented in sections 2.6 and 2.7, the critical repetition rate range depends on the seed pulse energy.

Thus, the picture of regenerative amplification is still incomplete and it is time to proceed to consideration of system optimization taking into account influence of the seed pulse energy.

Stability diagrams and pulse duration effects

It is possible to present data which allow evaluation of regenerative amplification of different seed pulse energies by considering a single diagram. The necessary premises for doing that have been formulated in the previous sections. The condition where bifurcations are absent in the whole parameter space (Table 1) gives a general understanding of the seed energy influence; however this condition is too strong from a practical point of view. In order to thoroughly utilize power capabilities of the regenerative amplifier at a certain repetition rate, the round trip number should be set equaled to τ_{max}. Stable operation requires having the point τ_{max} outside the instability zone (delimited by the separatrix curve on the parameter space). The inter-positioning of the separatrix and the τ_{max} curve contains sufficient information to determine the critical repetition rate range for a certain seed energy and the possible optimum operation points (τ_{opt}) within this range (which lie, as we have known, along one of the separatrix branches). The diagrams consisting of τ_{max} curves and separatrixes (further referred to as stability diagrams) for selected pulse seed energies are presented in Fig. 9.

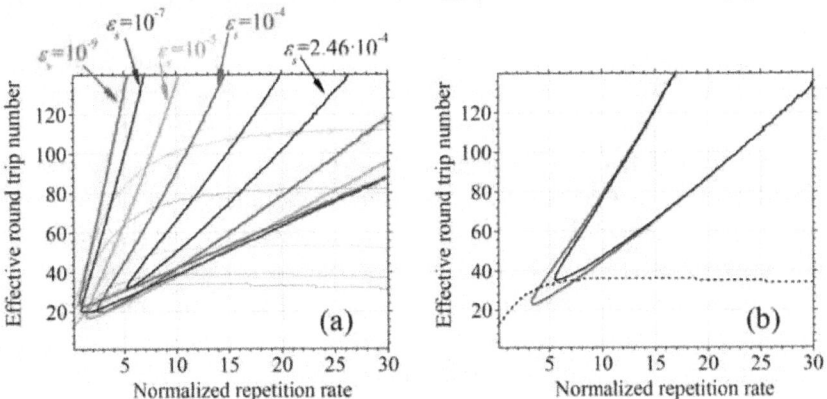

Figure. 9: Stability diagrams for the threshold gain $g_t=0.028$; the curves couples of the same color (the separatrix and τ_{max} curve) correspond to certain seed energies (a). The separatrixes are shown for the short pulse (black solid line) and long pulse approximations (red solid line) and τ_{max} curve (dash line) at g_t

$=0.028$ and $\varepsilon_s=1.7\times10^{-4}$ (b).

The approach of stability diagrams forms a more systematic concept of the system behavior and specifically allows estimation of the seed level which may enable one to avoid instability effects at the required pulse repetition rate. Also we can see that the critical range shrinks as the seed energy is increasing. The most critical repetition rate is $T/T_1=5.5$, the point of the "worst stability", requiring the highest seed energy for optimal operation. Finally one can determine the seed pulse energy at which τ_{max} curve does not pass instability zone at all. For the specific parasitic losses that we consider ($g_t=0.028$) this energy is equal to 2.46×10^{-4}. Consequently, a regenerative amplifier seeded with pulse energy higher than that value (further referred to as "ample") allows theoretically attainable average power and stable operation over the whole repetition rate range. Also we can mention here another advantage of the "ample" seed operation: As soon as the critical range has disappeared there is no need to operate at the upper separatrix branch and to suffer from large intracavity energy dissipation peculiar to this regime. The diagram of dissipated energy for the ample pulse seed energy, $\varepsilon_s=2.46 \times 10^{-4}$ is depicted in Fig. 8(b).

The approach of stability diagrams is a straightforward way of regenerative amplifiers optimization. However in real systems there are specific effects which influence dynamic behavior and output parameters and they cannot be elaborated by using only approximation of simplified rate equations. Nevertheless it is possible to understand some important contributions staying basically within present approach. The major application of regenerative amplifiers is amplification of short (even more common term is ultrashort) optical pulses. Now we shall consider the influence of pulse duration on regenerative amplification. The theory described above is based on rate equations formulated for idealized four-level system. One of the positions in the definition of that "ideality" is instant depopulation of the lower laser level (which is also called terminal level). In reality we can assume that the lower laser level is virtually unpopulated, only provided that the amplified optical pulse is much longer than the terminal-level lifetime (long pulse approximation). Otherwise, if the length of the optical pulse is much less than the terminal lifetime then the terminal level will remain populated resulting in "faster" decay of population inversion (indeed during amplification of a single pulse). Comprehensive evaluation of actual terminal-level lifetime provided by (Bibeau, et al., 1995) for different neodymium doped laser media gives actual values well exceeding 100 ps. So we can conclude that amplification of pulses shorter than 100ps virtually for all neodymium based systems is more appropriate to analyze within the short pulse approximation that is assuming negligible terminal-level depopulation during singe pulse amplification. This, in fact, constrains applicability of presently described approach and leaves beyond the scope functionally important ultrashort pulses. We can note in advance that net contribution of the terminal-lifetime effect to regenerative amplifier behavior is rather weak at high repetition rates. However it gives some quantitative refinements to the picture presented above. In order to elaborate regenerative amplification in the short pulse approximation we need to re-calculate the fundamental relations of final and initial gains $g_{af} = \hat{g}_a(g_{ai})$. In this approximation the amplification of a single pulse was regarded as that in truly three-level gain media with initially empty ground state. After single pass amplification the ground state becomes partially populated, but by the second pass it is empty again and consequently the gain defect, which appeared due to "instant" three-levelness, is recovered. The latter is the case since we assume the terminal-level lifetime to be much shorter than the round trip time, T_0. Typical T_0 value is in the range between ten and a few tens of nanoseconds, so this is good approximation for most of neodymium media [actually except some fluoride crystals and glasses in which the neodymium terminal lifetime is of 10 ns order (Bibeau, et al., 1995)]. By this means and by using sequential procedure the basic relations $g_{af} = \hat{g}_a(g_{ai})$ and corresponding output energies were determined. The subsequent procedures (evaluation of the fixed points and their stability analysis) stay unchanged for the short pulse approximation. We revealed that the influence of a terminal level appears as follows. There is noticeable deviation between the τ_{max} curves at low repetition rates. However both "short pulse" and "long pulse" curves virtually coincide at the repetition rates $T_1/T > 1.0$, indeed in the range that we are studying. The separatrix curves practically coincide at low seed energy levels, although filling of the instability zone differs from what have been seen earlier [e.g. in Fig 6(a)] towards not so wide variety of regimes. These peculiarities add little in practical essence. Therefore we have come to nothing more than qualitative description and statement of that fact.

The only noticeable influence of the terminal lifetime was found at high seed energy, approaching to the ample level. The stability diagram, depicted in Fig.9 (b), shows what the difference is. In comparing two separatrixes we can point out that the tip of the "short pulse" separatrix is somewhat shifted towards higher round trip number and higher repetition rate. Such a shrink of the instability zone gives certain improvement of general system stability. The ample seed pulse energy determined for short pulses is almost 1.5 times less than that found within long pulse approximation ($\varepsilon_s = 1.7 \times 10^{-4}$ against 2.46×10^{-4}). This improvement is rather unexpected result since terminal-level "bottleneck" in some respect hampers the ideal four-level amplification. Thus at least partially and shortly populated terminal level does not act as additional losses as one might imagine. Another phenomenon, substantial at high intensities and therefore requiring intent attention while ultrashort optical pulses are amplifying, is the Kerr effect. This nonlinearity makes an impact on amplification process by intensity dependent refractive index change in volume intracavity components. The influence occurs in both spatial and spectral domains and commonly is described as Kerr lensing and self-phase modulation. Reduction of this negative influence usually implies decrease of effective optical pulse intensity which can be quantified in terms of so-called B integral, in essence representing nonlinear on-axis phase shift (Brawn, 1981). In case of regenerative amplification multiple passes should be taken into account in order to evaluate total B integral accumulated during amplification stage:

$$B \approx B_1 \int_0^r \varepsilon(t^-) dt^-,$$

(16)

where B_1 is single pass B integral calculated for the intensity which is equal to the gain media saturation fluence divided by the pulse duration (see the Appendix section for details). The energy integral in this formula exhibits a factor of effective impact of multiple passes. We have already met such an integral when evaluated total lost part of the energy (see Eq. 14) and the value proportional to this factor has been depicted in previous section in Fig. 8. We can see that operation at the upper branch of the separatrix with short pulses is strongly unadvisable since the B integral is increased by several times. The cause of that is obviously multiple passes of intense pulses peculiar in a regime behind the instability zone. That concerns the B_1 value, its reduction, in a condition of fixed pulse duration, simply implies standard methods of the mode area increase and shortening of the volume intracavity components. However these possibilities are rather limited. Even so the thin disc geometry allows tremendous reduction of gain media length (Speiser & Giesen, 2008) but the Pockels cell still can exhibit a real challenge. Among known to date Pockels cell materials only the BBO crystal is suitable for high average power (due to low absorption) and for high repetition rates (thanks to relatively low acoustic ringing). However the transverse electro-optic effect, the only functional for this crystal, permits shortening the optical pass or aperture increasing only in limited extent until driving voltage becomes unacceptably high (Nickel et al., 2005). Thus, since the Kerr effect often restricts the capabilities of regenerative amplifiers it is important at least to select correctly the operation regime in order to minimize its impact. We can note that operation at higher seed energy (ε_s larger than ample seed pulse energy) is beneficial in this respect too. At that condition, operation at maximum output (τ_{max} curve) does not suffer from too high multiple pass factor of the B integral [see Fig. 8(b)]

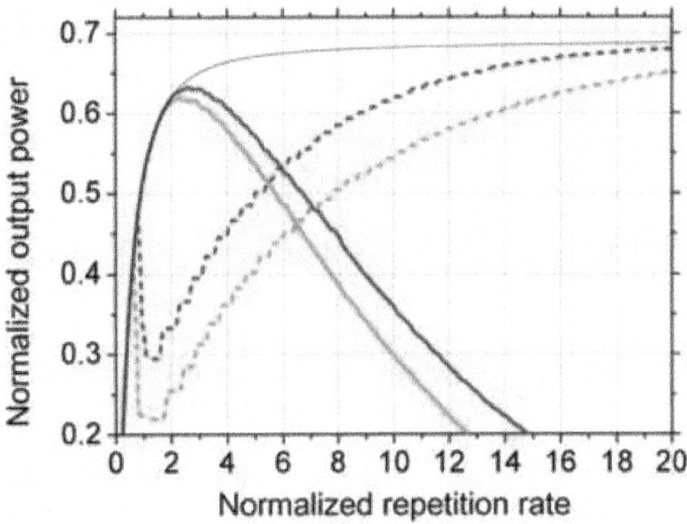

Figure. 10: Theoretical output power versus repetition rate for the threshold gain of 0.028 and normalized seed pulse energy of $\varepsilon_s =7.7\times10^{-7}$ (red line) and $\varepsilon_s =7.7\times10^{-9}$ (green line). The data of high and low separatrix branches correspond to solid and dashed lines respectively. The black line is the reference curve of achievable power.

At the end of the theoretical part of the chapter we are giving diagrams of the basic performance parameter – maximum output power which can be achieved in a condition of stability maintaining ($\tau=\tau_{opt}$), Fig. 10. Presented normalized average power is defined as a product of normalized energy and normalized repetition rate. The power curve, calculated for τ_{max} round trip number, can be regarded as the reference curve corresponding to theoretical efficiency limit. Really, in accordance with the τ_{max} definition, the corresponding output power is the highest obtainable average power in assumption of period doubling absence. The calculations show that this power is invariant under the seed energy, although the τ_{max} itself is not. Achieving the calculated maximum average power implies the best possible utilization of the stored pump energy, i.e. it assures the highest power efficiency. Obviously the same top performance we can reach operating at the ample seed energy provided that the round trip number is set equal to τ_{max}. This attribute is the main benefit of the high seed pulse energy at high repetition rates. The power defect with respect to the reference power curve observed for lower seed energies is essentially caused by instability effects. We present curves related to both branches of the separatrix within critical repetition rate range. The data related to the upper branch are not always reliable because as we have shown earlier this regime suffers from indirect effects related to multiple passes of intense optical pulse such as the Kerr effect and excessive heating of optical components. In concluding this theoretical part, now we are ready to proceed to experimental verification of the ideas developed above.

AMPLIFICATION EXPERIMENTS

Experimental setup

The amplification experiments were carried out in order to demonstrate basic features of amplification dynamics and to verify theoretical results presented above. The knowledge of potential capabilities and of general limitations makes it possible to provide the best regime selection and deliberate optimization of control parameters. In practicality, this means to maximize extraction of the given stored population inversion as a stable train of output pulses. We leave outside the scope of present consideration optimization of the pump characteristics and the geometry of the optical resonator allowing more power in TEM_{00} mode, as these do not relate directly to the amplification dynamics. The parasitic intracavity losses, although formally a governing parameter, are not an object of consideration; they should simply be reduced as much as technically possible. Since the repetition rate is usually imposed by the specifications it appears as a variable but a given parameter. There are two adjustable parameters which can be used for the system optimization – the number of roundtrips and the seed pulse energy. We performed experiments with a system based on $Nd:YVO_4$ crystals, a gain medium with truly four-level nature (except terminal-level nuances). The schematic diagram of the experimental setup that we used for investigation of regenerative amplification is shown in Fig. 11. In essence the system consists of the seed source and the regenerative amplifier itself.

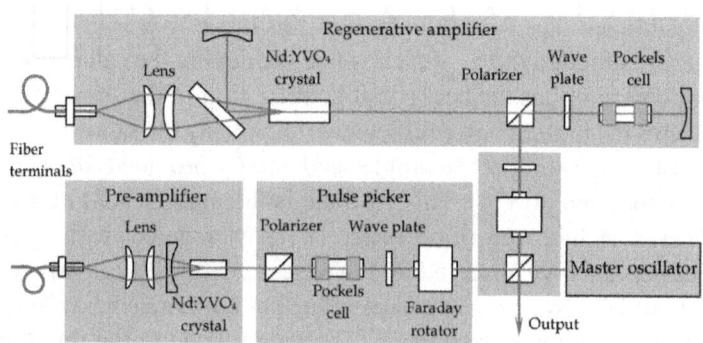

Figure. 11: Schematic diagram of the experimental setup.

The seed pulse source for regenerative amplification experiments was based on a diodepumped passively mode locked picosecond Nd:YVO$_4$ laser of moderate power. It generated a continuous pulse train with repetition frequency of 82 MHz and average power of 300 mW. The laser was able to produce optical pulses with duration as short as 6 ps. The short pulses were used in experiments where dynamics peculiar to high peak intensities were of interest. The initial investigations were focused on the "pure" dynamics not disturbed by optical nonlinearities. These experiments were carried out with 58 ps duration pulses obtained by installing an etalon in the oscillator cavity (the etalon narrows the bandwidth, thus widening the pulse duration).

A pulse picker was used to select pulses for further amplification and in this way to control the effective repetition frequency of the seed source. This part of the seed source system is important for high repetition rate operation, especially based on high gain laser media like neodymium doped vanadates. If the pulse picker is not used then two negative effects caused by unwanted pulses have place. These pulses continuously pass the optical resonator of the regenerative amplifier during pump stage and go out spatially coinciding with useful output signal. This leads first to reduced pulse contrast and second to parasitic consumption of stored energy. One can decrease the seed pulse energy and compensate for that by increasing number of round trips. This simple approach may often avoid bad influence of the unwanted seed background while operating at low repetition rates.

The reduction of the seed energy, as we have already seen (theoretically), is not a good idea when turning operation to high repetition rates. In our setup the pulse picker was an electro-optic switch based on an RTP Pockels cell. Selected pulses formed an input signal for the preamplifier. The remaining pulses of the master oscillator train were directed to the fast photodiode for synchronization of electro optic components of the system including the pulse picker itself. A double pass Nd:YVO$_4$ preamplifier installed behind the pulse picker was used to increase the seed pulse energy to required ample energy. High emission cross section of the Nd:YVO$_4$ crystal make this system efficient at relatively low input average power. Only 2 W of pumping was sufficient to achieve a gain coefficient of more than two orders of magnitude. The seed pulse energy was 3.2 nJ when pumping of the preamplifier was switched off. The energy of the pre-amplified pulse reached 1.1 μJ at 10 kHz and steadily decreased with the repetition rate to 370 nJ at 200 kHz. The calculation presented in the next section will show that the obtained energy is sufficient to ensure stable operation. Simple estimations show that the preamplifier is a good alternative in comparison with a more straightforward seed source scheme based on a powerful master oscillator. In order to provide 370 nJ pulses the master oscillator operating in a CW mode locking regime with a reasonable repetition frequency of 50 MHz should generate average power of 18.5 W. This way is really prodigal since the useful part of this power is much lower, e.g. only 74 mW even operating at 200 kHz. Obviously the preamplifier is a much more energy-efficient solution.

The regenerative amplifier was comprised of an optical resonator containing the gain medium (Nd:YVO$_4$ crystals) and an electro-optic switch. The electro-optic switch consisted of a BBO Pockels cell, a quarter-wave plate and a thin-film polarizer. The total multi-pass gain of the regenerative amplifier depends on the number of cavity round trips which is determined by the amplification-stage duration. This important parameter is easily controlled by setting the time interval during which the high voltage is applied to the Pockels cell. The laser crystal was continuously pumped by the fiber coupled laser diode module with fiber core diameter of 400 μm and numerical aperture of 0.22. Optimal pump power, providing maximum output in TEM$_{00}$ mode, was set to be 44 W. This optimization was performed in CW generation mode. Provided that no voltage was applied to the Pockels cell and the quarter-wave plate was adjusted for maximum output (optimal output coupling conditions), 12.5 W of average power was obtained. The output radiation was diverted from the input signal path by standard optical circulator based on Faraday rotator. The repetition rate of the system was limited to 200 kHz by electronics driving the electro-optic switches.

Application of stability diagrams to amplification experiments

Now we can apply the concepts developed in the theoretical part to estimating behavior of a real system. At the beginning the basic system properties should be evaluated in order to perform reciprocal transformation of normalized parameters to dimensional ones corresponding to real operation conditions. The steady state small signal gain, determined by the pump intensity, was found directly as the ratio of the seed energies measured right before and behind the active element of the regenerative amplifier.

The parasitic intracavity losses were derived from the specifications of the optical components. The laser characteristics of Nd:YVO$_4$ crystal (emission cross section and gain relaxation time) were taken from (Peterson et al., 2002). We explored amplification of three seed pulse energies differing by about two orders of magnitude: pre-amplified seed, unamplified seed and attenuated seed. This set of input signals covers the functionally important range. The value of the unamplified seed, 3.2 nJ, is of the same order of magnitude as the pulse energy of commonly used moderate power solid-state picosecond lasers. Operation with the seed energy intentionally attenuated to 32 pJ provides opportunity to evaluate typical behavior of the regenerative amplifier seeded with potentially attractive low power sources, e.g. with ultrafast laser diodes, which would substantially reduce system size and complexity. The seed energy obtained with the pre-amplifier was expected to be high enough to reach ample level. These seed energies were measured at the output of the seed formation system. However we observed that during further amplification the seed energy was not completely exploited. Mode mismatching reduces the effective seed energy. In general, it is difficult to avoid mode mismatching between a seed laser and the optical cavity of the regenerative amplifier in both spatial and spectral domains. Spectral mismatching can exist even with identical gain media because of e.g. different temperatures of laser crystals in those devices (Murray & Lowdermilk, 1980).

Table 2. Parameters used for stability diagrams calculation

Parameter	Value
Wavelength	1064 nm
Emission cross section	11.4×10^{-19} cm^2
Gain relaxation time	83 μs
Effective mode diameter in the laser rod	1 mm
Steady state small signal gain	2.94
Seed pulse energies	11 pJ (low); 1.1 nJ (medium); 240 nJ (high)

In order to have appropriate data for theory verification we experimentally estimated the overall level of mismatching. We measured real double pass small signal gain of the whole regenerative amplifier (output/input energy ratio while the Pockels cell was disabled) and compared this value with "pure" steady state gain of the laser element. Thus we determined the effective seed value to be about three times less than the measured one. The primary parameters of the regenerative amplifier which we used for calculations are summarized in Table 2.

Before proceeding to stability diagrams describing our particular experimental conditions, we note that we utilized a linear cavity that establishes double pass through the gain media, while the modeling has implied single pass for one cavity round trip. This factor of two was taken into account so that the y-axis of stability diagrams corresponds to real round trip number $t/T_0 = \frac{1}{2}\tau/G_0$. The other relations between dimensional and normalized parameters remain unchanged as they have been given in the section 2.2.

The stability diagrams describing amplification experiments are presented in Fig. 12. These data were obtained in the approximation of short pulses (accounting for terminal-level lifetime effect). As we have already known, the system capabilities are completely exploited when the τ_{max} curve is outside the instability region. The diagrams show that at low seed energy (11 pJ) the appropriate repetition rates should be less than 20 kHz. For the medium seed level (1.1 nJ) this range increases to 25 kHz. At higher repetition rates the τ_{max} curve enters the instability zone. It is important that for 240 nJ or higher seed energies the τ_{max} curve does not enter the instability zone in the whole range of repetition rates. So, for our laser system this energy corresponds to the ample seed value, sufficient to eliminate negative features of amplification dynamics, and thus to completely exploit the system capabilities.

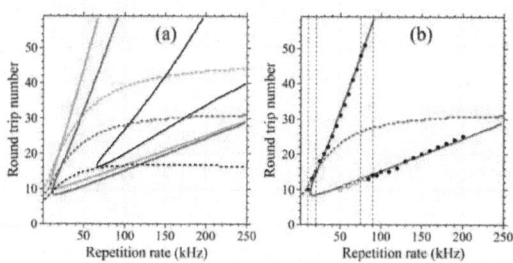

Figure. 12: (a) Separatrix curves (solid lines) and τ_{max} curves (dotted lines) in parameter space. Black, red and green lines correspond to seed pulse energies of 240 nJ, 1.1 nJ and 11 pJ, respectively. (b) Operating point trajectories (vertical dashed lines) and measured number of optimal round trips for the pulse durations of 58 ps (solid circles) and 9 ps (open circles) with respect to stability diagram for the seed energy of 1.1 nJ.

Experimental bifurcation diagrams

The initial experiments were carried out with medium seed pulse energy (the preamplifier was disabled). Various dynamic regimes, depending on set of control parameters, were observed. As an illustration, Fig. 13 shows oscilloscope screen shots of the output pulse train in typical single-energy and period doubling regimes.

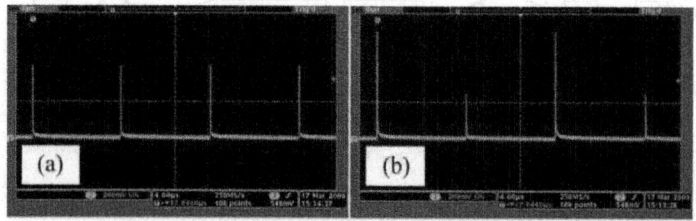

Figure. 13: Screenshots of typical pulse trains at 90 kHz. Stable energy output and the 2T period doubling regime were obtained at number of round trips equal to 14 (a) and 16 (b) respectively.

Experimentally obtained diagrams of the average output power and pulse energy versus number of round trips demonstrating system behavior at different repetition rates are presented in Fig. 14. The specific repetition rates were chosen to describe the most relevant cases of regenerative amplifier dynamics in respect of system optimization. The singlepeaked dependence inherent to low repetition rates appears at 10 kHz [Fig. 14(a)]. The average power and the pulse energy reach the maximum values simultaneously, when the round trip number is equal to ten. At 20 kHz the situation is different [Fig. 14(b)]. The shape of the energy curve shows that the system undergoes bifurcation in the 9–13 range of the round trip numbers. However, in this case the period doubling does not affect the system performance because the output power reaches its maximum value in a single-energy regime. This repetition rate is still not critical.

Instability effects become more pronounced at higher pulse repetition rates. The period doubling not only breaks the energy stability but also distorts the curve of the average power (as described in section 2.8). This curve now has two explicit peaks [Figs. 14(c) and 15(d)]. The first peak, corresponding to the maximum power, is located in a period doubling zone, whereas the second one is just over the instability edge. The optimal regime is obtained in the vicinity of the bifurcation point. At 75 kHz the optimal number round trips is equal to 48. This point is close to the second power peak, on the right side of the period doubling zone [Fig. 14(c)]. For 90 kHz repetition rate the optimal number round trips is equal to 13 and is situated right before the first bifurcation point [Fig. 14(d)].

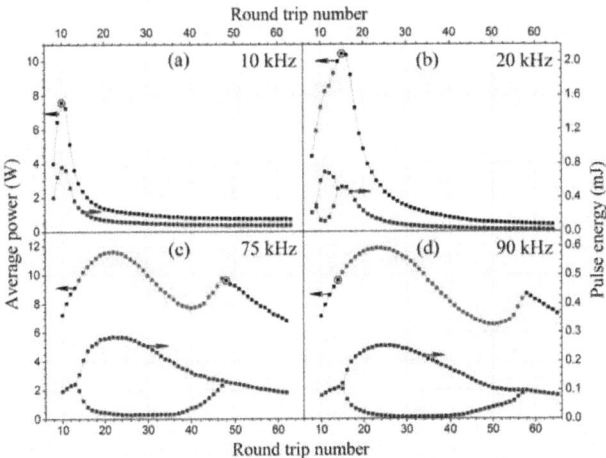

Figure. 14: Experimental average power (black and red dots correspond to stable and unstable regimes respectively) and pulse energy (blue dots) versus number of cavity round trips for the selected repetition rates. The encircled points correspond to the maximum power at stable operation.

The trajectories of the operating points corresponding to variation round trip number at a constant repetition rate are presented in Fig. 12(b). This trajectory at 10 kHz does not pass the instability zone. At 20 kHz the optimal operating point is above the instability zone. Both repetition rates 75 kHz and 90 kHz are critical – the optimal number of round trips is on the stability edge and is rather far from the point of the highest attainable power. The experimentally observed results confirm theoretical predictions that for critical repetition rates: (i) the output energy exhibits unacceptable fluctuations when the amplifier produces the highest average power, (ii) the highest stable pulse energy is reached close to the instability edge.

Some effects caused by nonzero terminal-level lifetime were also observed in the experiment. The presented bifurcation diagrams at 75 kHz and 90 kHz theoretically should exhibit not only 2T but also 4T regimes [like in Fig. 6(e)], if the long pulse approximation is applied. The regime of 4T period doubling was not observed experimentally and it should not exist theoretically, provided that contribution of the terminal-level lifetime is accounted. At the same time, as the theory has predicted, terminal-level lifetime effect does not influence the system performance at such low seed level.

Real deviation from theory is observed at the lowest repetition rate (10 kHz). The output energy decays too fast behind the peak point in comparison with theoretical expectations [see Fig. 6(b)]. This occurred because of the Kerr effect influence was substantial at low repetition rates even for initial experiments performed with relatively long 58 ps pulse duration.

Performance evaluation and discussion

It has been shown in the theoretical part that variation of only roundtrip number does not solve the stability problem; increase in the seed pulse energy is required in order to avoid bifurcations and corresponding instability at high repetition rates. So, we proceed to experimental verification of the seed energy influence. We compared operation for three cases: "low", "medium", and "ample" pulse seed energy.

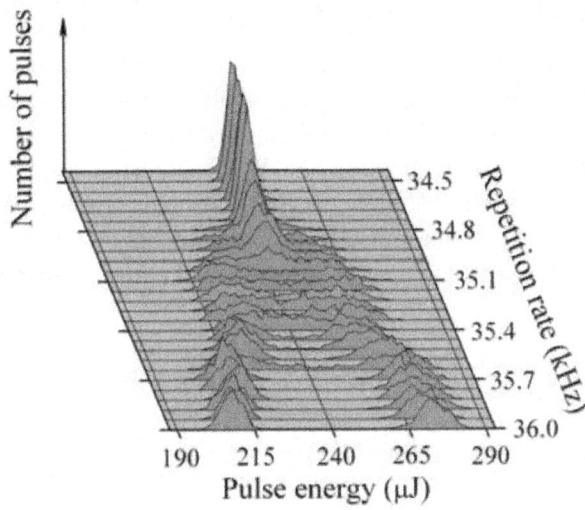

Figure. 15: Experimental energy histogram in the vicinity of the bifurcation point.

The measurements were performed at optimal round trip numbers. They were set for maximum average power while maintaining stable operation for every repetition rate. The formulation of stability criteria in real experimental environments requires certain attention. Experimental discrimination of amplification regimes to some extent suffers from uncertainty because of technical instabilities, not related to fundamental system properties. Technical noises, in essence slight modulation of governing parameters usually with random distribution, limit system stability. The most typical of these are pumping source noises, seed pulse energy fluctuation, synchronization jitter, resonator disturbance by mechanical vibrations and by air flow. In our setup the technical, not disturbed by period doubling, standard deviation of output energy was less than 0.7%. In proximity to a bifurcation point the deviation increased. The diagram of typical transition from single energy to period doubling regime "under magnification" of the repetition rate scale is shown in Fig. 15; the bifurcation point looks rather as a spot than as a point. The uncertainty takes place in the range where the deviation of energy is clearly higher than the technically conditional level but where the two peaks are still not distinguishable. In the repetition rate scale this range does not exceed 1 kHz (typically the value is 0.5 kHz). In the round trips scale measurements with such fine steps are not possible (in contrast to theoretically grounded terms the real experimental round trips are discrete), and besides, the analysis of the energy histogram is a time consuming procedure. So, we used simple phenomenological criterion formulated for the particular setup allowing real time measurements. The operation was considered stable when the standard deviation of the pulse energy did not exceed 1%. Apart from that, special care was taken regarding origin and spectrum of disturbing noises to be sure that they are virtually non-periodic. In particular the pulse picker driving electronics had to be improved in order to eliminate seed train modulation at the frequency equal to half the system repetition rate. We can note that even barely perceptible presence of "resonant" components (sub-harmonics of dumping frequency) in the spectrum of technical noises may tremendously enhance the influence of period doubling and can change the dynamical pattern beyond recognition.

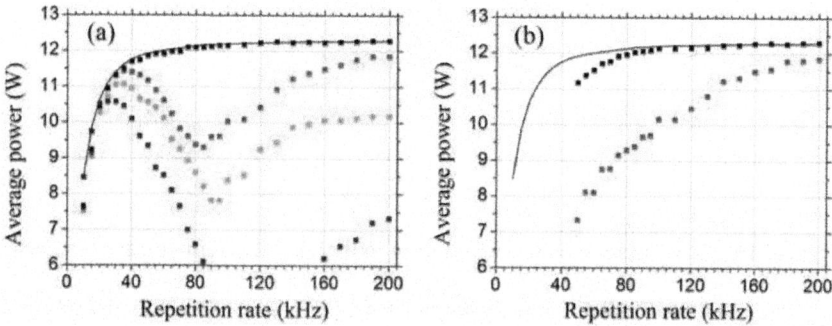

Figure. 16: Experimental output power versus repetition rate for 58 ps (a) and 9 ps (b) pulses. Black, red and green dots correspond to measured seed pulse energies of 700 nJ, 3.2 nJ and 32 pJ, respectively. Theoretical curve of achievable power is solid line in both diagrams. Blue dots in Fig. 16(a) are Q-switch experiment results.

Experimental dependences of the average output power versus repetition rate are presented in Fig. 16(a). The output is virtually independent of the seed level for low repetition rates. At higher rates there is a drop in power for low and medium seed levels in compliance with theoretical notions. The most significant power decrease appears in the 80–95 kHz range, then the output power steadily grows as the repetition rate increases. Non-monotonic behavior of the power curve originates from the specific location of optimal operation points in the parameter space. Corresponding experimental data with respect to the theoretical stability diagram for medium seed energy is presented in Fig. 12(b). The operating points coincide with the theoretical curve of τ_{max} until the latter enters into the instability zone (at 25 kHz). Then operation at the upper branch of the separatrix becomes optimal.

However as the repetition rate increases the stable operating point moves further away from the τ_{max} position, resulting in lower output energy. Consequently, starting from 85 kHz the optimal operation point switches to the lower border of the instability region. In this regime the operating point gradually comes closer to the τ_{max} position, and consequently, the output power steadily increases with increasing repetition rate. The pre-amplified seed pulse has sufficient energy to maintain stable operation at maximum power, in accordance with ample seed properties. The power curve obtained with the preamplifier shows no signs of downward excursion with respect to the reference curve (the curve of theoretically attainable power corresponding to τ_{max} round trip number). Some slight deviation observed at low repetition rates was assumed to be caused by the Kerr effect. This nonlinearity can cause additional intensity-dependent intracavity losses (more pronounced at higher pulse energies), and so its influence is more pronounced at lower repetition rates. In order to differentiate the Kerr effect influence we performed experiments with nanosecond pulses in the same setup. The seed source was disabled so that the regenerative amplifier was transformed to a Q-switched laser with cavity dumping.

Basically this device can be regarded as a regenerative amplifier seeded by spontaneous emission getting into the lasing mode. Thus, the seed energy becomes extremely low and additionally the laser system turns to the nanosecond domain (≈15 ns output pulse length was obtained in our setup). The former gives a large drop in power at critical repetition rates, the latter gives enhanced, with respect to the regular regenerative amplification, output energy at low repetition rate (the Kerr effect is negligible for such long pulses). The nanosecond pulse energy at low repetition rates exceeds the picosecond energy and very well agrees to the theoretical curve [Fig. 16(a)]. This demonstrates that experimental deviations appeared because of the Kerr effect and consequently general validates that the theoretical approach we have developed is wholly satisfactory. Heating of intracavity components was not observed to affect regenerative amplifier operation in optimal round trip number experiments. However, clear evidence of excessive heating was noticed during recording of bifurcation diagrams in the worse (from this point of view) regime.

The beam quality was observed to deteriorate for 90 kHz repetition rate and at round trip numbers between 46 and 56. From the other hand the average power defect reached maximum in this range [see Fig. 14(d)]. As we have concluded theoretically, this defect is indeed due to dissipated power, the absorbed fraction of which is heating intracavity optics. We also compared performances of the regenerative amplifier seeded by high and medium pulse energy for the functionally important case of shorter optical pulses. Typically about 9 ps pulse duration was obtained at the output of the amplifier seeded by the 6 ps pulse.

This duration is close to the minimum value supported by the gain bandwidth of the Nd:YVO$_4$ crystal in high total-gain applications such as regenerative amplification. The measurements were constrained to dumping rates above 50 kHz. Nevertheless, the intensities were substantial, and the Kerr effect influence was so strong that it eventually resulted in decrease of the output power [Fig. 16(b)]. The average power obtained with the pre-amplified seed was slightly lower than that theoretically predicted below 80 kHz and the difference reached 6.7% at 50 kHz. However, the comparison of these characteristics with those obtained at a medium seed level shows that the benefit of the preamplifier is even more pronounced in case of shorter pulses.

The difference is related to a large decrease of the average power below 85 kHz for the case of the medium seed. This is direct consequence of inefficient operation at the upper separatrix branch in critical repetition rate range under the Kerr effect influence. Experimental optimal operation points for both 58 ps and 9 ps pulses with respect to theoretical stability diagram are represented in Fig. 12(b). The optimal operating points for the short pulse experiments were always settled along the less efficient lower branch of the parameter separatrix. Attempts to operate at the upper branch (optimum for long pulses) resulted in an even larger decrease of the output. In order to quantify this difference we estimated the multi-pass B integral of the system (Eq. 22 in the Appendix). The B integral calculated at 50 kHz repetition rate gave values of 1.3 (acceptable) and 7.6 (problematic) at the transition from low to upper separatrix branch respectively. Thus the high seed energy gives additional advantage at shorter pulses due to a significantly lower value of the optimal number of round trips. The amplification experiments which were performed with $Nd:YVO_4$ regenerative amplifier have shown that the developed theoretical approach accurately agrees with experimental data and can be used for practical system-design guidelines.

CONCLUSIONS

Continuously pumped regenerative amplifiers are subject to energy instability at high pulse repetition rates due to period doubling bifurcation. Theoretical concepts representing a generalized picture of operation features have been in-detail worked out in order to differentiate and understand instability effects. Experimental data for $Nd:YVO_4$ regenerative amplifier have been presented; and possible techniques for performance optimization have been analyzed. An increase in the seed pulse energy has been demonstrated to improve amplification dynamics. Addition of a preamplifier is shown to be not only a convenient means for regenerative amplification investigation at a wide range of seed energies but also an efficient way to get top performance in practice. The $Nd:YVO_4$ preamplifier delivered seed energy high enough to provide stable operation at repetition rates up to 200 kHz with average output power near the theoretical limit.

We have not performed appropriate experiments to elaborate amplification dynamics for ytterbium based regenerative amplifiers. However, it would be interesting to verify theoretical predictions for ytterbium doped crystals as for media exhibiting pronounced quasi-three-level behavior. Especially, because similar types of crystals may exhibit different sensitivity to bifurcations in the same setup, as reported by (Buenting et al., 2009) and (Sayinc et al., 2009). In addition, bulk ytterbium doped materials are low gain materials and therefore the preamplifier technique is not so easily applicable to such systems. There is demand to create another method to keep regenerative amplifiers stable at maximum output power, possibly some kind of feedback. Such an idea was formulated by (Dörring etal., 2004) but as far as we know neither theoretical modeling nor practical realization of this approach has been reported so far. Essentially, this can be an attempt to stabilize the inherently unstable balance between pumping and inversion depletion – that is constant forced return of the system state to the originally repulsive fixed point. Actually the initial gain of amplification phase should be kept constant by the feedback. However, straightforward engineering does not work in this case because output energy signal of the current operation cycle does not contain sufficient information to adequately control pump power of the subsequent cycle. So elaboration of this problem can be a goal for further theoretical research and experimental work.

APPENDIX. MULTI-PASS B INTEGRAL FOR REGENERATIVE AMPLIFIER

The conventional quantitative gauge of the Kerr effect in laser systems is the B integral, nonlinear on-axis phase shift which light waves with wavelength λ undergo propagating through the media:

$$B_{\varphi} = \frac{2\pi}{\lambda} \int n_2(z) I(z) dz.$$

$$(17)$$

The terms $I(z)$ and $n_2(z)$ are distributions of on-axis intensity and nonlinear refractive index along current coordinate z. In order to evaluate the Kerr effect accumulated during regenerative amplification the integration should be performed over all the roundtrips of the optical cavity (multi-pass B integral). Then the full integration length is a product of the optical cavity pass length and round trip number. In an approximation of relatively low single pass gain (the integral within the gain medium can be replaced with the average) and also for moderate Kerr effect influence (iteration of the intensity profile in the optical resonator is not disturbed too much by self focusing) we can replace the overall integral with a sum of single pass integrals:

$$B \approx \frac{2\pi}{\lambda} \sum_{NRT} \int n_2(z) I_{NRT}(z) dz,$$

(18)

where index NRT implies summation over round trips. The sum of integrals is equal to the integral of the sum and also $n_2(z)$ is independent of round trip number function, then we obtain:

$$B \approx \frac{2\pi}{\lambda} \int n_2(z) \sum_{NRT} I_{NRT}(z) dz.$$

(19)

In an assumption of Gaussian beam shape for which the peak intensity is equal to $2P/(\pi w^2)$ (the term P is the optical power, the term w is the Gaussian beam radius) we can calculate the intensity for the pulse duration Δt in terms of pulse energy:

$$\sum_{NRT} I_{NRT}(z) \approx \frac{2}{\pi w^2(z) \Delta t} \sum_{NRT} E_{NRT}.$$

(20)

Summation of energies can be rewritten as time integration, and then dimensional current energy can be represented in terms of normalized energy which we have introduced in section 2.2:

$$\sum_{NRT} E_{NRT} \approx \frac{1}{T_0} \int_0^{NRT \cdot T_0} E(t) dt = \frac{A_a F_{sat} G_0}{T_0} \int_0^{NRT \cdot T_0} \varepsilon(t) dt = A_a F_{sat} \int_0^{\tau} \varepsilon(t^{\sim}) dt^{\sim}.$$

At that, the limit of integration NRT·T0 (overall pulse propagation time) is reducing to the effective round trip number τ when the integration variable, current dimensional time t, has been transformed to the normalized time $t^{\sim}$. Also, the laser material properties, responsible for stimulated emission, are combined to a conventional macroscopic term, saturation fluence, the ratio of photon energy and stimulated emission cross section $F_{sat} = \hbar \omega / \sigma$. The beam area in the active medium obviously can be expressed through the Gaussian beam radius: $A_a = \pi w_a^2$. And finally, using straightforward transformation and normalized energy integral calculated in section 2.8 (Eq. 14), we derive explicit expression of B integral, the diminished form of which has been used in section 2.9 (Eq. 16):

$$B = B_1 \int_0^{\tau} \varepsilon(t^{\sim}) dt^{\sim}, \text{ where } B_1 = \frac{4\pi}{\lambda} \frac{F_{sat}}{\Delta t} \int n_2(z) \left[\frac{w_a}{w(z)} \right]^2 dz \text{ and } \int_0^{\tau} \varepsilon(t^{\sim}) dt^{\sim} = \frac{g_{ef} - g_{ai} - \varepsilon_f}{g_1}.$$

(22)

We have obtained the multi-pass B integral for evaluation of the Kerr effect in regenerative amplifiers as the product of two factors. The first factor, B_1 represents attributes of the system geometry, material parameters and optical pulse duration. Essentially, this term is single-pass B integral calculated for the Gaussian beam in given optical cavity and for pulse energy fluence equal to the gain medium saturation fluence. The second factor is a function of regenerative amplifier regime represented in normalized terms (the dissipated energy divided by the threshold gain). This is convenient for practical application form in which functional physical contributions are separated.

ACKNOWLEDGEMENTS

The authors wish to acknowledge the technical assistance of Juozas Verseckas from EKSPLA UAB in preparation of the experimental setup, Vidmantas Gulbinas from Institute of Physics and Lucian Hand from Altos Photonics Inc. for fruitful discussions of the manuscript. This work was partially financed by the Eurostars Project E!4335-UPLIT.

REFERENCES

1.   K. Alligood, T. Sauer, J. Yorke, 1996 Chaos. An Introduction to Dynamical Systems, Springer-Verlag, 0-37894-677-2York

2.   F. Arecchi, R. Meucci, G. . Puccioni, J. Tredicce, 1982 Experimental Evidence of Subharmonic Bifurcations, Multistability, and Turbulence in a Q-Switched Gas Laser. Phys. Rev. Lett., 49 17 October 1982), 1217 1220 , 0031-9007

3.   C. Bibeau, S. Payne, H. Powell, 1995 Direct measurements of the terminal laser level lifetime in neodymium-doped crystals and glasses. J. Opt. Soc. Am. B, 12 10 October 1995), 1981 1992 , 0740-3224

4.   S. Biswal, J. Itatani, J. Nees, G. Mourou, 1998 Efficient energy extraction below the saturation fluence in a low-gain low-loss regenerative chirped-pulse amplifier. IEEE J. Sel. Top. Quantum Electron., 4 2 March 1998), 421 425 , 1077-260X

5.   D. Brawn, 1981 High Peak Power Nd:Glass Laser Systems, Springer-Verlag, 0-38710-516-6York

6.   U. Buenting, H. Sayinc, D. Wandt, U. Morgner, D. Kracht, 2009 Regenerative thin disk amplifier with combined gain spectra producing 500 µJ sub 200 fs pulses. Opt. Express, 17 10 May 2009), 8046-8050, 1094-4087

7.   D. Clubley, A. Bell, G. Friel, 2008 High average power Nd:YVO based pico-second regenerative amplifier. Proc. SPIE, 6871 February 2008), 68711D, 0277-786X

8.   J. Dörring, A. Killi, U. Morgner, A. Lang, M. Lederer, D. Kopf, 2004 Period doubling and deterministic chaos in continuously pumped regenerative amplifiers. Opt. Express, 12 8 April 2004), 1759 1768 ,

1094-4087

9.   M. Fermann, A. Galvanauskas, G. Sucha, 2002 Ultrafast Lasers: Technology and Applications, Marcel Dekker, 0-20391-020-6 York

10.   S. Forget, F. Balembois, P. Georges, P. Devilder, 2002 A new 3D multipass amplifier based on Nd:YAG or Nd:YVO$_4$ crystals. Appl. Phys. B, 75 4-5 (October 2002), 481-485, 0946-2171

11.   M. Grishin, V. Gulbinas, A. Michailovas, 2007 Dynamics of high repetition rate regenerative amplifiers. Opt. Express, 15 15 July 2007), 9434 9443 4-9443, 1094-4087

12.   M. Grishin, V. Gulbinas, A. Michailovas, J. Verseckas, 2008 Operation Features of Regenerative Amplifiers at High Repetition Rate, Technical digest in CD, paper CFB7, Conference on Lasers and Electro-Optics, CLEO-08, San-Chose, CA, USA, May 4-9, 2008, OSA

13.   M. Grishin, V. Gulbinas, A. Michailovas, 2009 Bifurcation suppression for stability improvement in Nd:YVO$_4$ regenerative amplifier. Opt. Express, 17 18 August 2009), 15700-15708, 1094-4087

14.   H. Haken, 1975 Analogy between higher instabilities in fluids and lasers. Physics Letters A, 53 1 (May 1975) 77 78 , 0375-9601

15.   Y. Jeong, J. Sahu, D. Payne, J. Nilsson, 2004 Ytterbium-doped large-core fiber laser with 1.36 kW continuous-wave output power. Opt. Express, 12 25 December 2004), 6088 6092, 1094-4087

16.   J. Kawanaka, K. Yamakawa, H. Nishioka, K. Ueda, 2003 30-mJ, diode-pumped, chirped-pulse Yb:YLF regenerative amplifier. Opt. Lett., 28 21 November 2003), 2121 213 , 0146-9592

17.   J. Kleinbauer, R. Knappe, R. Wallenstein, 2005 13-W picoseconds Nd:GdVO4 regenerative amplifier with 200-kHz repetition rate. Appl. Phys. B, 81 2-3 (July 2005), 163-166, 0946-2171

18.   J. Kleinbauer, D. Eckert, S. Weiler, D. Sutter, 2008 80 W ultrafast CPA-free disk laser, Proc. SPIE, 6871 February 2008), 68711B, 0277-786X

19.   W. Koechner, 2006 Solid-State Laser Engineering, Springer, 978-0-38729-094-2 USA

20.   C. Liu, T. Riesbeck, X. Wang, J. Ge, Z. Xiang, J. Chen, H. Eichler, 2008 Influence of spherical aberrations on the performance of dynamically stable resonators, Optics Communications, 281 20 October, 2008) 5222 5228 0030-4018.

21.   H. Liu, C. Gao, J. Tao, W. Zhao, Y. Wang, 2008 Compact tunable high power picosecond source based on Yb-doped fiber amplification of gain switch laser diode. Opt. Express, 16 11 May 2008), 7888-7893, 1094-

4087

22. E. Lorenz, 1963 Deterministic nonperiodic flow. Journal of the Atmospheric Sciences, 20 2 March 1963), 130 141 , 0022-4928

23. W. Lowdermilk, J. Murray, 1980 The multipass amplifier: theory and numerical analysis. J. Appl. Phys., 51 5 May 1980), 2436 2444 , 0021-8979

24. V. Magni, 1986 Resonators for solid-state lasers with large-volume fundamental mode and high alignment stability. Applied Optics, 25 1 January 1986), 107 118 0003-6935

25. Matsushima, H. Yashiro, T. Tomie, 2006 10 kHz 40 W Ti:sapphire regenerative ring amplifier. Opt. Lett., 31 13 July 2006), 2066 2068, 0146-9592

26. J. Meijer, K. Dub, A. Gillner, D. Hoffmann, V. Kovalenko, T. Masuzawa, A. Ostendorf, R. Poprawe, W. Schulz, 2002 Laser Machining by short and ultrashort pulses, state of the art and new opportunities in the age of the photons. CIRP Annals- Manufacturing Technology, 51 2 February 2002), 531 550 , 0007-8506

27. T. Metzger, A. Schwarz, C. Teisset, D. Sutter, A. Killi, R. Kienberger, F. Krausz, 2009 High-repetition-rate picosecond pump laser based on a Yb:YAG disk amplifier for optical parametric amplification. Opt. Lett., 34 14 July 2009), 2123-2125, 0146-9592

28. G. Mourou, D. Umstadter, 1992 Development and Applications of Compact High-Intensity Lasers. Phys. Fluids B, 4 7 July 1992), 2317-2325, 0899-8213

29. D. Müller, A. Giesen, H. Hügel, 2003 Picosecond thin-disk regenerative amplifier. Proceedings of SPIE, 5120 November 2003), 281 286 , 0277-786X

30. J. Murray, W. Lowdermilk, 1980 Nd:YAG regenerative amplifier. J. Appl. Phys., 51 7 July 1980), 3548-3555, 0021-8979

31. D. Nickel, C. Stolzenburg, A. Bevertt, A. Geisen, J. Haüssermann, F. Butze, M. Leitner, 2005 200 kHz electro-optic switch for ultrafast laser systems. Rev. Sci. Instrum., 76 3 March 2005), 033111 033117 0034-6748

32. T. Norris, 1992 Femtosecond pulse amplification at 250 kHz with a Ti:sapphire regenerative amplifier and application to continuum generation. Opt. Lett., 17 14 July 1992), 1009 1011 , 0146-9592

33. R. Peterson, H. Jenssen, A. Cassanho, 2002 Investigation of the spectroscopic properties of Nd:YVO$_4$. In: Proc. OSA TOPS, Advanced

Solid-State Lasers, M.E. Fermann and L.R. Marshall, (Ed.), 68 294-298, OSA

34. Pugžlys, G. Andriukaitis, A. Baltuška, L. Su, J. Xu, H. Li, R. Li, W. Lai, P. Phua, A. Marcinkevičius, M. Fermann, L. Giniūnas, R. Danielius, S. Ališauskas, 2009 Multi-mJ, 200-fs, CW-pumped, cryogenically cooled, Yb,Na:CaF2 amplifier. Opt. Lett., 34 13 July 2009), 2075-2077, 0146-9592

35. G. Raciukaitis, M. Grishin, R. Danielius, J. Pocius, L. Giniūnas, 2006 High repetition rate ps- and fs- DPSS lasers for micromachining. Congress Proceedings (in CD), 99 Paper M1001, 0-91203-585-4 Congress on Applications of Lasers & Electro- Optics, ICALEO 2006, Scottsdale, USA, October 30- November 2, 2006, Laser Institute of America

36. F. Röser, T. Eidam, J. Rothhardt, O. Schmidt, D. Schimpf, J. Limpert, A. Tünnermann, 2007 Millijoule pulse energy high repetition rate femtosecond fiber chirped-pulse amplification system. Opt. Lett., 32 12 December 2007), 3495 3497 , 0146-9592

37. Ross, G. New, P. Bates, 2007 Contrast limitation due to pump noise in an optical parametric chirped pulse amplification system. Optics Communications, 273 2 May 2007), 510 514 , 0030-4018

38. H. Sayinc, U. Buenting, D. Wandt, J. Neumann, D. Kracht, 2009 Ultrafast high power Yb:KLuW regenerative amplifier. Opt. Express, 17 17 August 2009), 15068 15071 , 1094-4087

39. M. Siebold, M. Hornung, J. Hein, G. Paunescu, R. Sauerbrey, T. Bergmann, G. Hollemann, 2004 A high-average-power diode-pumped Nd:YVO$_4$ regenerative laser amplifier for picosecond-pulses. Applied Physics B, 78 3-4 (February 2004), 287-290, 0946-2171

40. M. Siebold, J. Hein, M. Hornung, S. Podleska, M. Kaluza, S. Bock, R. Sauerbrey, 2008 Diode-pumped lasers for ultra-high peak power. Appl. Phys. B, 90 3-4 (March 2008), 431 437 , 0946-2171

41. J. Speiser, A. Giesen, 2008 Scaling of thin disk pulse amplifiers. Proc. SPIE, 6871 February 2008), 68710J, 0277-786X

42. D. Strickland, G. Mourou, 1985 Compression of amplified chirped optical pulses. Opt. Comm., 56 3 March 1985), 219 221 , 0030-4018

43. O. Svelto, 1998 Principles of Lasers, Plenum Press, 0-30645-748-2York

44. D. Tang, S. Ng, L. Qin, X. Meng, 2003 Deterministic chaos in a diode-pumped Nd:YAG laser passively Q switched by a Cr 4+ YAG crystal. Opt. Lett., 28 5 March 2003), 325-327, 0146-9592

45. S. Valling, T. Fordell, A. Lindberg, 2005 Experimental and numerical intensity time series of an optically injected solid state laser. Opt. Commun., 254 4-6 (October 2005), 282 289 -, 0030-4018

46. B. Walker, C. Toth, D. Fittinghoff, T. Guo, D. Kim, C. Rose-Petruck, J. Squier, K. Yamakawa, K. Wilson, B. Barty, 1999 A 50 EW/cm2 Ti:sapphire laser system for studying relativistic light-matter interactions. Opt. Express, 5 10 November 1999), 196-202, 1094-4087

Chapter 5

LOW-VOLTAGE FULLY DIFFERENTIAL CMOS SWITCHED-CAPACITOR AMPLIFIERS

Tsung-Sum Lee[1]

[1]National Yunlin University of Science and Technology, Taiwan (R.O.C.)

INTRODUCTION

Analog signal amplification in discrete-time system can be performed by switched-capacitor amplifiers (Martin et al., 1987). Switched-capacitor amplifier has been used in the design of digital-to-analog converter (Yang & Martin, 1989). The schematic for the switched-capacitor amplifier is shown inFigure 1.

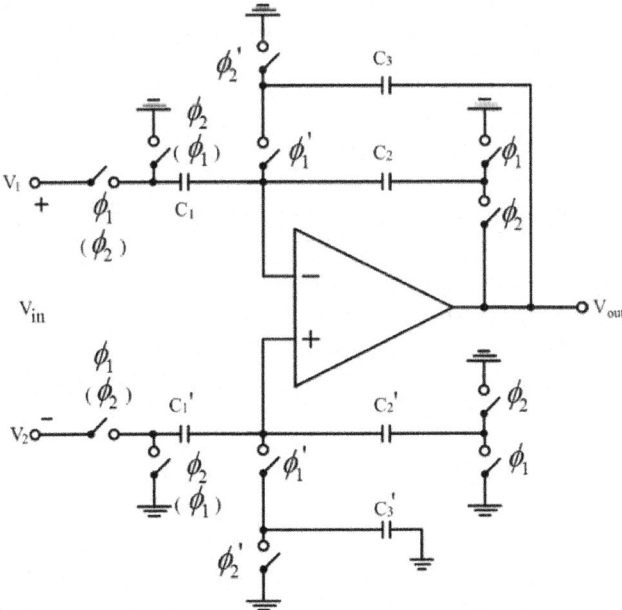

Figure 1: A differential-to-single-ended CMOS switched-capacitor amplifier. Depending on the input-stage clock signals, the amplifier can be either noninverting (as shown) or inverting (input-stage clocks shown in parentheses).

Assuming an infinite op amp gain, the output voltage at end of ϕ_2 is given by

$$V_{out}(nT) = \frac{C_1}{C_2}V_{in}(nT - \frac{T}{2}),$$

(1)

irrespective of the op amp offset voltage. If the clock waveforms shown in parentheses are used, then an inverting function is realized, and

$$V_{out}(nT) = -\frac{C_1}{C_2}V_{in}(nT),$$

(2)

again independent of the op amp input offset voltage. During the reset phase (ϕ_1) C_3 is connected in feedback around the op amp which causes the output change only by the op amp input offset voltage. The switches are realized as CMOS transmission gate. For low supply voltages, a conductance gap begins to appear around the middle of the supply range (Crols & Steyaert, 1994). This means that under low-voltage operation, this configuration no longer works. Existing solutions of low-voltage operation of switched-capacitor circuits include using low threshold voltage process (Matsuya & Yamada, 1994), switched-opamp technique (Baschirotto & Castello, 1997, Cheung et al., 2001,Cheung et al., 2002, Cheung et al., 2003, Crols & Steyaert, 1994, Peluso et al., 1997, Peluso et al., 1998, Sauerbrey et al., 2002, Waltari & Halonen, 2001, Wu et al., 2007), opamp-reset switching technique (Chang, & Moon, 2003, Keskin et al., 2002, Wang &. Embabi, 2003), voltage multiplier (charge pump) technique (Nicollini et al., 1996, Rombouts et al., 2001), clock multiplier (clock booster) technique (Au & Leung, 1997, Rabii & Wooley, 1997), and bootstrapping switch technique (Abo & Gray, 1999, Dessouky & Kaiser, 2001, Park et al., 2004). First, the use of low-threshold transistors involves special and high-cost technology (Matsuya & Yamada, 1994). The switched-opamp technique (Baschirotto & Castello, 1997, Cheung et al., 2001, Cheung et al., 2002, Cheung et al., 2003, Crols & Steyaert, 1994, Peluso et al., 1997, Peluso et al., 1998, Sauerbrey et al., 2002,Waltari & Halonen,

2001, Wu et al., 2007) and opamp-reset switching technique (Chang, & Moon, 2003, Keskin et al., 2002, Wang &. Embabi, 2003) can only be applicable to filters, delta-sigma modulators, and pipelined analog-to-digital converters. The main limitations of voltage multiplier (charge pump) technique (Nicollini et al., 1996, Rombouts et al., 2001) regards: the gate-oxide breakdown reliability, the need to supply a dc current to the op amps from the multiplied supply (this necessitates the use of an external capacitor, with additional cost), and the conversion efficiency of the charge pump (which is lower than 100%). The clock multiplier (clock booster) technique (Au & Leung, 1997, Rabii & Wooley, 1997) suffers from the technology limitation associated with the gate oxide breakdown. Device reliability can be assured in the bootstrapped switch technique (Abo & Gray, 1999, Dessouky & Kaiser, 2001, Park et al., 2004), owing to keeping the terminal-to-terminal voltages of the MOSFET devices within the rated operating supply voltage of the technology. The bootstrapped switch provides a small, nearly constant input resistance. The switch linearity is also improved, and signal-dependent charge injections is reduced.

To improve the overall linearity, minimize the effect of common-mode interference and noise, the fully differential approach has obtained wider acceptance for accurate and/or high-speed signal processing. The switched-capacitor amplifier in (Martin et al., 1987) is a differential-to-single-ended design. A fully differential switched–capacitor amplifier using series compensation MOSFET capacitors has been presented in (Yoshizawa et al., 1999). However its operating voltage is ±2.5-V. Consequently there is an increasing demand to extend these improvements to this circuit.

This chapter describes the design of two 1V fully differential CMOS switched-capacitor amplifiers in a standard CMOS technology using improved bootstrapped switches. In section 2, the circuit realization of these two switched-capacitor amplifiers is addressed. In section 3 the circuit design of low-voltage building blocks is described. Experimental results are presented in section 4 to support the ideas put forth in paper. Finally conclusion is given.

CIRCUIT DESCRIPTION

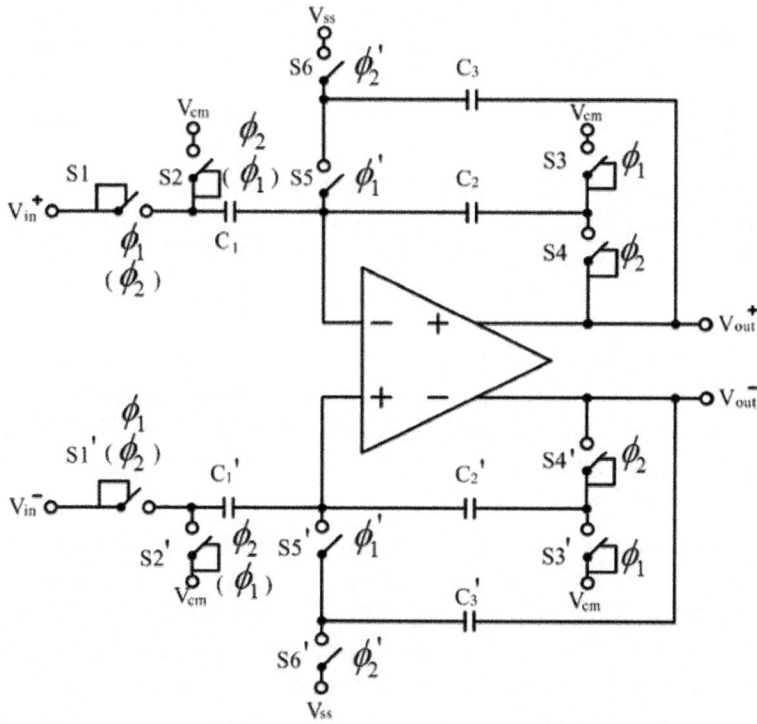

Figure 2: First low-voltage fully differential CMOS switched-capacitor amplifier. Depending on the input-stage clock signals, the amplifier can be either noninverting (as shown) or inverting (input-stage clocks shown in parentheses).

Figure 2 shows the first low-voltage fully differential CMOS switched-capacitor amplifier based on improved bootstrapped switches described in section 3.2, where switches S1-S4 and S1'-S4' are matched improved bootstrapped switch pairs and switches S5-S6 and S5'-S6' are NMOS matched switch pairs. In order to minimize the number of improved bootstrapped switches, two analog reference voltages are used: VSS at the op amp input where a normal NMOS switch can be used to switch the lowest supply voltage, and a $\frac{V_{DD} + V_{SS}}{2}$ common-mode voltage at the op amp output and the circuit input to maximize the signal swing. The improved bootstrapped switch is used to switch signals at this voltage level. Figure 3 is the single-ended version of Figure 2.

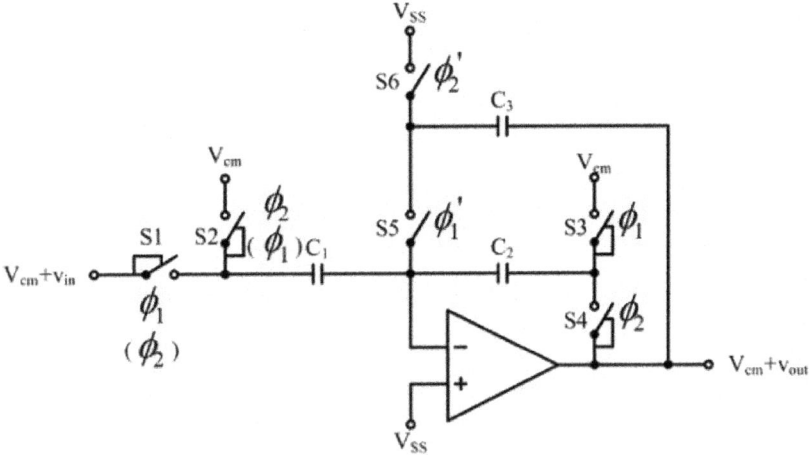

Figure 3: Single-ended version of Figure 2.

To see how this circuit operates, consider the inverting circuit during the reset phase (ϕ_1) and during valid output phase (ϕ_2), as shown in Figure 4. Then based on charge conservation principle we can write:

$$C_1(V_{SS} + V_{off} - V_{cm}) + C_2(V_{SS} + V_{off} - V_{cm})$$

$$= C_1[V_{SS} + V_{off} - V_{cm} - v_{in}(nT)] + C_2[V_{SS} + V_{off} - V_{cm} - v_{out}(nT)],$$

$$\text{or } v_{out}(nT) = \frac{C_1}{C_2} v_{in}(nT).$$

$$(3)$$

It should be noted that the clock waveforms with the primed superscripts change before the nonprimed waveforms in order to reduce nonlinearities due to charge injection.

Another technique to further reduce the number of improved bootstrapped switches is shown in Figure 5, where switches S1 and S4 and S1' and S4' are matched improved bootstrapped switch pairs. Those switches connected to VSS are realized with NMOS transistors, while those switches connected to VDD are realized with PMOS transistors. In Figure 5 a single reference voltage at VSS is used. However, the signal still varies around VDD+VSS at the circuit input as well as at the op amp output to preserve the maximum swing. The difference between the two reference voltages is compensated by injecting a fixed amount of charge at the op amp input using extra capacitor pairs $C_{M1} = \frac{C_1}{2}$ and $C_{M2} = \frac{C_2}{2}$ ($C'_{M1} = \frac{C'_1}{2}$ and $C'_{M2} = \frac{C'_2}{2}$) and (Baschirotto & Castello, 1997). Figure 6 is the single-ended version of Figure 5.

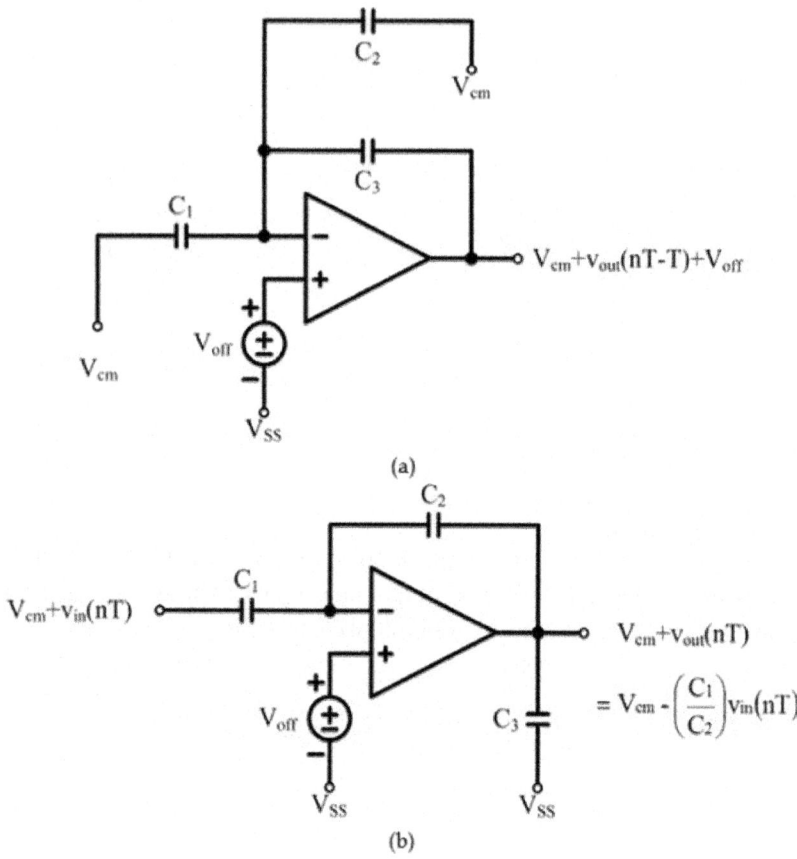

Figure 4: Single-ended CMOS switched-capacitor amplifier, (a) during reset phase (ϕ1), (b) during valid output phase (ϕ2).

To see how this circuit operates, consider the inverting circuit during the reset phase (ϕ1) and during valid output phase (ϕ2), as shown in Figure 7.

Then based on charge conservation principle we can write:

$$C_1(V_{SS}+V_{off}-V_{SS})+C_2(V_{SS}+V_{off}-V_{SS})+(C_{M1}+C_{M2})(V_{SS}+V_{off}-V_{DD})$$

$$=C_1[V_{SS}+V_{off}-V_{cm}-v_{in}(nT)]+C_2[V_{SS}+V_{off}-V_{cm}-v_{out}(nT)]$$
$$+(C_{M1}+C_{M2})(V_{SS}+V_{off}-V_{SS})$$

$$\text{or } v_{out}(nT)=-\frac{C_1}{C_2}v_{in}(nT).$$

(4)

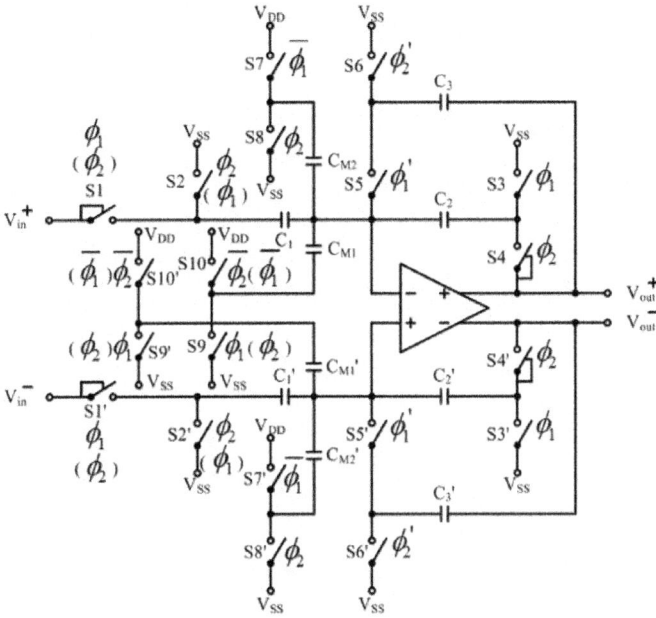

Figure 5: Second low-voltage fully differential CMOS switched-capacitor amplifier. Depending on the input-stage clock signals, the amplifier can be either noninverting (as shown) or inverting (input-stage clocks shown in parentheses).

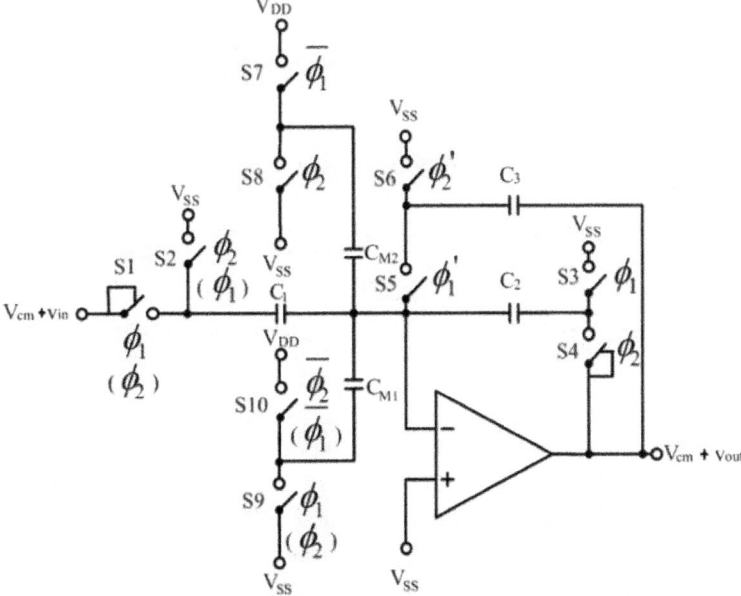

Figure 6: Single-ended version of Figure 5.

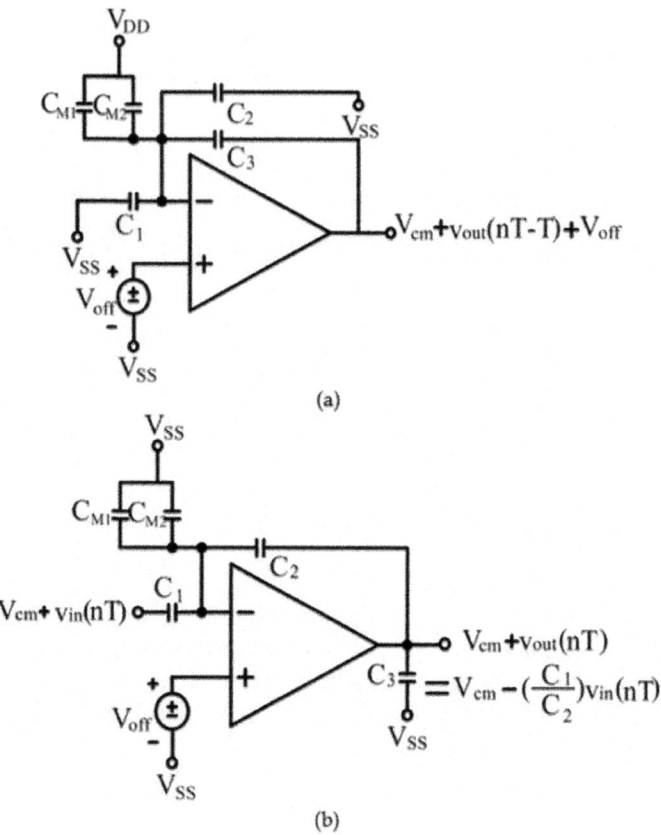

Figure 7: Single-ended CMOS switched-capacitor amplifier, (a) during reset phase (ϕ1), (b) during valid output phase (ϕ2)

LOW-VOLTAGE BUILDING BLOCKS

In this section, the low-voltage circuit building blocks used in the two fully differential CMOS switched-capacitor amplifiers are discussed

OP AMP

Figure 8 shows the used op amp. It is based on a fully differential folded-cascode p-type two-stage Miller-compensated configuration. The second stage is a common-source amplifier with active load which also allows a large output swing. In order to avoid the common-mode feedback (CMFB) circuit for the first stage, transistors $M51$, $M52$, $M61$, and $M62$ are used, which is similar to (Waltari & Halonen, 1998). For the second stage, a simple passive switched-capacitor CMFB circuit, shown inFigure 9, is used. The improved bootstrapped switches are used to connect and disconnect the common-mode sensing capacitor.

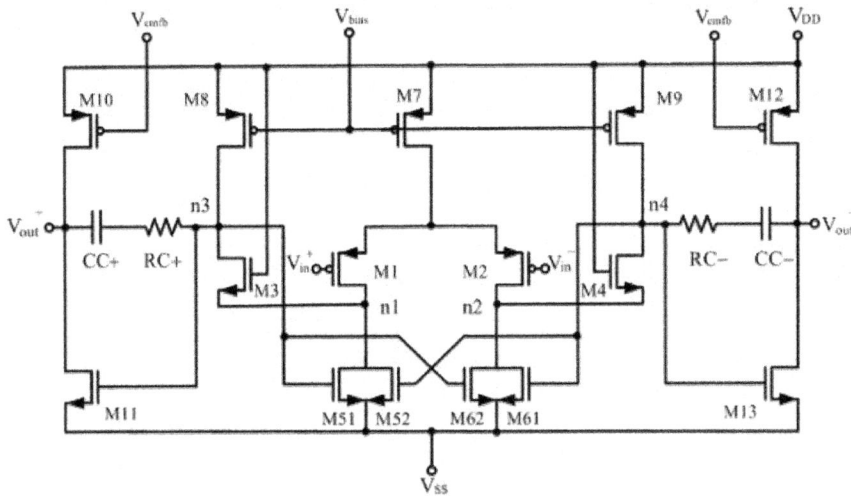

Figure 8: Low-voltage op amp.

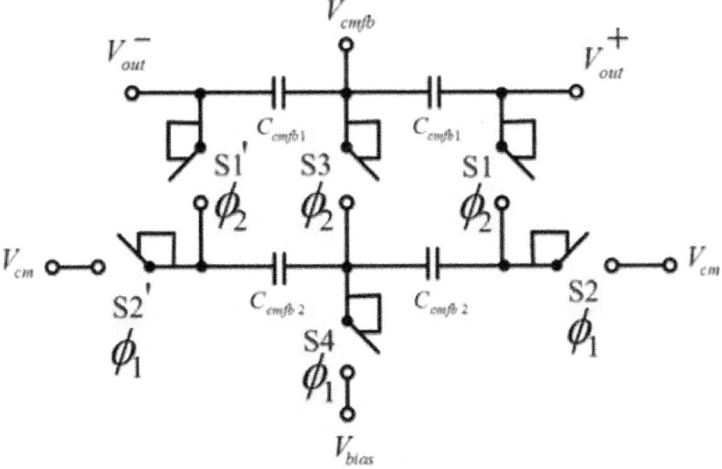

Figure 9: Common-mode feedback circuit for the low-voltage op amp.

Improved Bootstrapped Switch

The improved bootstrapped switch shown in Figure 10 is utilized in the proposed circuit. The circuitry is improved version of that presented in (Abo

& Gray, 1999). In the circuit presented in (Abo & Gray, 1999), the voltage at the drain side of the main switch *M11* must be always higher than that at the source side at the switching moment to prevent the gate-drain voltage from exceeding VDD during the turn-on transient. In order to overcome this limitation, an additional transistor *M14* has been added on the drain side, such that the switch *M11* becomes completely symmetrical. This bootstrapping circuit thus allows switch operation (transistor *M11*) from rail-to-rail while limiting all gate-source/drain voltages to VDD avoiding any oxide overstress.

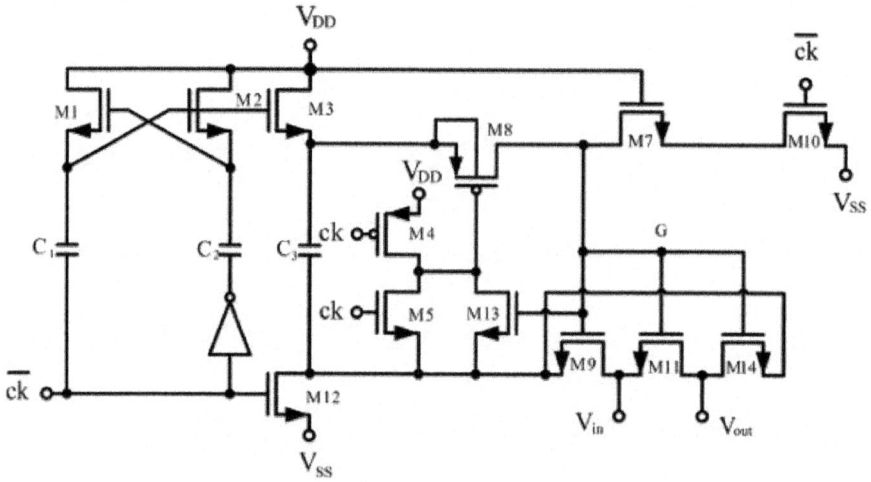

Figure 10: Improved bootstrapped switch.

EXPERIMENTAL RESULTS

Based on the principles presented earlier, we have designed two 1-V fully differential CMOS switched-capacitor amplifiers. These two switched-capacitor amplifiers were operated with ±0.5-V. The capacitor sizes used were C_1 =1.25-pF, C_2 =0.25-pF, and C_3 =0.25-pF, for a nominal gain of -5. The circuits of Figure 2 and Figure 5 were fabricated using a TSMC 0.35-μm double-poly four-metal CMOS technology. Figure 11 and Figure 12 show the photomicrographs of Figure 2 and Figure 5, respectively. The chip areas of Figure 2 and Figure 5 excluding bonding pads are 414×278-μm² and 460×330-μm², respectively.

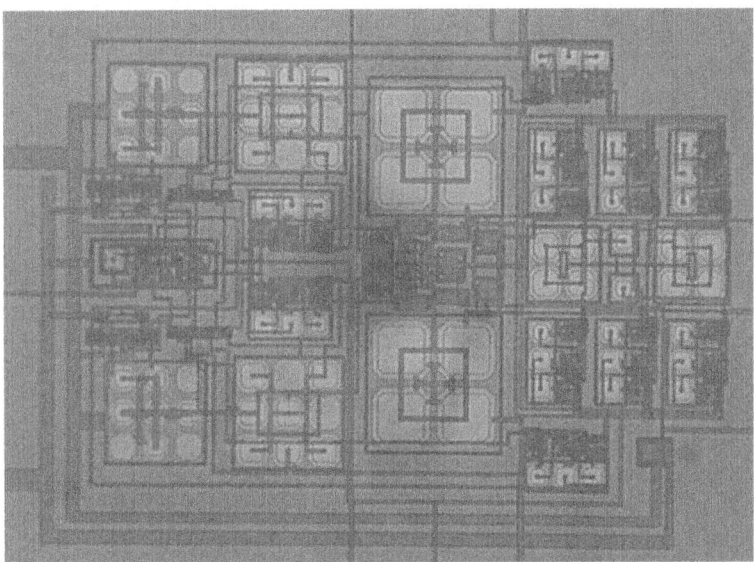

Figure 11: Photomicrograph of Figure 2.

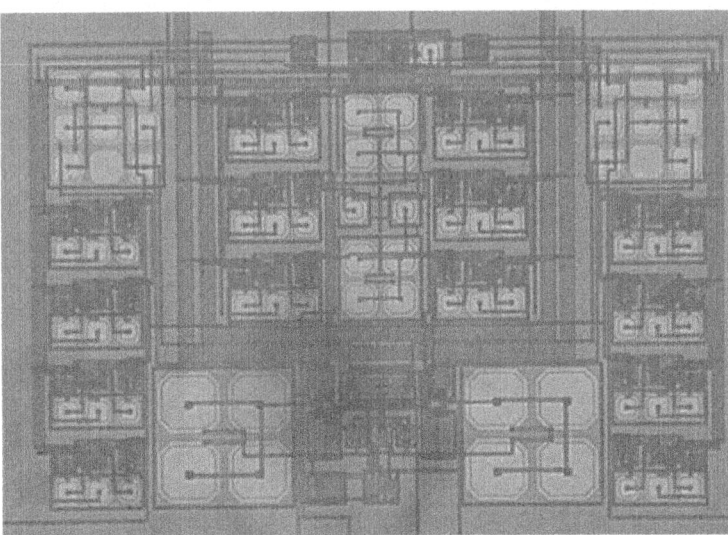

Figure 12: Photomicrograph of Figure 5.

Two figures of the measured input/output waveforms for 0.2V peak-to-peak sinusoidal differential input signal are shown in Figure 13 and Figure 14, respectively. The input signal was at 10kHz whereas the clock signal was at 1MHz. It can be seen that the gain is very close to the nominal value of -5.

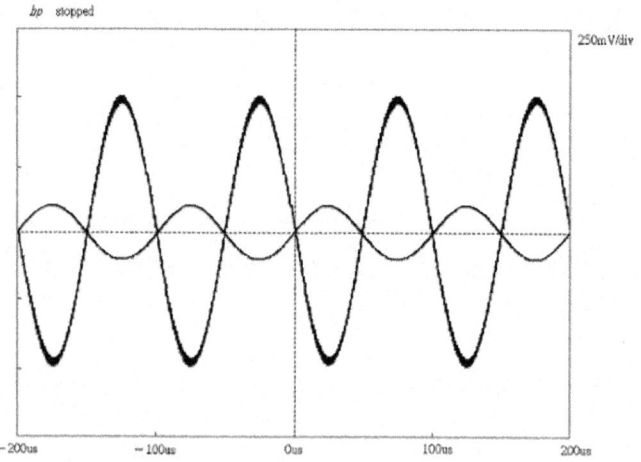

Figure 13. Measured differential input and output waveforms of Figure 2 (f_{clk}=1-MHz, f_{in}=10-kHz, sinusoidal differential input voltage=0.2-V_{pp}).

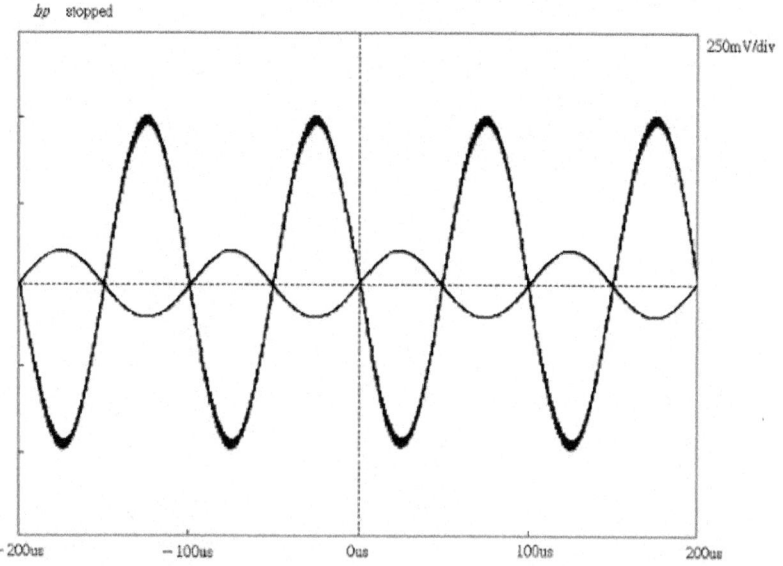

Figure 14: Measured differential input and output waveforms of Figure 5 (f_{clk}=1-MHz, f_{in}=10-kHz, sinusoidal differential input voltage=0.2-V_{pp}).

Figure 15 and Figure 16 show the resulting output spectrum. As shown in Figure 15 and Figure 16, the even-order harmonics have been largely attenuated by the fully differential topology and 59dB and 52dB spurious-free dynamic range (SFDR) are exhibited, respectively. The circuits of Figure

2 andFigure 5 dissipate 206.5μW and 206.6μW, respectively with a 1V power supply.

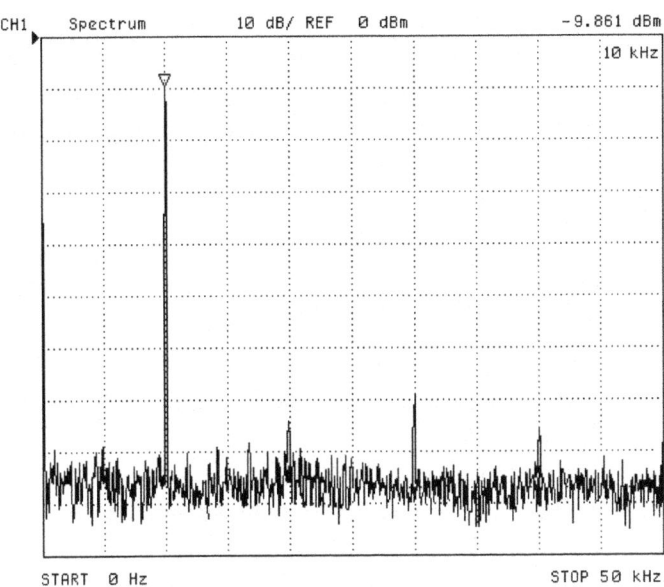

Figure 15: Measured output spectrum of Figure 2.

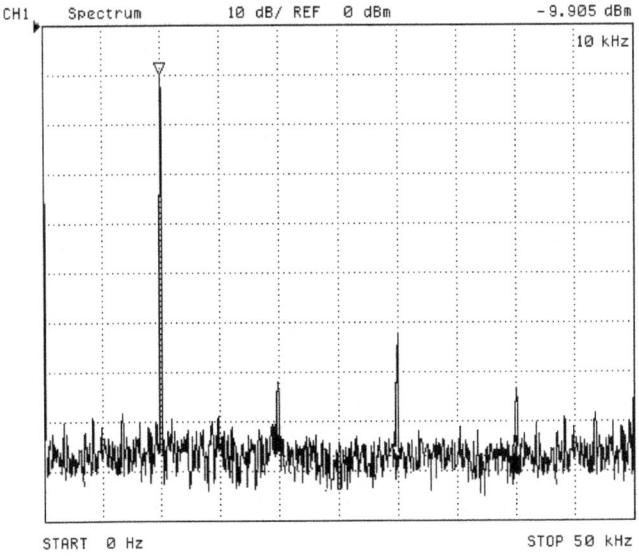

Figure 16: Measured output spectrum of Figure 5.

CONCLUSION

Two fully differential CMOS 1-V switched-capacitor amplifiers have been described. Rail-to-rail operation of improved bootstrapped switches allows very low voltage robust switched-capacitor designs in standard CMOS technologies while avoiding transistor gate oxide overstress. The circuits have been fabricated and all aspects of their performance have been confirmed.

REFERENCES

1.   A. M. Abo, P. R. Gray, 1999 A 1.5-V, 10-bit, 14.3-MS/s CMOS pipeline analog-to-digital converter., IEEE J. Solid-State Circuits, May, 34 599 606 ,0018-9200

2.   S. Au, B. H. Leung, 1997 A 1.95-V, 0.34-mW, 12-b sigma-delta modulator stabilized by local feedback loops., IEEE J. Solid-State Circuits, March, 32 321 328 , 0018-9200

3.   A. Baschirotto, R. Castello, 1997 A 1-V 1.8-MHz CMOS switched-opamp SC filter with rail-to-rail output swing., IEEE J. Solid-State Circuits, December, 32 1979 1986 , 0018-9200.

4.   D. Y. Chang, U. K. Moon, 2003 A 1.4-V 10-bit 25-MS/s pipelined ADC using opamp-reset switching technique., IEEE J. Solid-State Circuits, August, 38 1401 1404 , 0018-9200

5.   V. S. L. Cheung, et al. 2001 A 1-V CMOS switched-opamp switched-capacitor pseudo-2-path filter., IEEE J. Solid-State Circuits, Jan.2001, 36 14 22 , 0018-9200

6.   V. S. L. Cheung, et al. 2002 A 1-V 10.7-MHz switched-opamp bandpass$\sum\Delta$ modulator using double-sampling finite-gain-compensation technique., IEEE J. Solid-State Circuits, October, 37 1215 1225 , 0018-9200.

7.   V. S. L. Cheung, et al. 2003 A 1-V 3.5-mW CMOS switched-opamp quadrature IF circuitry for Bluetooth receivers, IEEE J. Solid-State Circuits, May., 38 805 816 , 0018-9200.

8.   J. Crols, M. Steyaert, 1994 Switched-opamp: an approach to realize full CMOS switched-capacitor circuits at very low power supply voltage, IEEE J. Solid-State Circuits, August, 29 936 942 , 0018-9200.

9.   M. Dessouky, A. Kaiser, 2001 Very low-voltage digital-audio $\sum\Delta$ modulator with 88-dB dynamic range using local switch bootstrapping, IEEE J. Solid-State Circuits, March, 36 349 355 , 0018-9200.

10.   M. Keskin, et al. 2002 A 1-V 10-MHz Clock-Rate 13-Bit CMOS $\sum\Delta$ modulator using unity-gain-reset opamps, IEEE J. Solid-State Circuits, July, 37 817 824 , 0018-9200.

11. K. Martin, et al. 1987 A differential switched-capacitor amplifier, IEEE J. Solid-State Circuits, February, 22 104 106 , 0018-9200.

12. Y. Matsuya, J. Yamada, 1994 1-V power supply, low-power consumption A/D conversion technique with swing-suppression noise shaping, IEEE J. Solid-State Circuits, December, 29 1524 1530 , 0018-9200.

13. G. A. Nicollini, et al. 1996 A-80dB THD, 4-Vpp switched capacitor filter for 1.5-V battery-operated systems., IEEE J. Solid-State Circuits, August, 31 1214 1219 , 0018-9200.

14. J. B. Park, et al. 2004 A 10-b 150-MSample/s 1.8-V 123-mW CMOS A/D converter with 400-MHz input bandwidth., IEEE J. Solid-State Circuits, August, 39 1335 1337 , 0018-9200..

15. V. Peluso, et al. 1997 A 1.5-V 100-μW $\sum\Delta$ modulator with 12-b dynamic range using the switched-opamp technique., IEEE J. Solid-State Circuits, July, 32 943 952 , 0018-9200

16. V. Peluso, et al. 1998 A 900-mV low-power $\sum\Delta$ A/D converter with 77-dB dynamic range.," IEEE J. Solid-State Circuits, December, 33 1887 1897 , 0018-9200.

17. S. Rabii, B. A. Wooley, 1997 A 1.8-V digital-audio sigma-delta modulator in 0.8-μm CMOS., IEEE J. Solid-State Circuits, June, 32 783 796 , 0018-9200.

18. P. Rombouts, et al. 2001 A 13.5-b 1.2-V micropower extended counting A/D converter., "IEEE J. Solid-State Circuits, February, 36 176 183 , 0018-9200.

19. J. Sauerbrey, et al. 2002 A 0.7-V MOSFET-only switched-opamp $\sum\Delta$ modulators in standard digital CMOS technology., IEEE J. Solid-State Circuits, December, 37 1662 1669 , 0018-9200

20. M. Waltari, K. A. I. Halonen, 2001 1-V 9-Bit pipelined switched-opamp ADC," IEEE J. Solid-State Circuits, January, 36 129 134 , 0018-9200.

21. M. Waltari, K. Halonen, 1998 Fully differential switched opamp with enhanced common-mode feedback., Electron. Lett., November, 34 23 2181 2182 , 0013-5194..

22. L. . Wang, S. H. K. Embabi, 2003 Low-voltage high-speed switched-capacitor circuits without voltage bootstrapper, IEEE J. Solid-State Circuits, August, 38 1411 1415 , 0013-5194.

23. P. Y. Wu, et al. 2007 A 1-V 100-MHS/s 8-bit CMOS Switched-Opamp Pipelined ADC Using Loading-Free Architecture, IEEE J. Solid-State Circuits, April, 42 730 738 , 0013-5194.

24. J. W. Yang, K. W. Martin, 1989 High-resolution low-power D/A converter, IEEE J. Solid-State Circuits, October, 24 1458 1461 , 0013-5194.

25. H. Yoshizawa, et al. 1999 MOSFET-only switched-capacitor circuits in digital CMOS technology," IEEE J. Solid-State Circuits, June, 34 734 747 , 0013-5194.

Chapter 6

FLEXIBLE POWER AMPLIFIER ARCHITECTURES FOR SPECTRUM EFFICIENT WIRELESS APPLICATIONS

Alessandro Cidronali, Iacopo Magrini and Gianfranco Manes

Department of Electronics and Telecommunications, University of Firenze, Italy

INTRODUCTION

The wireless systems evolution known as "beyond the 3rd generation" (B3G) will make use of dynamic spectrum access techniques to provide wide bandwidth to mobile users via heterogeneous wireless networks. A consistent step toward this scenario is represented by the outcome of the last World Radio communication Conference [1] which established new primary frequency bands allocation spanning from the UHF band to low microwaves and thus reflecting the increasing demands for broadband mobile and cellular systems. We have become used to the doubling of processing power of chips based on Moore's law, but the progress in radio interface technologies still poses significant challenges. High spectrum efficiency performance becomes therefore another major requirement of the design, along with the more consolidated ones: energy efficiency, integration, cost and reliability.

While the IMT-advanced roadmap foresees a 100 Mbps data rate for mobile users and a peak of 1 Gbps is expected for nomadic users, the available spectrum for legacy wireless communications is fragmented and reaches the amount of 750 MHz in the S-C band. A radio technology that is expected to interact with a multi-services network should be able to change between different operating bands and adapt its features according with the different available standard and requirements. Most of the research efforts performed during the last years dealt with issues related to the physical layer of the communication stack [2]; however, despite the growing interest in multi-standard operation, less attention has been devoted to the radio-frequency front-end, which therefore remains one of the most challenging parts of a multi-band radio. One main reason for the delay in effectively implementing multi-standard transceivers can be attributed to the implementation of the RF transmit power amplifier (PA). Today, dedicated, single standard PAs achieve very good power added efficiency (PAE) and, in this way, long battery lifetime. Any multi-standard PA, needed for the support of different, not always predefined, communication systems, should compete with such dedicated solutions. A conceptual framework to this is provided by the so-called software-defined radio (SDR), i.e. a radio communication system, using software for the recon Figure ration of the digital and analog parts in order to perform the modulation and demodulation of radio signals, [3]. In practice, however, due to the difficulty of implementing the fast signal processing implied in the SDR approach, most of the systems on the market, based on more traditional approaches, are still supporting only a very limited number of standards (e.g. 4 GSM frequencies, UMTS and, possibly, Bluetooth). In the near future, further standards will have to be supported, and more could have to be added during the handset lifetime, hopefully without hardware reconFigureuration. This will determine the need of multiband PAs capable to transmit efficiently more than one service with variable radio access schemes.

Example of realizations in different technologies are provided in this Chapter as demonstrators of the discussed multiband design methods. The flexibility of the operative frequency is thus introduced by analyzing new PA architectures and design methodologies which consider the inclusion of tunable and switching components to enable a change in the operative frequency. A review of the most promising circuit topologies suitable to design reconFigureurable matching networks is given in this Chapter. Varactor diodes based and MOS switched based topologies are compared, highlighting their point of strength and weakness. It is shown as a concurrent dual-band PA implemented by the proper combination of frequency-dedicated PAs, each of them optimized to work in a given bandwidth would be an easy approach, it becomes unsuited due to the complexity of the power combiners. For this reason the true concurrent dual band PA presented in this chapter is to be considered as an enabler components for high efficiency multiband systems

Spectrum efficiency is just one of the challenges a wireless system designer faces, further come from linearity and energy efficiency resulting from the use of multicarrier and complex envelope modulation schemes. As the spectrum efficiency increases a more demanding requirement in term of PA linearity faces to the designers. Energy efficiency and linearity are conventionally traded-off considering that increasing the power back-off increases the linearity at the expenses of lower energy efficiency. To maintain signal integrity, the resulting waveforms in turn require linear transmission paths for their successful deployment. A way to match signal integrity and energy efficiency consists in the use of digital pre distortion algorithm applied at base-band and implemented in the digital section of the transmitter. In spite of their large development in frequency dedicated PA architectures, the development of a technique suitable for multi-band applications is not yet completely available. In this Chapter a comprehensive treatment a novel technique for Dual Band Digital Pre distortion (DB-DP) is discussed. The DB-PD is based on the simultaneous pre distortion of both channels at intermediate frequency (IF), it uses a single band memory polynomial DP for linearization, while the feedback path is based on a subsampling receiver. The memory polynomial DB-DP system is presented by simulation with Matlab Simulink® for a deep understanding of performance.

A POSSIBLE APPLICATIVE SCENARIO FOR MULTI-BAND TRANSMITTERS

Extending the scenario to already experienced 3G voice/data systems, users may be moving while simultaneously operating in a broadband data access or multimedia streaming session. The need to support low latency and low packet loss handovers of data streams as users transition from one access point to another may require the concurrent use of more than one frequency band at the time. For full-mobile data services, no user interaction will be required to adapt their service expectations because of environmental limitations that are technically challenging but not directly relevant to the user (such as being stationary or moving). The enabling front-end of future mobile unit thus will accommodate more than one system in a effective and efficient way to make possible the connectivity capabilities depicted in Figure. 1.

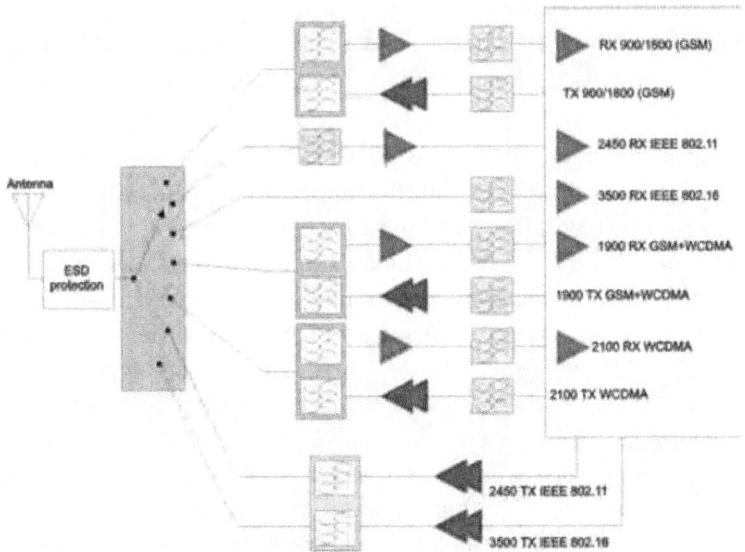

Figure. 1: Concept of a multi-band transmitter.

The Wireless Local Area Network (WLAN) industry has become one of the fastest growing segments of the communications industry. This growth is due, in large part, to the introduction of standards-based WLAN products, regulated by the IEEE 802.11. The expectation of the WLAN's continuing growth stems from the promise of new standardized WLAN technologies, from improved cost/performance of WLAN systems, and from the growing availability of WLAN solutions that consolidate voice, data, and mobility functions. This, combined with market forecasts reporting that WLAN will experience a continuous growth in the next years, show that WLAN technologies will play a significant role in the future and will have a significant impact on our business and personal life styles. The WiMAX is an alternative and complementing standard for high data rate transmission, which will transform the world of mobile broadband by enabling the cost-effective deployment of metropolitan area networks based on the IEEE 802.16 standard to support notebook PC and mobile users on move. There are many advantages of systems based on 802.16, e.g. the ability to provide service even in areas that are difficult for wired infrastructure to reach and the ability to overcome the physical limitations of traditional wired infrastructure. The standard will offer wireless connectivity of up to 30 miles. The major capabilities of the standard are its widespread reach, which can be used to set up a metropolitan area network, and its data capacity of 75 Mbps. This high-speed wireless broadband technology promises to open new, economically viable market opportunities for operators, wireless Internet service providers and equipment manufacturers. The flexibility of wireless technology, combined with high throughput, scalability and long-range features of the IEEE 802.16 standard helps to fill the broadband coverage gaps and reach millions of new residential and business customers worldwide.

With WLAN 802.11 and now WiMAX 802.16, there has been a growing interest in technologies that allow delivery of higher data rates over large geographical areas. The IEEE 802.16 family of standards (802.16-2004 and 802.16e) are intended to provide high bandwidth wireless voice and data for residential and enterprise use. The modulation used to achieve these high data rates is orthogonal frequency-division multiplexing (OFDM). WiMAX OFDM features a minimum of 256 subcarriers up to 2048 subcarriers, each modulated with either BPSK, QPSK, 16 QAM or 64 QAM modulation. Having these carriers orthogonal to each other minimizes self-interference. This standard also supports different signal bandwidths, from 1.25 MHz to 20 MHz to facilitate transmission over longer ranges and to accommodate different multipath environments. This represents a significant increase in system profile complexity as compared to the 802.11 standard, mostly to guarantee a wider, more efficient, more robust network. More subcarriers and variablelength guard intervals contribute to this enhancement.

The ability to develop and manufacture a single recon Figureurable terminal, which can be conFigureured at the final stage of manufacture to tailor it to a particular market, clearly presents immense benefits to equipment manufacturers. With the design, components used, and hardware manufacturing processes all being identical for all terminals worldwide, the economy of scale would be huge. This has the potential to offset the additional hardware costs which would be inevitable in the realisation of such a generic device. Based on this, the scenario adopted reflects in the request for transceiver architectures capable to support cellular phone, WLAN and WiMAX in an 'always and everywhere connected' solution. The transceiver performance in this multi-standard operation, however, comes at the expense of RF specifications that are more difficult to achieve. Furthermore, the choice and definition of the proper transceiver architecture becomes a difficult task, since several parameters - as now imposed by two standards - must be taken into account.

SUITABLE ARCHITECTURES FOR MULTIBAND-MULTI-MODE TRANSMITTERS

The concept of a multiband or general coverage terminal is, strictly speaking, an extension of the basic SDR concept into that of a broadband flexible architecture radio, since the basic recon Figureurability and adaptability aspects of operation do not depend upon multiband coverage. It would be possible, for example, to construct a useful SDR which operated in the 800-900 MHz area of spectrum and which could adapt between AMPS, GSM, DAMPS, PDC, and CDMA. It is now normal, however, for a handset to have multi-frequency operation and hence the extension of this principle to a SDR is a natural one. The international business traveler market is still seen as both large and lucrative, particularly in terms of call charges, hence making this type of handset attractive to both manufacturers and network providers. An ideal SDR is shown in Figure. 2; note that the A/D converter is assumed to have a built-in anti-alias filter and that the D/A is assumed to have a built-in reconstruction filter. The ideal software defined radio has the following features [4]:

- The radio access scheme (i.e. modulation scheme, channelization, coding) and equalization for transmitter and receiver are all determined in software within the digital processing subsystem. This is shown containing a DSP in Figure. 2

- The ideal circulator is used to separate the transmit and receive path signals, without the usual frequency restrictions placed upon this function when using filter-based solutions (e.g., a conventional diplexer). This component relies on ideal matching between itself and the antenna and power amplifier impedances and so is unrealistic in practice over a broad frequency band. Since the primary alternative, a diplexer, is very much a frequency-dedicated component, its elimination is a key element in a multiband or even multimode transceiver.

- The linear, or linearised, PA ensures an ideal transfer of the RF modulation from the DAC to a high-power signal suitable for transmission, with ideally no adjacent channel emissions. Note that this function could also be provided by an RF synthesis technique, in which case the DAC and power amplifier functions would effectively be combined into a single high-power RF synthesis block.

- Anti-alias and reconstruction filtering is clearly required in this architecture (not shown in Figure. 2

- It should, however, be relatively straightforward to implement, assuming that the ADC and DAC have sampling rates of many gigahertz. Current transmit, receive, and duplex filtering can achieve excellent roll-off rates in both handportable and (especially) base-station designs. The main change would be in transforming them from bandpass (where relevant) to lowpass designs.

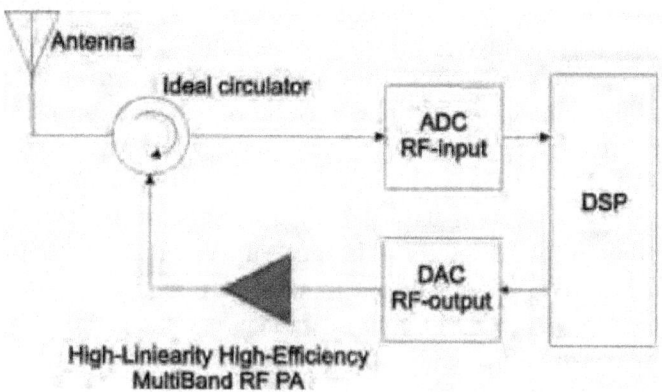

Figure. 2: Ideal software defined radio architecture

Possibly the most important element of any SDR system, whether in a base station or handset, is the linear or linearised multiband transmitter. Receiver systems have always required a high degree of linearity, as they must possess a good signal handling capability, in addition to good low-noise performance. In the case of transmitters, however, a high degree of linearity is a relatively recent requirement, arising predominantly from the widespread adoption of multi symbols envelope-varying digital modulations. This follows from the fact that most modern modulation formats incorporate some degree of envelope variation, the only significant exception at present being GSM and its derivatives (DCS and PCS). The basic architecture of a SDR transmitter revolves around the creation of a baseband version of the desired RF spectrum, followed by a linear path translating that spectrum to a high-power RF signal. Nevertheless the implementation of a true SDR poses a further very critical issues, i.e. the power consumption of the analogue-digital converter. Let's consider for instance the use of a flash converter, largely available in the market with a maximum number of bit about 18 preceded by a sample and hold circuit. Carrying out a simplified calculation, given the converter dynamic range, Dc, the power consumption of this systems is:

$$P_{dc} \frac{kT}{t_s} 10^{Dc/10},$$

$$(1)$$

Where the k is the Boltzmann's constant $= 1.38 \times 10^{-23}$ J/K, T is the device temperature and t_s the sampling time. Furthermore the dynamic range of the converter is given by

$$D_c = 6.02N + 1.76 - PAR + 10\log_{10}[2OSR],$$

Where number of bit, N, a peak-average ratio for the signal, PAR, and an oversampling ratio, OSR. From this easy calculation we can straightforwardly estimate the AD power consumption Pdc in a significant scenario for SDR. Assuming to digitize a frequency band from 800 MHz to 5.5 GHz with a 11GS/s ADC and assuming that the receiver dynamic range is from -20 dBm to -120 dBm, with a SNR of 12 dB at minimum sensitivity, the average PAR of 4, the required N is 20; it results that a such ADC consumes hundred of watt, thus preventing the use of the ideal architecture in Figure. 2. in practical implementation.

RECONFIGUREURABLE MATCHING NETWORKS

The multiband-multimode demands of today's wireless market, is fulfilled by implementations based on parallel line-ups completed by antenna diplexers and switches to meet the specific requirements of each communication standard, (c.f. Figure. 1). Utilizing only one adaptive transmit path to replace the parallel path concept is conceptually simple, but practical design considerations place severe design constraints and technology. Major challenges consists in creating the tunable filters and PAs [5]. Addressing these challenges means to develop flexible PAs capable to maintain the power-added efficiency (PAE) and linearity while moving among different operating frequencies. In conventional PA implementations, the linearity requirement typically results in the use of class-AB operation for the output, which provides a workable compromise between linearity and efficiency. When considering linearity, the class-AB output stage must be dimensioned in such a way that it can provide its peak output power without saturation. As a result, for a given peak output power and battery voltage, the load impedance for a class-AB stage at the fundamental frequency is fixed to $R_L \approx 0.5 \cdot V_{cc}^2 / P_{Peak}$ Unfortunately, class-AB operation provides its highest efficiency only under maximum drive conditions. When operated at the required back-off level, due to linearity reasons for a given communication standard, a rather dramatic loss in efficiency occurs. For these reasons improving amplifier efficiency, while maintaining linearity, is currently a major research topic in wireless communications. In linearity-focused researches, the circuit is designed so that the resulting overall linearity performance of the PA module is improved. In this way, the active device can be operated closer to its peak-power capabilities and still be able to meet the linearity requirements. Techniques that address the efficiency in the back-off mode are dynamic biasing or regulation of the supply voltage of the output stage

[6]. Dynamic biasing provides only modest improvements in efficiency, and supply voltage regulation requires an efficient DC-to-DC conversion, increasing system cost and complexity and operative bandwidth. Nevertheless this techniques appear very promising for future transmitter architectures. An alternative for improved class-AB efficiency is load-line adjustment as a function of output power using an adaptive or reconFigureurable output matching network. An ideal ReconFigureurable Matching Network has to provide:

- Low Loss
- High linearity
- High Tuning Speed
- Sufficient impedance coverage
- Low complexity
- Low area usage

Power handling of matching networks is a critical issue in PA applications. To reduce the losses in a matching network, the use of a limited number of reactive elements is mandatory, beside the choice of high Q tunable components. Typically, such a network is based on varactor diodes, PIN-diodes or FET switching of matching elements like inductors, transmission-lines or capacitors, also involving micro electromechanical systems to improve the power handling capability [7]. We can conclude that these integrated adaptive networks will play an important role for the realization of the next generation of adaptive transceivers and this paragraph is aimed to describe the ongoing basic researches on this subject.

Varactor based switching matching network

Varactor diodes, although characterized by a relatively low Q factor at microwave frequencies, can be a choice for enabling RF tuning. Unfortunately, because of their inherently non linear behavior, their use with modern communication standards (characterized by high peak-to-average power ratios), has to be carefully analyzed according to the specific case considered. In Figure. 3 are shown varactor diode based circuit topologies [5] suited to provide matching tuning overcoming the issue related to the linearity of the electron devices.

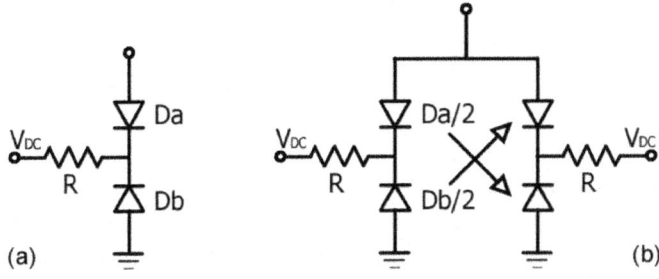

Figure 3: (a) Anti-series connection of varactor diodes to minimize third-order distortion, (b) Anti-series/anti-parallel connection of varactor diodes to minimize second and third-order distortion.

Basically, the capacitance of a single varactor diode can usually be expressed as:

$$C(V) = \frac{K}{(\varphi + V)^n}$$

$$(3)$$

where φ is the built-in potential of the diode, V is the applied voltage, n is the power law exponent of the diode capacitance, and K is the capacitance constant. The power law exponent can exhibit wide variation in different situations, from a value of n≈0.3 for an implanted junction to n≈0.5 for a uniformly doped junction to n≈1.5/2 for a hyper-abrupt junction. If the applied dc voltage is V_{DC}, then the incremental capacitance of a single varactor diode as a function of the incremental voltage v can be expressed as

$$C(v) = C_0 + C_1 v + C_2 v^2 + \dots$$

$$(4)$$

where the term C_1 gives rise to second-order distortion and the C_2 term gives rise to thirdorder distortion. The diode conFigureuration in Figure. 3a can be employed to realize a voltage variable capacitor with theoretically no distortion. Indeed, (referring to the circuit in Figure. 3a) evaluating the expressions for the linear and nonlinear terms of the capacitance, and setting to s the ratio of the diode areas Db/Da, it follows that:

$$C_0 = \frac{sK}{(1+s)(\varphi + V_{DC})^n}$$

$$C_1 = \frac{(1-s)nC_0}{(1+s)(\varphi + V_{DC})}$$

$$C_2 = \frac{C_0\left[(s^2+1)(n+1) - s(4n+1)\right]}{2(\varphi + V_{DC})^2(1+s)^2}$$

(5)

It can be noted that for $n \geq 0.5$, C_2 can be made equal to zero, resulting in zero third-order distortion, by setting

$$s = \frac{4n + 1 + \sqrt{12n^2 - 3}}{2(n+1)}$$

(6)

It can be observed from eq. (6), that the particular case of constant doping profile in the diode (the abrupt junction case where n=0.5) results in a value of s=1. This case is particularly attractive because this set of conditions (n=0.5, s=1), sets both C_2 and C_1 equal to zero. A more elaborate analysis shows that all higher order distortion terms also vanish, yielding (in theory) a "distortion-free" operation for this unique case. When dealing with process technologies where n>0.5, eq. (6) provides a direct means of calculating the required diode area ratio to minimize C_2.

For example, in the case where n=1, the required area ratio is exactly two. In the case of n=2, which corresponds to the ideal hyper-abrupt junction, the required area ratio is 2.6. Although this approach can minimize C_2, it is clear from eq. (5), that a value of s≠1 will result in a finite value of C_1. In this case, a relatively high third-order distortion product will unfortunately still arise, resulting from the secondary mixing of the fundamental with the second-order non-linearity C_1. Fortunately, this distortion contribution can be eliminated, by placing an identical varactor stack in anti-parallel conFigureuration (see Figure. 3b). The linear capacitance of the circuits of Figure. 3a and b are identical, but the circuit of Figure. 3b has $C_1=C_2=0$ when the proper area ratio is set. It must be underlined that, in this conFigureuration (Figure. 3b), all the even-order coefficients are zero (C_1, C_3, C_5, ...) in this topology, but the higher coefficients that create odd-order distortion (C_4, C_6, C_8, ...) are not zero, although the IM3 contributions due to the 5th and higher order nonlinearities are very small. The implications of this analysis can be summarized as follows:

- The classical conFigureuration of Figure. 3a provides theoretically a "distortion-free" varactor stack when n=0.5, corresponding to a uniform doping profile of the varactors.
- The more generalized conFigureuration of Figure. 3b provides an ultra-low distortion varactor stack for any value n>0.5, by setting the proper ratio of the diode areas, which sets C1 and C2 to zero, providing more freedom for use in different process technologies.

It must be underlined that each of the circuits in Figure. 3 requires a very high tap impedance (R) for proper operation. A high tap impedance, limits the impact that forward biasing of one of the diodes by RF signal has on linearity. An effective way to implement the high impedance while keeping the RC time limited for the control signal is the anti-parallel diode pair depicted in Figure. 4.

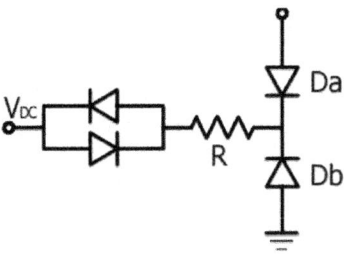

Figure. 4: – Anti-series connection of varactor diodes with modified center tap.

Analyzing, by a two tone test the described matching and comparing them with a single varactor diode we can observe the linearity of the different topologies. The IM2 and IM3 distortion of the circuit in Figure. 3a are comparably low, while the IM3 of the circuits in Figure. 3b is limited by fifth-order distortion due to complete cancellation of the third-order products. In this case indeed, a 1:5 slope dependence for the IM3 components can been found [8], confirming the elimination of the C_1 and C_2 contributions.

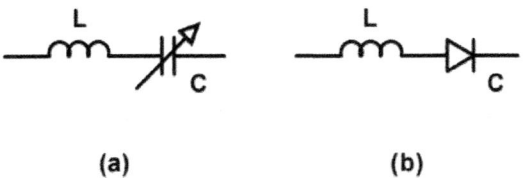

(a)   (b)

Figure. 5: (a) Equivalent circuit of the "tunable inductor", (b) Simplified circuit of a varactorbased "tunable inductor".

A different option for the implementation of recon Figure matching network is to act directly on the inductance value, rather than the value of the capacitors as discussed hereinabove. For this purpose varactor diodes can be exploited to create a "tunable inductor" [9]. By considering the circuits in Figure. 5, the equivalent impedance Z of the LC circuits can be written as:

$$Z = \frac{1 - \left(\omega/\omega_c\right)^2}{j\omega C}$$

(7)

Where $\omega_c = 1/\sqrt{LC}$ is the resonant frequency of the circuit. When the condition $\omega > \omega_c$ is achieved, the impedance Z represents an equivalent inductor L_{eq}, that, for the simplified circuits in Figure. 5, is analytically expressed as:

$$L_{eq} = \frac{\left(\omega/\omega_c\right)^2 - 1}{\omega C}$$

(8)

The resulting equivalent inductor is a "tunable inductor" whose value is related to the varactor capacitance C (and also ω_c). In Figure. 5b, the varactor parasitic elements have not been represented, however they have to be taken into account during the design of the "tunable inductor".

Switching matching networks

If the output power of the PA is so high to make the use of varactor diodes prohibitive, a possible solution could be to replace the varactor diodes with fixed value capacitors controlled by switching PIN diodes. By observing to the examples in Figure. 6, if the PIN diode connected to the capacitor is in the ON state, the capacitor adds its own capacity to the global circuit. On the other hand, if the PIN diode is in the OFF state, the capacitor does not affect the global circuit.

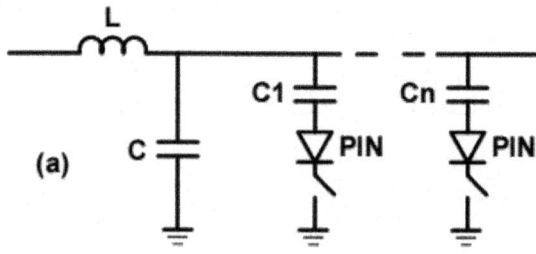

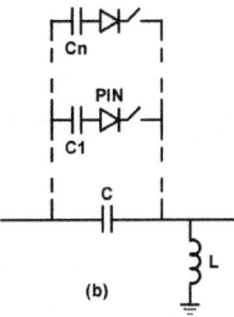

(b)

Figure. 6: (a) Low-Pass Matching Network example, (b) High-Pass Matching Network Example.

By using a set of digital signals that change the state of the PIN diodes to active or inverse condition, it is possible to generate $N=2^n$ impedances (where n is the number of PIN diodes used). Unfortunately, such a technique permits only a discrete set of tuning possibilities, resulting inappropriate where fine tuning is requested, unless a considerable number of components is used, thus increasing losses, area usage and costs. A different option of switched matching network is based on MOS devices . The use of MOS devices rather than PIN diodes makes them more suitable for IC designs. The operative principle of the network is discussed by a following example.

Let's consider a Si-Ge HBT device, the optimum fundamental loads at one dB compression point are 15 W at 2.45 GHz and 5 W and 3.5 GHz respectively, the reactive parts are negligible. These resistance values have been determinate accounting for the output power level to be supplied by the power amplifier in the two different bands. The aim of the output network consists in synthesizing these loads using a MOS switching network topology, schematically depicted in Figure. 7 and investigating the general features.

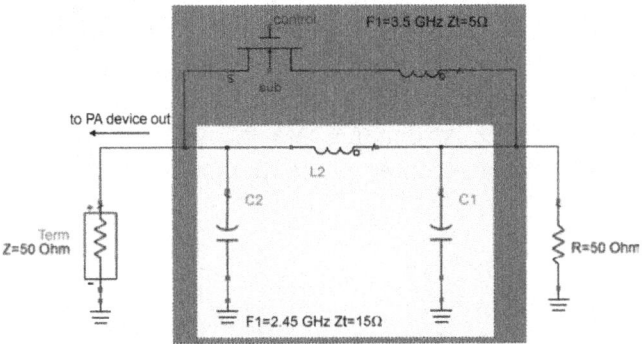

Figure. 7: MOS based switched matching network p topology

The basic simple network topology is based on a p-structure, with an additional branch composed by a NMOS device acting as a switch, with in series an inductor to change the network impedance when the switch is closed. The behavior of the network depends on the switch condition:

1.   when the switch is ON (i.e. its R_{ds} value is low, zero in the ideal case) the network should present at its input an impedance value equal to 15 W that is the optimum value for the device at 2.45 GHz.

2.   When the switch is OFF (i.e. its R_{ds} value is high, infinite in the ideal case) of course, the frequency behavior of the network change. In particular the equivalent value of the inductor between the two shunt capacitors becomes equal to the parallel of L_1 and L_2 and the input impedance of the network decreases to 5 W at 3.5 GHz.

Moving on the actual schematic, for the switch it has been necessary to introduce the biasing network in order to guarantee the right switching functionalities. The simulations have been performed biasing the NMOS device, with dc voltages on drain and source set to V_D=3V and V_S=3V respectively, to ensure that the NMOS device is properly biased in the origin of its output characteristics. The gate control voltage is raised to V_G=5V, when the NMOS switch has to realize a short circuit condition, and dropped to V_G=2V when the NMOS switch has to realize an open circuit condition. The feed lines for all three terminals have been realized using 4.8 KW resistor for each lines to guarantee the request isolation. Figure. 8 shows the small signal parameters S11 and S21 of the networks, as a function of frequency from 1 GHz to 4 GHz, respectively when the switch is OFF, e.g. when the network has to synthesize the load at 3.5 GHz, and when the switch is ON, e.g. when has to be synthesized the load at 2.45 GHz.

In particular in Figure. 8 left, the blue line in the right part of the Figureure represents the real part of the input impedance, which can be note is roughly equal to 15 ohm. Unfortunately, the input impedance shows a residual imaginary part due to the non ideal behavior of the inductor and capacitor elements present in the network. Similarly, Figure. 8 right reports the same features when the switch is OFF, e.g. when the network has to realize the load required at 3.5GHz. Also in this case, the network exhibits the requested real part of the input impedance, with a residual imaginary part of the input impedance.

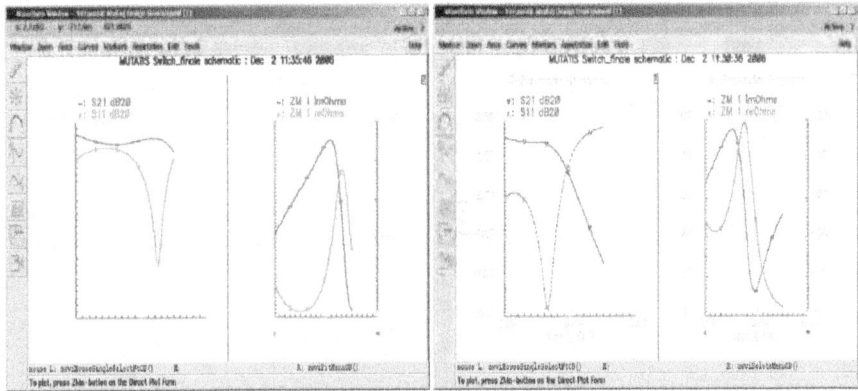

Figure. 8: S11 and S21 frequency behavior together with the real and imaginary parts of the input impedance when the switch is ON (left) and OFF (right).

DUAL-BAND RECONFIGUREURABLE SIGE HBT AMPLIFIER DESIGN

The above described matching network can been adopted to implement the two matching networks for a class-AB amplifier. In the following example of such design approach is presented. The design of the reconFigureurable PA is based on a power device composed of 17x8 elementary SiGe HBT with and emitter area of 8.49 μm². The bias circuitry of the power section is designed to provide other than the required base bias current, a circuit-level linearization. Let's start introducing this latter part, referencing to the Figure. 9. The size of the devices used for the bias circuitry and the values of passive components scale accordingly, so that accurate (i.e. matched) current mirroring can take place. In addition, resistor R4 introduces base ballasting and its value is selected to be 350 Ω per emitter. It helps in reducing the risks due to thermal runaway. A larger value would have better effect, however, that could result in high DC voltage drop at high power drive, and thus early gain compression of T4, the power transistor, would take place.

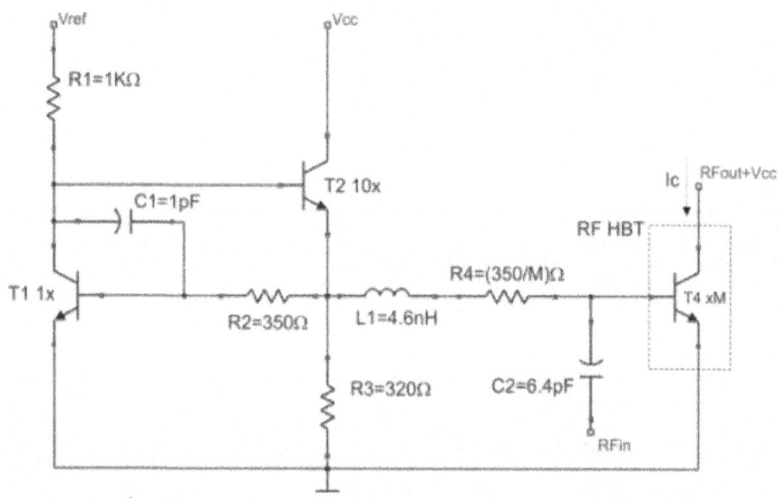

Figure. 9: PA core schematic including the power transistor and the bias network

The power device, which is shown in Figure. 9 within the dotted outline, is biased through the current mirror. By inspection of this Figureure the current through the reference transistor T1 is:

$$I_{REF} = \frac{(V_{REF} - 2V_{be})}{R1} \, ,$$

(9)

with Vbe ~ 0.8 V. If the ratio between the reference device and the RF device is M, in this case M = 17x8, then the current that will flow through the RF transistor is M*I$_{ref}$. in addition, the current mirror behaves like an ideal voltage source, since its output impedance is

$$R_{out} \approx \frac{1}{g_m A_{loop}} \, ,$$

(10)

where g$_m$ is the transconductance of the emitter follower device, and A$_{loop}$ is the loop gain of the loop formed by the reference device and the emitter follower (buffer) transistor. At high frequencies, R$_{out}$ becomes inductive and its value should be kept low throughout the frequency band of interest and could be further reduced if resistor R3 was further reduced in value, at the expenses of an higher current trough this resistor.

At 1 dB compression point, the power transistor should be biased at a collector current which has a value near the optimum value that guarantees that the maximum transition frequency (f$_t$) of the transistor is achieved. Due to self bias effect, which forces the DC average collector current to increase with increasing input power, the value of the quiescent bias current is chosen to be much lower compared to the value that it reaches under full power drive conditions. So, when the transistor T_4 is self biased it starts to draw more current. Since the current flowing through R_2 is constant, as the base-emitter voltage of T_1 is constant, this additional current is supplied by transistor T_2. The capacitor C_1 in the bias network helps to stabilize the loop formed by the reference transistor, T1, and the emitter follower transistor, T_2. The stability analysis of the loop formed by T_1 and T_2 shown that for unity loop gain the phase margin is approximately 78 degrees, which guarantees the loop unconditional stability. Figure. 10 reports the DC current flowing in the collector of the power device, T_4, as a function of the input power at the frequency of 2.45 GHz. From the curve is seen that the quiescent current increases according with the description given above and reaches the value related to the peak f_T, at the 108 mA for an input power of 6 dBm. The value of the bias current for low input power is defined according with the eq. (9), adjusting the value of either R_1 or V_{ref}, in this case to the $R_1 = 1$ kW, corresponds a $V_{ref} = 3.3$ V.

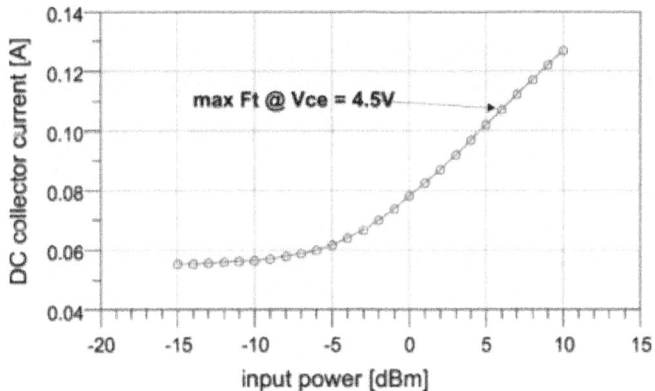

Figure. 10: Behavior of the collector current for the power transistor, T4 in Figure. 9, as a function of the input power at 2.45 GHz.

The layout of the recon Figureurable PA is illustrated in Figure. 11 (left), where are clearly visible the inductors adopted for the matching and the area for the active part of circuit. The total size of the layout with the bonding pads is 1x1mm. While in operation, the voltage in the node between the inductors swings between positive and negative values. This causes the control MOS drain-bulk np junction to be forwarded biased, which degrades switch performance and may results in latchup. In order to overcome the specific drawback, we set an offset voltage (pin Vofs) which shifts the voltage in the abovementioned node in positive values.

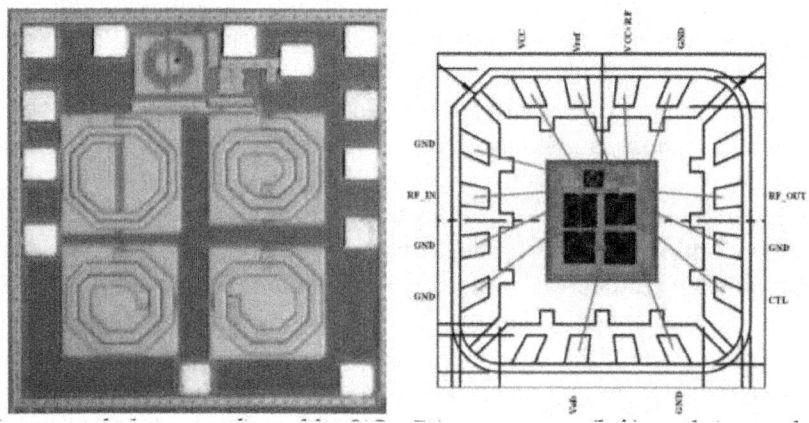

Figure. 11: Layout of the reconFigureurable SiGe–PA prototype (left) and its packed wiring diagram (right)

For the design of the inductive part of the matching networks a planar electromagnetic simulator should be used, this permits to calculate any mutual inductances, between the inductors that are in proximity along with the losses in the silicon. Due to the large number of controls and dc-supply for the operation of the components a package is normally required and has to be taken into account during the design. The selected solution consists in the QFN package which allows up to 16 leads. The bonding diagram for this component is reported in the Figure. 11 (right), along with the pin description. At the time of this report editing, the packaged components were not yet available, for the reason the measurement results that will follow consider only the on-wafer device.

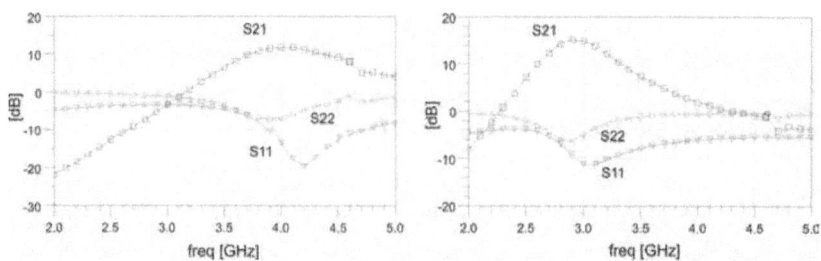

Figure. 12: S-parameters in the state corresponding to the lower and higher bands.

The first test of measured data considers s-parameter in the range 2 to 5 GHz. The S21-S11 and S22 in dB for the state corresponding to the lower and higher frequency band are reported in Figure. 12. From the Figureures is clearly observed the shift in frequency due to the above mentioned reasons, which is estimated in the range of 500 MHz. While the smallsignal gain maximum is 2.5 dB lower than the simulated in the lower frequency band and 1 dB less in the higher band. This reduced matching have to be attributed to the matching which, being lower than estimated for the discussed reasons, introduces and matching loss consistent with the reduced gain observed during the characterization phase. We exclude problems related to the measurement set and calibration, although a problem related to measurements, the ringing in the 4.7GHz, was observed. The characterization of the sample in the large-signal regime is reported in the below Figure.

It is carried out at the frequencies where the device exhibits the maximum gain and matching in the two states that are respectively at 2.9GHz and 3.9GHz, which correspond to 500MHz frequency shift from the design target as discussed above. The data related to the large signal gain up to the compression are reported in the Figure. 13, respectively for the state corresponding to the lower and higher frequency. In this Figure the CW single tone signal is applied in the two states and the Pout-Pin curves are recorded. The value of the gain is consistent with the small-signal gain while the compression point is estimated to be about 16 dBm for the lower band and 15 dBm for the higher frequency.

These Figure are 1 dB lower that estimated during the simulations. All this data are consistent with the simulation and again supports the hypothesis that introducing the additional inductive parts due to the packaging the proper frequency behavior can be reached. An intermodulation product characterization considering a two-tone signal with a center frequency at the selected frequencies of 2.9GHz and 3.9GHz and 1MHz offset, was applied to the device. The input power was swept from -10dBm to 13dBm. The results of the characterization, in terms of the higher IM3 are reported in the Figure. 13. It is observed that any consistent change in slope is observed in the traces. This allows concluding that the MOS involved in the switched matching networks doesn't introduce any additional nonlinearities. In fact, if the additional nonlinearity required changing the state of the device, was effectively excited we would have observed an addition component in the IM3. It is also worth to observe that the input power dealt by the device is below the threshold for which this effects become evident. This threshold from simulation is estimated in the range of 28-30 dBm.

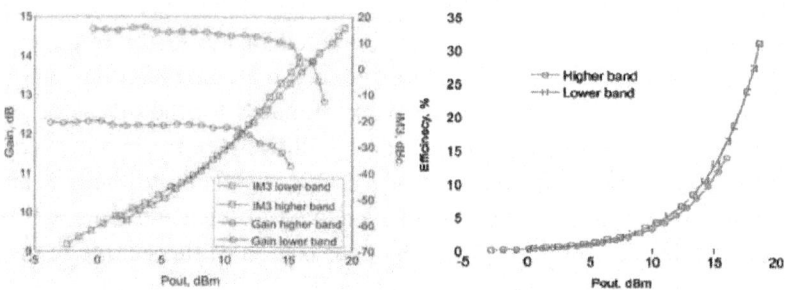

Figure. 13: single tone large-signal gain at 2.9GHz and 3.9GHz and two-tone large-signal intermodulation offset frequency 1MHz, data reported for the higher IM3 (left), PAE

DUAL-BAND POWER AMPLIFIER ARCHITECTURES

The main objective of this paragraph is about the consistent and quantitative evaluation of a two possible architectures of dual-band PA both suitable for their involvement in the concurrent dual-band systems, [12]. The first is based on two dedicated PAs combined by a frequency diplexer while the second is specifically designed to be operated in dual-band state. For the sake of the comparison the operative frequency are defined as 1.98 and 3.42 GHz respectively suitable for WCDMA and OFDM radio access technologies. The two dual-band architectures considered in this paragraph are based on the schematic representations reported in Figure. 14. In the first (Figure. 14, left) the PA is implemented by making use of two dedicated PAs combined by a frequency diplexer. This latter device has to be designed to combine the two PAs introducing band-pass and band-stop behavior in each of the two branches. This is implement by the most innovative technique and technology and it still represents a very critical part of the entire PA structure. Indeed, this component must guarantee an almost lossless behavior in the two transmission paths and as much as possible isolation between them; without sacrificing the matching. In particular the transmission loss characteristic is required to preserve the combined efficiency of the entire structure, while the isolation is a required feature to avoid the cross-modulation between the two dedicated PAs. The constraints on the diplexer become more critical in the case of closer operative spectrum bands. During the two dedicated PAs design, the eventual combination with the diplexer implies a specific additional PA design consideration related to an accurate evaluation of the out-of band termination, which might degrade the output power and efficiency of the two units. The treatment of the harmonic termination of the due to the diplexer is out of the scope for the present treatment.

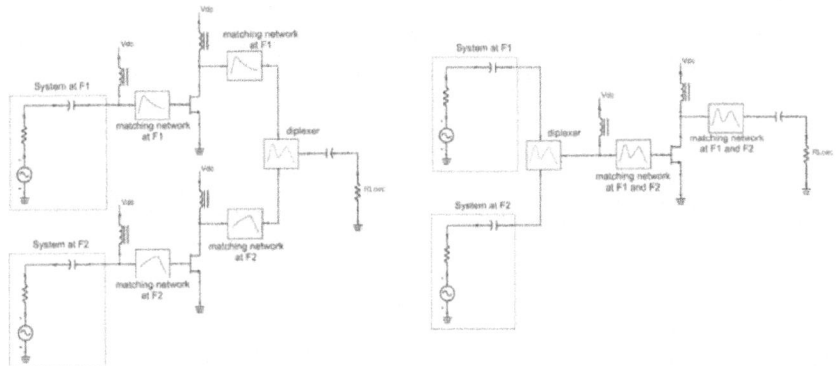

Figure. 14: Schematic of the concurrent dual-band PA implemented by two combined dedicated PAs (left), and by a dual-band PA (right).

Either the dedicated frequency PAs and concurrent PA topology consist of conventional class-AB PA designs, where the tuned matching networks can be synthesized by either passive and\or distributed elements properly dimensioned, without external tuning controls. For the concurrent PA (Figure. 14, right), the two signal sources are combined prior to be applied to the PA input.

The design method of concurrent dual-band PA based on multituned networks composed of lumped elements is discussed in [10]. The investigation carried out in this chapter relies on prototypes designed and fabricated using low cost off the shelf active devices along with discrete SMD passive components assembled on FR4 0.8 mm thick evaluation printed circuit board designed with microstrip technology. For this specific test several electron device technology can be considered, eider bipolar or FET fabricated using several different material, spanning from Si to GaAs and possibly GaN. In the present case we choose as active device a GaAs FET, namely the ATF50189 from AVAGO Technologies, a medium power enhanced mode p-HEMT with a cut off frequency of 6 GHz and a 1-dB compression point of 29 dBm at 2 GHz. Optimum bias point for efficiency, linearity and gain can be fund either from manufacturer specifications or CAD simulations, on the basis of nonlinear model analysis. For the specific device possible bias point is found to be 4.5V drain supply voltage with a corresponding quiescent current of 200 mA. The chosen bias point drives the ATF50189 transistor in the AB class operation. The design was based on load and source pull simulations carried out at the two fundamental frequencies of 1.98 and 3.42 GHz, adopting a nonlinear device model which included the package parasitic. Simulations provided saturated output power of 28 dBm and 26 dBm respectively in the lower and higher frequency bands with a power added efficiency of approximately 40% and 35% at the 1-dB gain compression point. The resulting load and source constant power contours are shown respectively in Figure. 15.

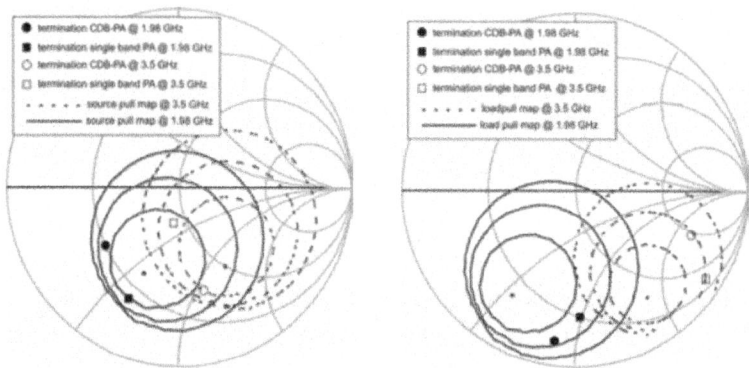

Figure. 15: Simulated source-pull (left) load-pull (right) contours at 1.98 GHz and 3.42GHz, 1 dB steps, and terminations at fundamentals for the single band and dual-band prototypes

The implementation of the source and load terminations defined by the source- and loadpull analysis was obtained by using lumped elements matching networks. This technique, by employing a different approach with respect to standard microstrip technology, enabled the achievement of highly compact prototypes. All the designed PAs adopt the same general topology for the input and output matching networks. Different nominal values and the absence of some of the components determine the difference between the prototypes. In addition, the input network accommodates a stabilizing network which has bee implemented by all three prototypes. The presence of shunt capacitors at both the gate and drain terminals, provide a short circuit to the second harmonic.

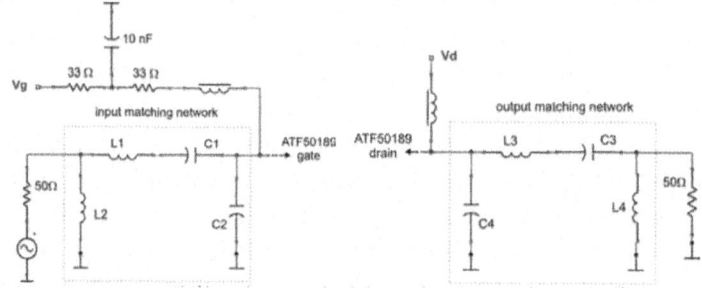

Figure. 16: Prototypes input (left) and output (right) matching network circuit schematic

The selected matching networks pi-topology, exhibits high out of band frequency roll off and a null in the transfer characteristic between the two fundamental frequency bands at 1.98 GHz and 3.42 GHz so enhancing isolation between frequency bands. In order to properly define the networks the additional conditions for maximum efficiency and 1-dB compression output power under large signal excitations were taken into account. To simplify the description we can consider that the three prototypes adopt the same general topology for the input and output matching networks, whose schematics are represented in Figure. 16. As far as the networks dimensioning is of concern, it is possible to show that the 4 unknown (2 inductors and 2 capacitors for each network) are calculated solving a nonlinear system of four equations, where the number of equations results directly equating the two real parts and the two imaginary parts of the equivalent impedance, in symbolic format, with the required optimum impedances. Different nominal values and the absence of some of the components will determine the difference between the prototypes. In addition to matching purpose the input network accommodates a stabilizing network which is common for all the three prototypes. The resulting matching networks result very compact and capable to satisfy the conditions for the optimization of both the output power and gain. The presence of shunt capacitors at both the gate and drain terminals, C2 and C4 in the Figureures, provide a short circuits at the second harmonics. The resulting best values for the SMD capacitances and inductances are those indicated in Table 1 and Table 1. Input matching network L-C values respectively for the input and output matching networks. The achieved impedance are reported in the Figure. 16 where the mismatch between the actual values and the optimum impedances for power level take into account for the additional condition of gain and commercial availability of the nominal values

Table 1: Input matching network L-C values

	C1	L1	C2	L2
concurrent dual band	1.7 nH	0.33 pF	7.6 nH	0.3 pF
single band at 1.98GHz	0.6 nH	1.47 pF	n.a.	n.a.
single band 3.42GHz	1.6 nH	0.26 pF	n.a.	n.a.

Table 2: Output matching network L-C values

	C4	L4	C3	L3
concurrent dual band	0.6 pF	3.2 nH	0.77 pF	2.88 nH
single band at 1.98GHz	n.a.	n.a.	1.7 nH	1.9 pF
single band 3.42GHz	n.a.	n.a.	1.1 nH	0.5 pF

The three PA modules were fabricated using FR4 PCB technology and then adopted to implement the two dual-band PA conFigureurations, namely the combined PAs and the dualband PA. The diplexer used in the large-signal test benches it is realized in microstrip technology and provides an insertion loss of 0.6 dB and 0.8 dB respectively at 1.9 GHz and 3.4 GHz, and isolation between the two channels better than 30 dB and a return loss higher than 20 dB.

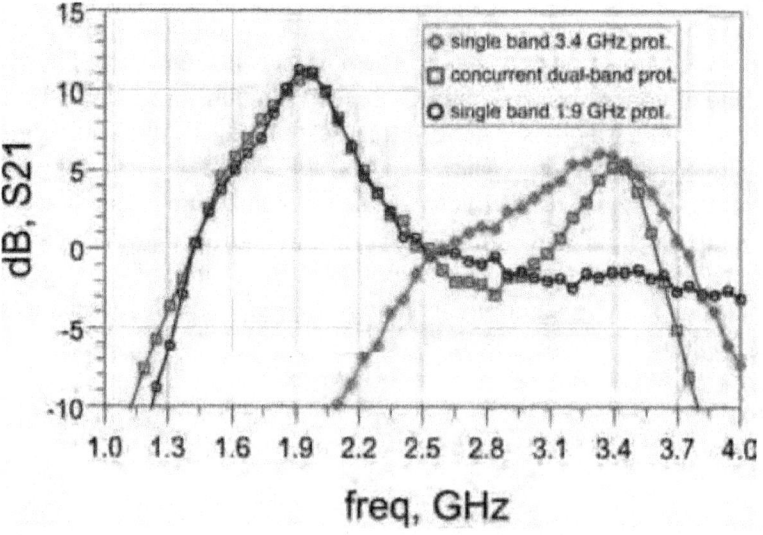

Figure. 17: Measured small-signal gain for the concurrent dual-band and the two single band prototypes

The preliminary test was performed in small signal regime to assess the prototypes performance and to verify the consistency of the comparison. The measured small signal gains associated with the concurrent dual-band PA and with the two single-band PAs prototypes are compared in Figure. 17. The Figure indicates that at 1.98 GHz the maximum linear gain is approximately 11dB for both single band and dual band circuits. In the 3.4 GHz band the PA prototypes exhibit a maximum linear gain of 6 dB at 3.42 GHz, and present a 0.5dB gain bandwidth of approximately 60 MHz. Input and output return losses are not reported but are below −15 dB in the respective frequency bands for all the PAs. Small signal characterizations have indicated a very close correspondence between the single band circuits and the concurrent dual band PA in terms of both input/output return loss and gain. This results show that the design of a concurrent dual-band PA using compact lumped elements is feasible without loss of performance at small-signal and makes the characterization and comparison with large and modulated signals meaningful. The first set of large signal measurements test bench is deployed to fully characterise the PA prototypes with CW large signal excitations. The data are useful to compare the maximum linear power, the gain and the efficiency of the two architectures. In particular the comparison between the power gains as a function of the two CW signals at 1.98 GHz (namely F1) and 3.4 GHz (namely F2), reported in Figure. 18, shows that the combined PA architecture is capable to maintain the two channels mostly separated producing a gain compression which is insensible from the concurrent signal at the side band.

In fact the contour plots show a linear behaviour which is independent of the power level in the other channel. At the contrary for the dual-band PA, the mutual interaction between signals at the two frequencies is evidenced by bended constant gain loci, see Figure. 19. Nevertheless, although the effect of the side band is quite evident, the effect of the output diplexer is such that the maximum output power was slightly higher. For example, we can notice that at 1.98 GHz for input power equal to 16 dBm the gain is 8.7 dB for the combined PAs while 9.5 dB for the dual band. This latter condition is maintained only when the input signal at 3.4 GHz is lower than 10 dBm. Similar behaviour is observed in respect to the power gain calculated at 3.4 GHz. From the contour plots it is observed that the loci corresponding to 8.5 dB and 4 dB correspond to the linear gain that the dual-band PA can provide and, consequently, the input ranges which guarantee the linear operation of the PA. Differently, in the case of the combined PAs, the linearity in one frequency band does not depend of the side band.

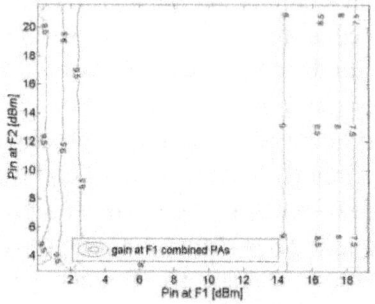 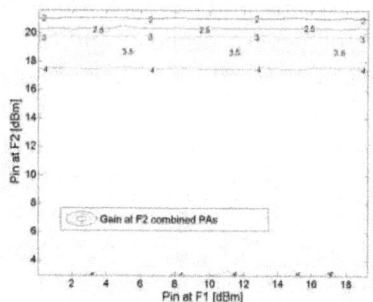

Figure. 18: Power gain [dB] in large signal regime evaluated at the frequency of 1.98 GHz (left) 3.4 GHz (right) as function of input power at 1.98 GHz (F1) and 3.45 GHz (F2), for the combined PAs architecture.

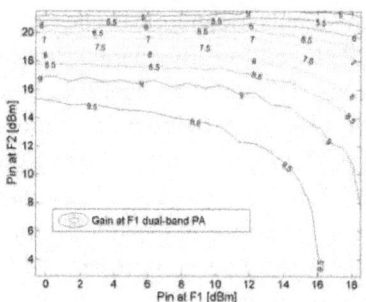 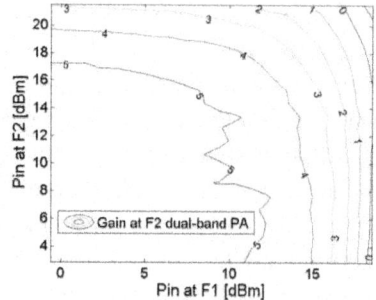

Figure. 19: Power gain [dB] in large signal regime evaluated at the frequency of 1.98 GHz (left) 3.4 GHz (right) and as function of input power at 1.98 GHz (F1) and 3.45 GHz (F2), for the dual-band PA architecture.

From the contour plots it is observed in the case of the combined PAs and in the case of the dual band PA, the presence of the diplexer and the mutual interaction between carriers respectively, determine the output power in correspondence of a 1-dB compressed power gain reported in Table 3. This latter consideration leads to the conclusion that from the point of view of the CW output power there are not differences between the two architectures. This latter consideration leads to the conclusion that from the maximum output power in CW condition there are not differences between the two possible architectures.

Table 3: Simultaneous maximum linear output power (at 1 dB gain compression)

architecture	1.98 GHz	3.4 GHz
Combined PAs	25 dBm	22 dBm
Dual-band PA	24.5 dBm	23 dBm

A further very significant Figureure is represented by the Power Added Efficiency (PAE) for the two PA architectures. In the case of dual-band concurrent PA the PAE is calculated by:

$$PAE = \frac{\left(P_{load}^{F1} - P_{av}^{F1}\right) + \left(P_{load}^{F2} - P_{av}^{F2}\right)}{P_{dc}},$$
(11)

The above equation admits that the two signal are uncorrelated and where Pdc takes into account for the total current drawn by the PA modules. The PAE as a function of the input power at the two carrier frequencies, respectively for the combined PAs and the dual band PA architectures are reported in Figure. 20.

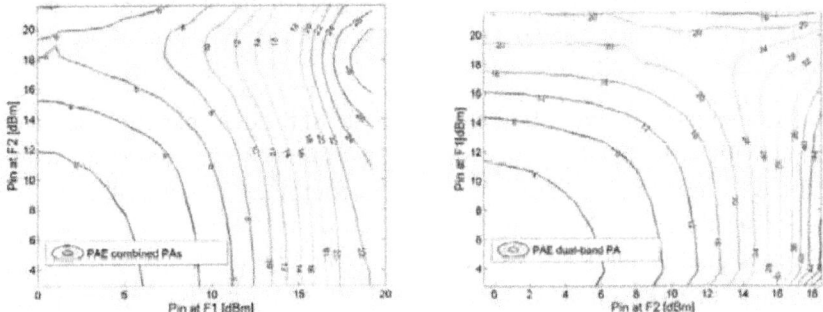

Figure. 20: Power added efficiency [%] in large signal regime as function of input power at 1.98 GHz (F1) and 3.45 GHz (F2), for the combined PAs architecture, (left), and the dual-band PA architecture, (right).

From these experimental verification it is confirmed the intuition that the dual-band PA PAE takes advantage from the current reuse which is inherent in the use of a single power device, when compared with the case of the combined PAs which need twice of the DC power to bias the two PAs. This determine an almost factor 2 in the PAE for the dual-band PA for almost the entire range of evaluation. In particular at 1-dB gain compression the PAE achieved with the combined PAs architecture is in the range of 20%, as evidenced in Figure. 20 left, while in the case of the dual-band PA it reaches 32 %, see Figure. 20 right. The maxima are 28 % and 44% respectively for the combined PAs and the dual band PA architectures. By this Figureure we can observe a significant improvement of the dual-band PA with respect to the combined PAs architecture.

The absolute maximum power for the two PAs are reported in the Figure. 21, calculated by summing the power level at the two carrier frequencies, assumed uncorrelated. Form the contour plots is observed that, regardless linearity concerns, the total power provided by the two systems are slightly the same.

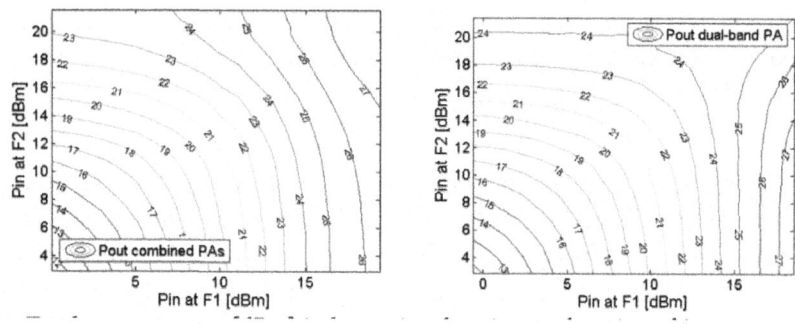

Figure. 21: Total output power [dBm] in large signal regime as function of input power at 1.98 GHz (F1) and 3.45 GHz (F2), for the combined PAs architecture (left) and for the dual-band PA architecture, (right).

The next test concerns about the capability of the dual-band PA architecture to deal with modulated signals and its performance are compared with the combined single band PAs; henceforth the combined PAs architecture is not longer considered. In this case, the baseband signals were down-loaded in the arbitrary signal generators (Agilent ESG 4438C) by using the tools available in the Agilent ADS2006A systems. Two different digitally modulated signals were employed to evaluate PA performance: a 3GPP up-link W-CDMA 3.84 MHz chip rate signal at 1.98 GHz and a 5MHz OFDM 16-QAM signal at 3.42GHz corresponding to one of the WiMAX modes. The output of the PA under test was connected to the VSA (Agilent N9020, 26MHz bandwidth) which was synchronized with the two arbitrary signal generators. The first set of data refers to the large signal gain plotted against the output power for the three PA modules; the comparisons between several operating conditions are shown in Figure. 22, left and right, for the lower and higher frequency bands respectively, which include also CW for the sake of a better comparison.

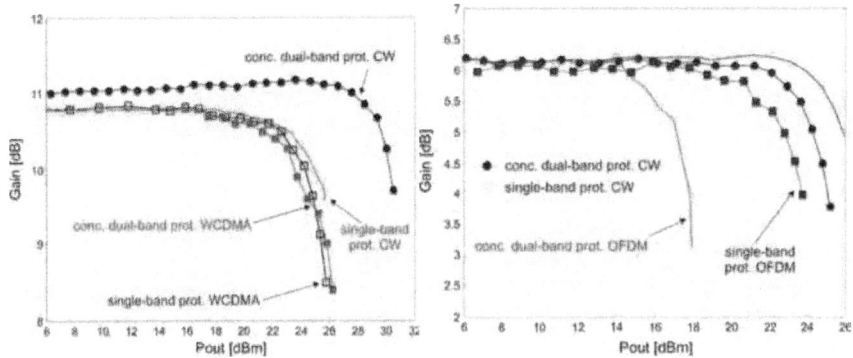

Figure. 22: Gain curve versus output power for CW and WCDMA modulated excitations, both with carrier at 1.98 GHz (left) and 3.42 GHz (right).

It is observed that when the amplifiers are driven by a single modulated signal peak power at 1dB gain compression point decreases: this effect is explained by the fact that gain compression in PAs driven by digitally modulated signals occurs at lower power levels than for 1-tone CW signals. In addition, load pull CAD analysis and successive design were performed based on a CW test signal, while experimental results show that the optimum load impedance for maximum linear output power as well as peak efficiency varies depending on the characteristics of the input signal, i.e. pulsed, modulated or CW. Concurrent mode was then operated by simultaneously feeding the dual-band PA with OFDM and WCDMA signals at the two center band frequencies. Reduction of peak output power with respect to single-channel excitations is mainly due to the simultaneous presence of two modulated signals in the same device which cause cross-modulation between the two time varying envelopes.

A resulting 4 dB and a 4.5 dB peak power reduction at 1.98 GHz and at 3.42 GHz respectively were measured with respect to the single channel cases. Moving on to system level Figureures, the 5.6% EVM WiMax standard limit and a minimum ACPR of 33dBc for a WCDMA signal as settled by the 3GPP specifications have been taken as a reference for power and efficiency values. The goal of the large signal characterisation has so being focussed on the evaluation of the peak output powers and the resulting PAE levels achievable in the two frequency bands with both concurrent dual-band and singleband excitations so as to satisfy EVM and ACPR constrains.

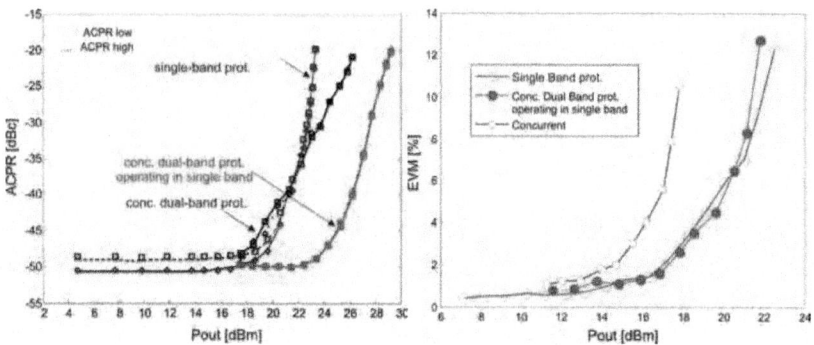

Figure. 23: (Left) Adjacent channel power ratio measured at 5 MHz offset and integrated over the bandwidth, for the single band and the dual-band prototypes with the WCDMA signal at 1.98 GHz; (Right) Error vector magnitude measured for the single band and the dualband prototypes with the OFDM signal at 3.42 GHz.

From Figure. 23 it is observed that at 1.98 GHz the maximum achievable output power, due to ACPR constrains, is 27.5 dBm when the dual-band PA is working in single-channel mode, while for the concurrent dual-band case this limit decreases to 23 dBm. Data in Figure. 23 show EVM versus the output power results, for single channel operation and dual band mode at 3.42 GHz: a maximum output power of 20 dBm is achieved in the first case while when the dual-band PA is working in concurrent mode, maximum output power settles to 17 dBm. The above data indicate that a significant change in performance arises when the PA is driven in the concurrent dual band mode, specifically resulting in a peak power back off of about 4.5 dB and 3dB respectively for the lower and higher frequency bands due to meet the EVM and ACPR restrictions. Envelope cross-modulation and inter-modulation explain the EVM and ACPR increased growth with input power when compared with single channel mode. Experimental data showed that a 2 dB back-off is necessary with concurrent operation to maintain the EVM at 4.1%. It can be concluded that the proposed solution is capable to provide the same system level performance of more conventional solutions while increasing the overall PAE and allowing a significant reduction of the system complexity. A further implementation of dual band PA in GaN technology can be found in [13].

DUAL-BAND POWER AMPLIFIER DIGITAL LINEARIZATION

As discussed in the above paragraphs, the application of digitally modulated signals to a PA, which is considered hereinafter as an nonlinear (NL) dynamic system, causes in-band distortion and spectrum spreading and finally determine a degradation of the signal quality. The most effective broadband linearization systems have usually been based upon the feed forward technique [14]. However, RF and baseband pre distortion linearization techniques have become an attractive solution owing to their reduced cost and complexity. For multicarrier PA applications, an effort has been placed to increase the bandwidth of pre distortion linearization to combat fast memory effects. Baseband Digital Pre distortion (DP) seems to be the most promising one. It works by the introduction, in the digital baseband, of an opposite NL of the PA's one, allowing for greater efficiency through a significant power back off reduction. The most of the available DP techniques deal with single-band operation, although recently an approach to deal with multi-carrier and potentially for multiband systems was presented in [15]. In this technique the modulation bandwidth in several bands and then a DP algorithm is applied selectively the in band and inter band third-order intermodulation distortion (IMD3) As the approach relies on third order Volterra model the accuracy of the DP depends upon the identification procedure and the frequency band spacing.

Here we discuss a novel method of Dual Band DP (DB-DP), based on the simultaneous pre distortion of both channels at intermediate frequency (IF), [16]. The proposed method uses a single band memory polynomial DP for linearization. As a feedback path we propose a subsampling receiver.

Basic principles of digital linearization

Let's start reviewing the basic concepts of the DP system. It consists basically in the introduction, at the baseband, of a subsystems which has a transfer function which is opposite to the one of the PA, as in Figure. 24.

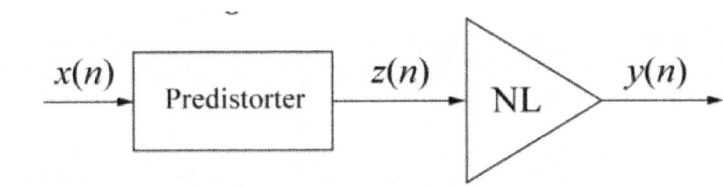

Figure. 24: Predistorsion Principle applied to a nonlinear system

The predistorter can be Look-Up Table (LUT) based or polynomial-based: in the first case, a LUT indexed by the input power is filled with complex coefficients, and the input $x(n)$ is multiplied with the corrisponding one; in the second, the complex coefficients of a k-order polynomial approximating the inverse of the PA's characteristics are found, and the DP output $z(n)$ is given by:

$$z(n) = a_1 x(n) + a_2 x(n) + a_3 x(n) + \cdots \tag{12}$$

The coefficients vector a can be found through a recursive algorithm based on the Indirect Learning architecture shown in Figure. 25. The name derives from the fact that the polynomial coefficients are found without passing by the determination of the PA's characteristics.

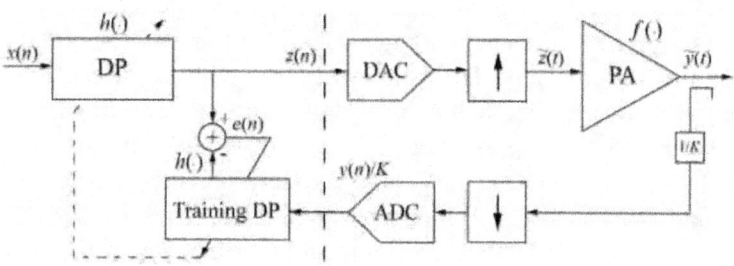

Figure. 25: Indirect Learning schematic principle for a DPD system

The indirect learning techniques works by two identical predistorters, the first – the actual one – in the transmission path and the second – the training one – in the feedback path. The outputs of both are compared to produce an error signal:

$$e(n) = \hat{z}(n) - z(\tilde{n}) \tag{13}$$

where $\hat{z}(n)$ is the output of the training DP. It can be demonstrated that when the error energy is zero the PA's baseband output y(t) is linear with the baseband input x(n), that is the cascade of pre distorter and amplifier becomes linear. If the PA's memory length is comparable to the envelope variations of the signal, the baseband model – which we call memory polynomial – can be adopted:

$$y(n) = \sum_{\substack{k=1, \\ k \text{ odd}}}^{K} \sum_{l=0}^{L-1} b_{k,l} \, z(n-l) |z(n-l)|^{k-1} \tag{14}$$

where K represent the order of the DP, while L is the number of memory samples; the baseband equivalent input z(n), the output y(n) and the coefficients k_{lb} of the model, are all complex valued in general. The pre distorter's output can be written the same way:

$$z(n) = \sum_{\substack{k=1, \\ k \text{ odd}}}^{K} \sum_{l=0}^{L-1} a_{k,l} \, x(n-l) |x(n-l)|^{k-1} \tag{15}$$

It has to be equal to the Training DP's output $\hat{z}(n)$ to minimize the error energy, that is:

$$z(n) = \sum_{\substack{k=1,\\ k \text{ odd}}}^{K} \sum_{l=0}^{L-1} a_{k,l} \frac{y(n-l)}{K} \left| \frac{y(n-l)}{K} \right|^{k-1}$$

$$(16)$$

The objective consists in finding the parameters $k_{l,a}$ that define the predistorter. Since $z(n)$ is linear in the $k_{l,a}$, the latter can be estimated by a simple least-squares method. By defining a new sequence:

$$u_{k,l}(n) = \frac{y(n-l)}{K} \left| \frac{y(n-l)}{K} \right|^{k-1}$$

$$(17)$$

we can rewrite $z(n)$ in matrix form as:

$$\mathbf{z} = \mathbf{Ua} \qquad (18)$$

where $\mathbf{z} = [z(0), \cdots, z(n-1)]^T$, $\mathbf{U} = [\mathbf{U}_0, \cdots, \mathbf{U}_L]$, $\mathbf{U}_l = [\mathbf{u}_{1l}, \cdots, \mathbf{u}_{Kl}]$, $\mathbf{u}_{kl} = [u_{kl}(0), \cdots, u_{kl}(N-1)]^T$ and

$\mathbf{a} = [a_{10}, \cdots, a_{K0}, \cdots, a_{1Q}, \cdots, a_{KQ}]^T$. The least-squares solution for a is given by:

$$\mathbf{a} = [\mathbf{U}^H \mathbf{U}]^{-1} \mathbf{U}^H \mathbf{z}$$

$$(18)$$

where $(\cdot)^H$ denotes complex conjugate transpose. A direct implementation of the polynomial predistorter is difficult, because it requires several sample-per-sample multiplications and power raisings. However, an efficient implementation is possible by observing that (15) is equivalent to:

$$z(n) = \left[\sum_{\substack{k=1,\\ k \text{ odd}}}^{K} a_{k,0} |x(n)|^{k-1} \right] x(n) + \left[\sum_{\substack{k=1,\\ k \text{ odd}}}^{K} a_{k,1} |x(n-1)|^{k-1} \right] x(n-1) + \cdots$$

$$+ \left[\sum_{\substack{k=1,\\ k \text{ odd}}}^{K} a_{k,L-1} |x(n-L+1)|^{k-1} \right] x(n-L+1)$$

$$(20)$$

The nonlinear polynomial can be implemented with a LUT indexed by the input magnitude, $|x(n-l)|$ [1]. This way, only L complex multiplications per sample are needed. LUT coefficients calculation is performed once the $k_{l,a}$ are found. The performance of the memory polynomial-LUT predistorter depends on the number of quantization points, on the memory length L and on the order of the polynomial, K.

Sub-sampling receiver

A key component for the DB-DP is the sub-sampling receiver, it operates on the principle of the band-pass sampling theorem, and it is used as feedback path of the DP system. If RF signals have a narrow bandwidth B, they can be sampled with a frequency:

$$f_s \geq 2B \qquad (21)$$

As a result of the sampling process, spectrum aliases are generated around all the multiples of s $_f$ as in Figure. 26. The image that falls in $[0;f_s/2]$ (first Nyquist zone) is the exact representation of the input signal, unless a potential phase inversion, and can be digitized. The same principle can also be used to convert two (or more) band-pass signals s_1 and s_2 'located at different carrier frequencies f_{c1} and f_{c2}', with band-widths B$_1$ and B$_2$. With a proper sampling frequency there will be replicas of the two signals located side-by-side in the first Nyquist zone with no overlap, as shown in Figure. 27. The proper sampling frequency respect the condition:

$$f_s \geq 2(B_1 + B_2) \qquad\qquad (22)$$

That is, a Nyquist Zone must be wider than the sum of the two bands.

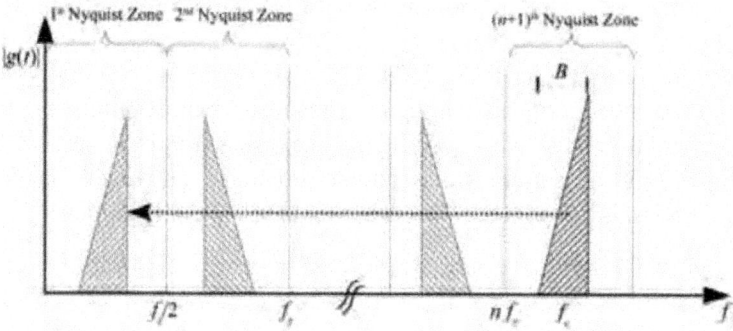

Figure. 26: Single band band-pass sub-sampling principle.

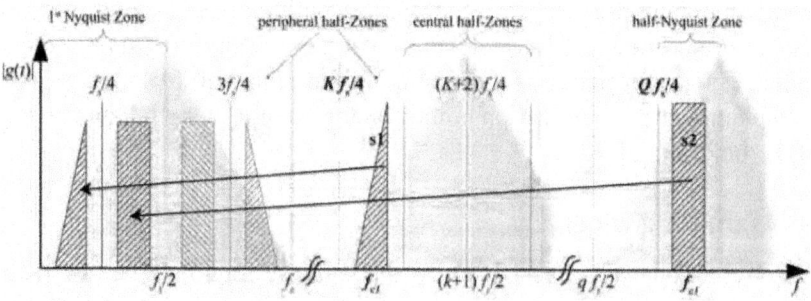

Figure. 27: Dual band sub-sampling principle.

The condition of no overlap consists of the both signals to be comprised in a single halfNyquist zone, i.e. $[nf_s/4;(n+1)f_s/4]$, where n is integer. If we define:

$$K = floor\left(\frac{f_{c1} - B_1 / 2}{B_1}\right) \quad Q = floor\left(\frac{f_{c2} - B_2 / 2}{B_2}\right)$$

(23)

where floor() is the operation of rounding to the lower integer, the conditions of no overlap are first given by

$$\begin{cases} kf_s / 4 + B_1 / 2 \le f_{c1} \le (k+1)f_s / 4 - B_1 / 2 \\ qf_s / 4 + B_2 / 2 \le f_{c2} \le (q+1)f_s / 4 - B_2 / 2 \\ k \le K \\ q \le Q \end{cases}$$

(24)

where k and q are integers identifying the order of the half-Zone in which the first and the second signals stand, respectively. The other condition, i.e. standing in central vs. peripheral half-zones, are given by:

$$\begin{cases} K = 4n \quad \Leftrightarrow \quad Q \neq 4n, 4n - 1 \\ K = 4n - 1 \quad \Leftrightarrow \quad Q \neq 4n, 4n - 1 \\ K = 4n + 1 \quad \Leftrightarrow \quad Q \neq 4n + 2, 4n + 1 \\ K = 4n + 2 \quad \Leftrightarrow \quad Q \neq 4n + 2, 4n + 1 \end{cases}$$

(25)

These conditions lead to a not closed form formulation which require an iterative approach for the solution. Once the suitable sampling frequency is found, the two signals replicas in the first Nyquist zone are located at the frequencies f_{bb1} and f_{bb2} which are given by:

$$f_{bb1} = \begin{cases} f_{c1} - floor(k / 4) \cdot f_s & k = 4n, k = 4n + 1 \\ (floor(k / 4) + 1) \cdot f_s - f_{c1} & k = 4n + 2, k = 4n - 1 \end{cases}$$

$$f_{bb2} = \begin{cases} f_{c2} - floor(q / 4) \cdot f_s & q = 4n, q = 4n + 1 \\ (floor(q / 4) + 1) \cdot f_s - f_{c2} & q = 4n + 2, q = 4n - 1 \end{cases}$$

(26)

The distortion introduced by a sub-sampling receiver is due in large part to the transfer function of the sampling device. In general, a T/H is preferred over a S/H, because of the lower distortion and higher sampling frequency reachable. The transfer function of a T/H is:

$$G_s(f) = \sum_{n=-\infty}^{\infty} G\left(f - \frac{n}{T_s}\right) \left[\frac{\tau}{T_s} \text{sinc}(\tau f) e^{-j\pi\tau f} + \frac{T_s - \tau}{T_s} \text{sinc}\left(\frac{n(T_s - \tau)}{T_s}\right) e^{-j\pi(T_s + \tau)\frac{n}{T_s}}\right],$$

(27)

where T_s is the sampling period and t is the length of the hold period. Due to the sinc()in order to avoid an amplitude distortion, t should be as low as possible to move at high frequency the first null. Also, the baseband aliases should be as near as possible to the zero. As regards the phase, different replicas have a different offset depending on the order n and the frequency of the alias. Replicas falling into the first Nyquist zone have a phase offset

Depending on k and f_{BB1}, or q and f_{BB2}. This offset must be compensated if a synchronism between the two signals is necessary, as in our proposed Dual Band DP method. This approach exhibits some critical points, [17]. The first ones to be considered are noise aliasing and aperture jitter; then out-of-bands signals and wideband noise must be filtered out before the sampler. That noise would otherwise, after sampling, translate and accumulate into the ⇒rst Nyquist zone. Besides, as even a perfect filter would reject the noise introduced by downstream circuits, low noise components have to be chosen. However, noise aliasing reduces with sampling frequency increase. Aperture jitter can be treated as a white noise if the jitter is low, and it doesn't depend on the sampling frequency. When designing a sub-sampling receiver, another important parameter to take care of is the analog bandwidth of the sampler, that must be greater than the highest frequency of the RF signals.

Dual Band Digital Predistortion

Architecture The DP-DP is achieved by a RF-level predistortion: a signal predistorter (as opposed to a data predistorter) is able to treat any kind of signal, that is it doesn't depend either on the bandwidth or the center frequency. Let's consider an input signal made of the superposition of two signals at different center frequencies, that is $x(n) = x_1(n) + x_2(n)$.

The input is pre distorted ($z(n)$), converted into analog ($\tilde{z}(t)$) and amplified ($\tilde{y}(t)$). A portion of $\tilde{y}(t)$ is drawn to have a feedback signal and to train the DP. A scheme is shown in Figure. 28. The main problem with this setup is the lack of sufficiently fast D/A and A/D converters, that will remain so in the foreseeable future because ADC dynamic range and conversion are known to progress at a rate much slower than Moore's law. Also, a RF predistortion is not possible at the moment, because it must be performed sample-per-sample and the sample rate is at least twice the maximum RF frequency (baseband sampling theorem).

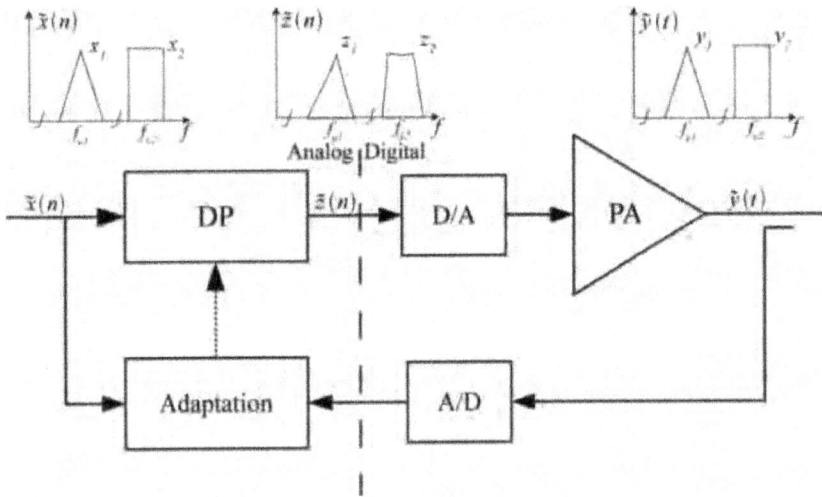

Figure. 28: RF DB-DP, principle of operation

Actually, the converters related problem can be easily overcome. The RF DAC can be replaced by two baseband DAC preceded by a proper digital filtering and digital frequency conversion system. In a similar way, the RF ADC can be replaced by two frequency converters and two baseband ADCs. There remains the sample rate problem. The last limit can be overcome by introducing a new architecture which is capable to lowering the sample rate, that is predistorting at intermediate frequency (IF). In this case the baseband digital signals $x_1(n)$ and $x_2(n)$ are shifted to f_{IF1} and f_{IF2} then summed, creating $\tilde{x}'(n)$. This IF signal is predistorted $(\tilde{z}'(n))$, and the two bands are separated and shifted to the baseband to be analog converted. The analog PA's input $\tilde{z}'(t)$ is built by those baseband signals, shifted to the RF frequencies f_{c1} and f_{c2} It is amplified $(\tilde{y}'(t))$ and a portion of it is drawn to create the feedback signals. As a feedback path we propose a subsampling receiver: the two bands composing $\tilde{y}'(t)$ are aliased side-by-side in the baseband, then digitized by a single ADC. In the digital domain, the bands are separated and shifted to IF, composing the signal $\tilde{y}''(n)$ that will be compared to $\tilde{x}'(n)$.

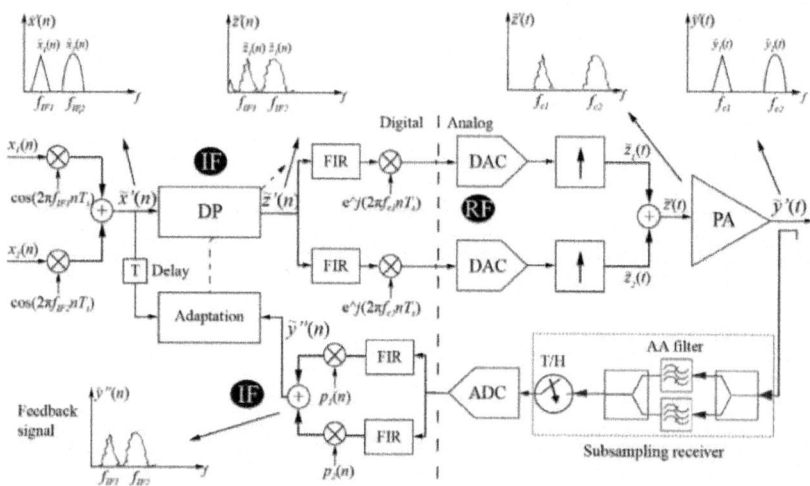

Figure. 29: DB-DP system with IF predistortion and subsampling feedback The block diagram of the whole system is shown in

When using a subsampling receiver, it is necessary to compensate the different phase offset applied to both bands. This may be done in the digital domain. If a T/H is used, the right phase shift can be calculated through eq. (27). Anti-aliasing filters must be carefully designed with in general out of band rejection. The IFs setting is a crucial point of the system design. They have to be far enough to leave room for out-of-band distortion and to simplify filtering; on the other side, they should be as low as possible to reduce computational constraints. As a rule, for the proposed DB-DP you may consider a sample rate at least four times higher than in a SB-DP system. The DB-DP was simulated by Matlab/Simulink®. We considered two 16 QAM signals, with amplitudesP = -10dBm and centre frequencies $f_{c1} = 2.1\,\text{GHz}$ and $f_{c2} = 3.5$ GHz; the sampling frequency was set to $f_s = 146.5\,\text{MHz}$. The PA was modeled with the WienerHammerstein model. LTI blocks preceeding and following the memoryless NL were set to have the following transfer functions:

$$H(z) = \frac{1 + 0.5z^{-2}}{1 - 0.4z^{-1}} \ , \ G(z) = \frac{1 - 0.3z^{-2}}{1 - 0.4z^{-1}}$$

(28)

It was chosen a tanh-shaped AM/AM NL, that has G=20dB, IP3=38dB and whose AM/PM is linear, with 5°/dB slope.

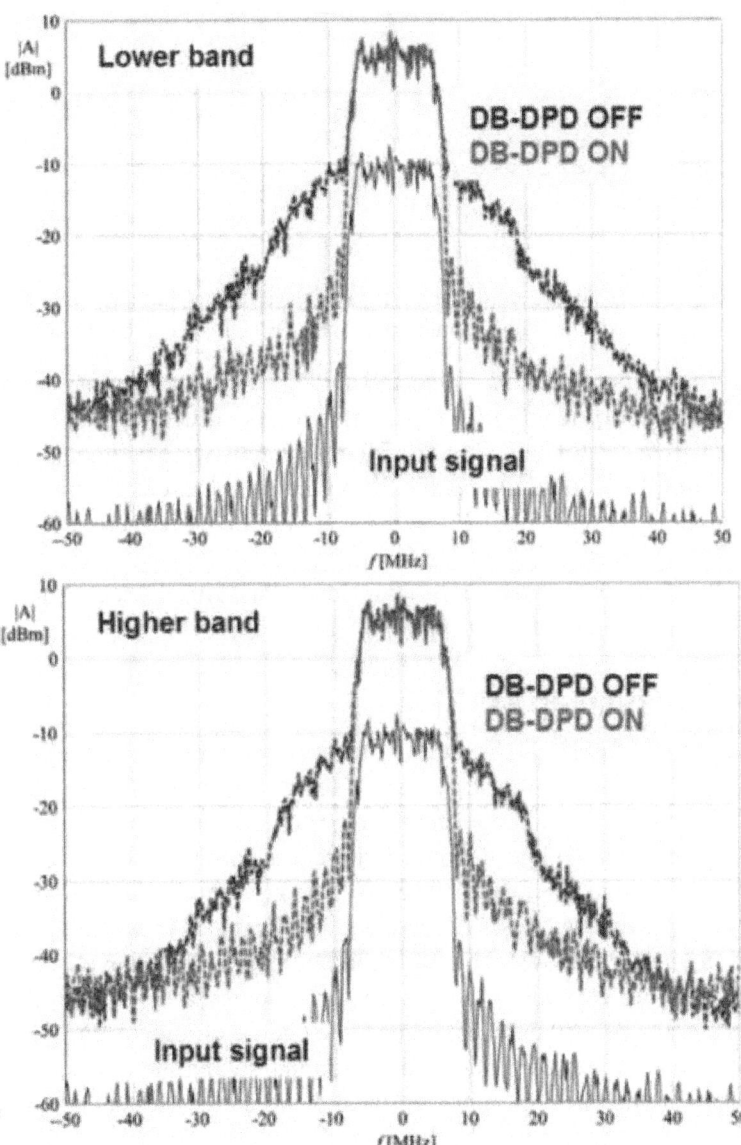

Figure. 30: Spectra comparison for lower and higher channels, between transmitted signal and input signal, with DB-DP OFF and DB-DP ON (left).

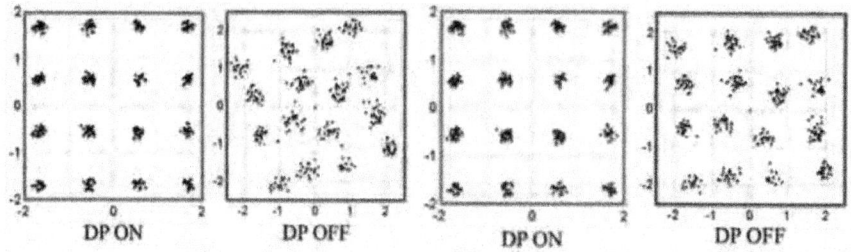

Figure. 31: Constellations comparison for lower (left) and higher (higher) channel, between transmitted signal and input signal, with DB-DP OFF and DB-DP ON.

For the implementation of the DB-DPD we used a memory polynomial DP, with a memory length of 4 taps, a polynomial order K=9 and a LUT pre distorter with a size of 512. Polynomial coefficients were estimated on a basis of 8192 samples. Simulation results for both channels are shown in Figure. 30 and Figure. 31, where an ACPR and EVM significant reduction is observed. The method proved to be able to correct most NLs, but it is not as good as a SB-DP. While in that case we obtained a Normalized Mean Square Error (NMSE) of 3e-4, in the DB-DP case we obtained an NMSE of 1e-3.

CONCLUDING REMARKS

The design of flexible PAs and multiband transmitter architectures is at a crucial stage; the number of research teams and projects that approached this field increased over the recent years. The number of special sessions and workshops in the main international conferences confirmed this interest. Some commercial products appeared recently, although they remain mainly based on very simple arrangements of frequency dedicated PAs with limited tuning control. Some technological and methodological problem have to be solved. The first set are related to the device technologies for both the RF power devices and the control devices. Indeed, the energy efficiency and peak power have to be maintained for wideband operation, making the device technology more challenging. Design approach have to take into account for multiband driving which reduce sensibly the power handling capability of the power device. Control devices, like switches and tuning elements have to cope with high peak power increasing the demand of linearity and efficiency, in this field MEMS appears a promising technology. An additional consideration is due for the architectures of multiband-multiband transmitters. Other than flexibility they have to provide excellent signal quality, which is much more threated by simultaneous concurrent signals. Polar transmitters versus Cartesian architectures are investigated as the two mainstreams for future transmitter architectures.

ACKNOWLEDGEMENT

The contents of this chapter are mainly based on the results of the research activities performed in the context of the project TARGET– "Top Amplifier Research Groups in a European Team" supported by the Information Society Technologies Programme of the EU under contract IST-1-507893-NOE, www. target-net.org.

REFERENCES

1.   Hashimoto, A.; Yoshino, H.; Atarashi, H., "Roadmap of IMT-advanced development," Microwave Magazine, IEEE , vol.9, no.4, pp.80-88, Aug. 2008

2.   F. K. Jondral, "Software-Defined Radio Basics and Evolution to Cognitive Radio", Journal on Wireless Communications and Networking, 2005, vol. 3, 275-283

3.   A. A. Abidi, "The Path to the Software-Defined Radio Receiver", IEEE Journal of SolidState Circuits, Vol. 42, no. 5, May 2007, pp. 954-966

4.   P. B. Kennington, RF and Baseband Techniques for Software Defined. Radio. Norwell, MA: Artech House, 2005.

5.   J. Laskar, R. Mukhopadhyay, Y. Hur, C. -H. Lee, and K. Lim, "ReconFigureurable RFICs and modules for cognitive radio", Digest of Topical Meeting on Silicon Monolithic Integrated Circuits in RF Systems, 2006. Jan. 2006 pp. 18-20

6.   F. Wang, D. F. Kimball, J. D. Popp, A. H. Yang, D. Y. Lie, P. M. Asbeck, L. E. Larson, "An Improved Power-Added Efficiency 19-dBm Hybrid Envelope Elimination and Restoration Power Amplifier for 802.11g WLAN Applications," Trans. On Microwave Theory and Techniques, Vol. 54, Dec. 2006, pp. 4086-4099

7.   Q. Shen, N. S. Barker "Distributed MEMS tunable matching network using minimalcontact RF-MEMS varactors," Microwave Theory and Techniques, IEEE Transactions on , vol.54, no.6, pp.2646-2658, June 2006

8.   K. Buisman, L.C.N. de Vreede, L.E. Larson, M. Spirito, A. Akhnoukh, T.L.M. Scholtes, L.K. Nanver "Distortion-free varactor diode topologies for RF adaptivity", Microwave Symposium Digest, 2005 IEEE MTT-S International,12-17 June 2005 pp. 157-160

9.   A. Jrad, A.-L. Perrier, R. Bourtoutian, J.-M. Duchamp, P. Ferrari, "Design of an ultra compact electronically tunable microwave impedance

transformer", Electronics Letters, Volume 41, Issue 12, 9 June 2005 pp. 707 – 709

10. P. Colantonio, F. Giannini, R. Giofrè, L. Piazzon, "Simultaneous Dual-Band High Efficiency Harmonic Tuned Power Amplifier in GaN Technology", European Microwave Conference Digest, Munich Oct., 2007

11. W.C.E. Neo, Yu Lin, Xiao-dong Liu, L.C.N. de Vreede, L.E. Larson, M. Spirito, M.J. Pelk, K. Buisman, A. Akhnoukh, A. de Graauw, L.K. Nanver, "Adaptive MultiBand Multi-Mode Power Amplifier Using Integrated Varactor-Based Tunable Matching Networks," Solid-State Circuits, IEEE Journal of , vol.41, no.9, pp.2166- 2176, Sept. 2006

12. A. Cidronali, I. Magrini, N. Giovannelli, M. Mercanti, G. Manes "Experimental system level analysis of a concurrent dual-band power amplifier for WiMAX and WCDMA applications"; International Journal of Microwave and Wireless Technologies, Cambridge University Press and the European Microwave Association, Vol.1 Special Issue 02, April 2009 pp 99-107

13. P. Colantonio, F. Giannini, R. Giofre, L. Piazzon, "Simultaneous dual-band high efficiency harmonic tuned power amplifier in GaN technology", European Microwave Integrated Circuit Conference, 8-10 Oct. 2007 pp.127 - 130

14. R. Meyer, R. Eschenback, and W. Edgerley, Jr., "A wideband feedforward amplifier," IEEE J. Solid-State Circuits, vol. SCC-9, no. 6, pp. 422–448, Jun. 1974.

15. P. Roblin, S. K. Myoung, D. Chaillot, Y. Gi Kim, A. Fathimulla, J. Strahler, S. Bibyk"Frequency-Selective Predistortion Linearization of RF Power Amplifiers" IEEE Trans on Microwave Theory and Tech., Vol. 56, Jan. 2008, pp 65-76

16. A. Cidronali, I. Magrini, R. Fagotti, G. Manes, "A new approach for concurrent DualBand IF Digital PreDistortion: System design and analysis," Workshop on Integrated Nonlinear Microwave and Millimetre-Wave Circuits, 2008. INMMIC 2008, pp.127-130, 24-25 Nov. 2008

17. G. Avitabile, A. Cidronali, G. Manes, 'A S-band digital down converter for radar applications based on GaAs MMIC fast sample&hold', IEE Proceedings-Circuits, Device and Systems, Vol.143, No.6 Dec. 1996 pp.337-342

Chapter 7

ACTIVE COMB FILTER USING OPERATIONAL TRANSCONDUCTANCE AMPLIFIER

Rajeev Kumar Ranjan, Surya Prasanna Yalla, Shubham Sorya, and Sajal K. Paul

Department of Electronics Engineering, Indian School of Mines, Dhanbad, Jharkhand 826004, India

ABSTRACT

A new approach for the design of an active comb filter is proposed to remove the selected frequencies of various signals. The proposed filter is based on only OTAs and capacitors, hence suitable for monolithic integrated circuit implementation. The workability of the circuit is tested using PSPICE for test signals of 60, 180, 300, and 420 Hz as in ECG signal. The results are given in the paper and found to agree well with theory.

INTRODUCTION

When harmonics of a certain frequency components are coupled into the circuit or the signal transmission line of an instrumentation system, the data acquired may suffer from harmonic interference. The power line interference is a common type of interference for various types of signals such as biomedical signals. There are basically two components in power line interferences, namely, electric field interference and magnetic field interference. Electric field interference generates spikes at 50/60 Hz frequency, whereas magnetic field which is generated due to the transformer in the power supply causes interference to generate harmonic frequencies of the fundamental. As an example, the source of these interferences is present in the entire clinic [1, 2], where a number of biomedical instruments run on AC power line. Hence physiological signal gets corrupted by power line frequency and its harmonics. The interference may be removed by both digital and analog filtering techniques [1–9].

Operational amplifiers are the most popular building block for analog circuit design. However, op-amp has limitations in bandwidth and slew rate which lead the analog designer to search for other possibilities [5–7,10–17]. Recently operational transconductance amplifier (OTA) has been found to be one of the most significant building blocks in analog signal processing. In high-frequency continuous-time filters, OTA-C filters have often been employed since OTAs provide high bandwidth, high slew rate, and a transconductance gain (G) which can be electronically controlled using a bias current. Hence the circuits developed using OTAs are most likely to possess intrinsic electronic control of parameters such as the cutoff frequency, quality factor, gain of a filter or frequency of oscillation, and the condition of oscillation of an oscillator.

In this paper a new analog comb filter based on notch filter is proposed. The presented filter is developed using all OTAs and capacitors. Hence it is an OTA-C comb filter and suitable for IC implementation. The parameters of the comb filter can be easily tuned electronically using bias current of OTAs.

CIRCUIT DESCRIPTION

The circuit of second order passive notch filter is shown in Figure 1.

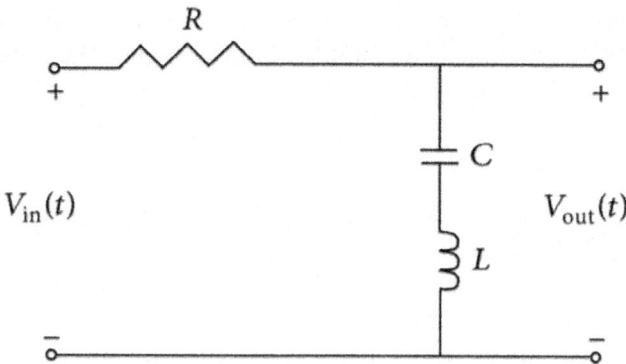

Figure 1: RLC circuit of notch filter.

The routine analysis gives voltage transfer function H(s) as

$$H(s) = \frac{s^2 LC + 1}{s^2 LC + sCR + 1}.$$

(1)

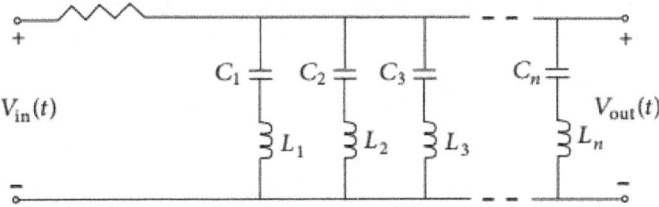

Figure 2: Comb filter using a basic RLC circuit.

The parameters of notch filters are obtained as

$$\omega_0 = \frac{1}{\sqrt{LC}},$$

$$Q_0 = \frac{1}{R}\sqrt{\frac{L}{C}},$$

$$\Delta f = \frac{R}{L}. \tag{2}$$

Proposed Active Comb Filter

The extension of L-C section of circuit in Figure 1 gives a comb filter as shown in Figure 2. It can remove n-number of harmonics of the power line interference, which corrupt the input signal $V_{in}(t)$. The routine analysis of the circuit in Figure 2 results in a voltage transfer function of the active comb filter as

$$H(s) = \frac{1}{R\sum_{k=1}^{n}\left(sC_k/\left(s^2 L_k C_k + 1\right)\right) + 1}. \tag{3}$$

The Kth notch filter is used to eliminate the Kth harmonic component from the input signal $V_{in}(t)$. The transfer function of the Kth notch filter is obtained as

$$H^k(s) = \frac{1}{\left(sC_k R/\left(s^2 L_k C_k + 1\right)\right) + 1}. \tag{4}$$

It is well known that the implementation of inductance in integrated circuits is very difficult or almost impossible. The passive resistance (R) is also not encouraged in integrated circuit implementation. Hence to ease the integrated circuit implementation, the proposed notch filter is implemented based on a new active element, namely, operational transconductance amplifier (OTA).

Operational transconductance amplifier (OTA) is an active current mode building block. It is widely used block in integrated circuit technique and suitable for various applications. It consists of an input differential pair and an output current mirror. The input of OTA is voltage and output is current.

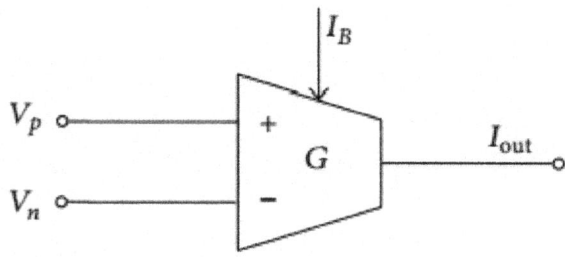

Figure 3: Symbol of OTA.

It has higher bandwidth and slew rate than op-amp. The transconductance gain (G) of OTA can be controlled electronically by bias current over a wide range. The symbolic representation of an OTA is given in Figure 3 and internal structure in Figure 4. The transconductance gain for CMOS based OTA is expressed as

$$G = B\sqrt{2\mu C_{ox}\left(\frac{w}{L}\right)I_B},$$

$$(5)$$

Where B is constant, μ is the mobility of electron, C_{ox} is oxide capacitance, I_B is biasing current which is controlled by biasing voltage V_{con}, W is channel width, and L is the channel length of the transistor. The circuit of notch filter using all OTAs and capacitors is shown in Figure 5. The resistance (R) and inductance (L) are expressed as

$$R = \frac{1}{G_R}, \qquad L = \frac{C_L}{G_L^2},$$

$$(6)$$

Where G_R and G_L are the transconductance of the OTAs, which implements R and L, respectively. It reveals that the value of resistance (R) and inductance (L) may be varied using bias current of respective OTAs. Hence, ω_o, Q_o, and bandwidth (Δf) can be tuned electronically with the bias current of OTAs. It is also evident that Q_o can be tuned independently of ω_o by R. The generalized proposed OTA-C comb filter which can absorb n number of unwanted frequencies is shown in Figure 6. The expressions of the characteristic parameters of each notch filter are modified as

$$\omega_{on} = \frac{1}{\sqrt{L_n C_n}} = \sqrt{\frac{G_{Ln}^2}{C_{Ln} C_n}} = \frac{G_{Ln}}{\sqrt{C_{Ln} C_n}},$$

$$Q_{on} = \frac{1}{R}\sqrt{\frac{L_n}{C_n}} = G_R \sqrt{\frac{C_{Ln}}{G_{Ln}^2 C_n}} = \frac{G_R}{G_{Ln}}\sqrt{\frac{C_{Ln}}{C_n}},$$

$$\Delta f_n = \frac{G_{Ln}^2}{G_R C_{Ln}}.$$

$$(7)$$

It is evident that once the values of C_n and CL_n are fixed as per requirement, the notch frequencies can still be tuned by varying GL_n using bias currents of nth set of OTAs. The same is also true for quality factor (Q_{on}) and bandwidth (Δf_n).

SIMULATION AND RESULT

The OTA in Figure 4 has been simulated using PSPICE in 0.5 μm CMOS Technology. The dimensions of the MOS are given in Table 1 [4]. The supply voltages used for simulation are V_{dd} = 5 V and V_{ss} = −5 V. The results of simulations are shown in Figures 7, 8, and 9.

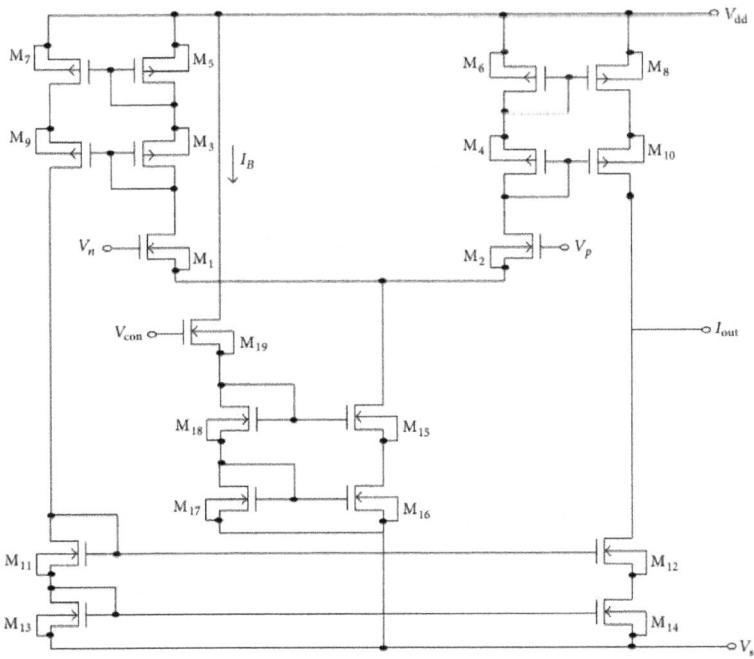

Figure 4: Internal structure of OTA [4].

Table 1: MOS dimensions [4]

MOS	Dimensions of MOS transistors		
		W (μm)	L (μm)
M_1, M_2, M_{11}, M_{12} M_{13}, M_{14}, M_{15}, M_{16}, M_{17}, M_{18}, and M_{19}		5	3
M_3, M_4, M_5, M_6, M_7, M_8, M_9, and M_{10}		10	3

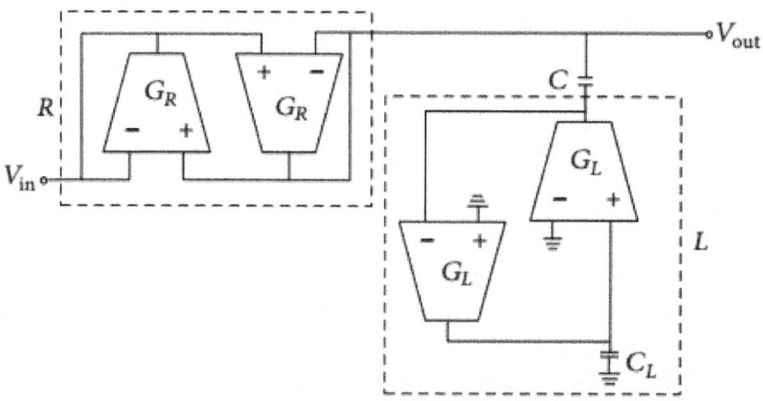

Figure 5: Realization of OTA-C notch filter

Physiological signal such as ECG signal is known to contain the power line frequency and its harmonics along with the actual physiological signal. Hence as an example, the proposed comb filter is designed for n=4 to show. its performance to remove undesired power line signals of fundamental frequency of 60 Hz and its odd harmonics 180, 300, and 420 Hz in ECG signal. As it is a low-frequency operation, a low noise and low distortion OTA as presented in Figure 4 suitable for low-frequency application [4] is used. The values of capacitors used are C_1 = 998.85 nF, C_2 = 110.98 nF, C_3 = 39.95 nF, C_4 = 20.38 nF, and C_{L1} = C_{L2} = C_{L3} = C_{L4} = 21 nF and R is adjusted to 100 kΩ by biasing current of corresponding OTAs. The frequency response of the output of the comb filter is shown in Figure 10. It verifies that the simulated response matches perfectly the theoretical result. The time responses of the input having frequencies 60, 180, 300, and 420 Hz signals and their corresponding outputs are shown in Figures 11, 12, 13, and 14, respectively. It reveals that the output is almost insignificant at the set (rejection) frequency of the comb filter. To know the quality of the output and dynamic range at the passband, the total harmonic distortion has been obtained for a signal at 100 Hz as shown in Figure 15. It indicates that % THD is very low up to 4.5 V (peak to peak). It indicates that the dynamic range is wide.

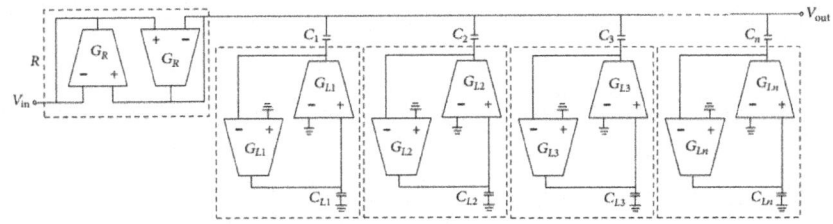

Figure 6: Generalized OTA-C comb filter.

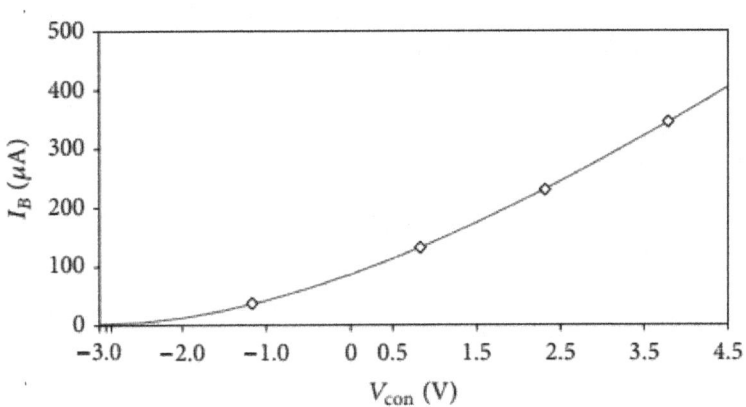

Figure 7: Variation of biasing current I_B with control voltage V_{con}.

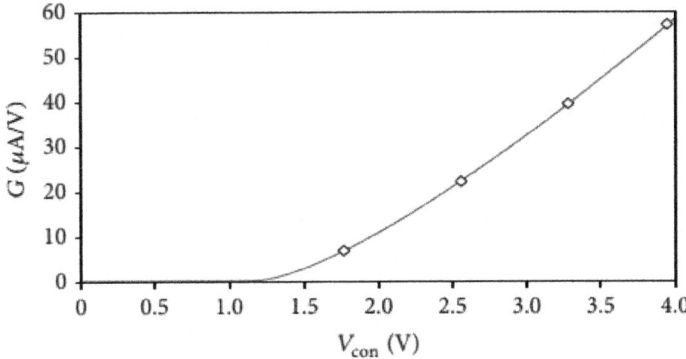

Figure 8: Variation of transconductance G with control voltage V_{con}.

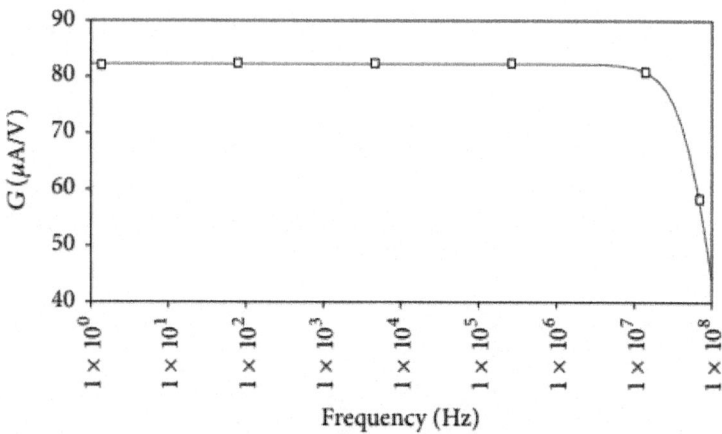

Figure 9: Frequency response of OTA transconductance for bias current $I_B = 206.3$ μA.

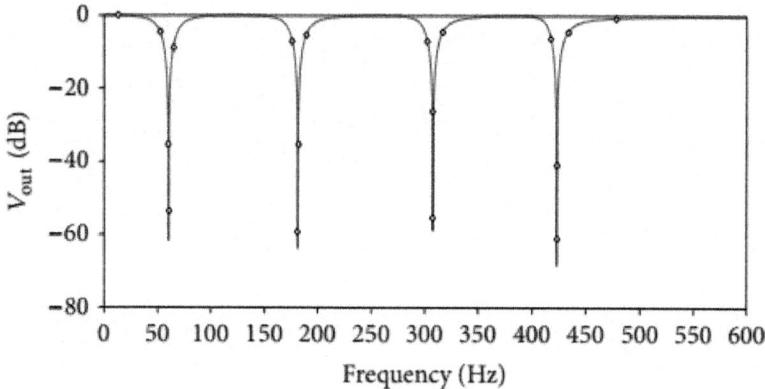

Figure 10: Simulated result of proposed comb filter.

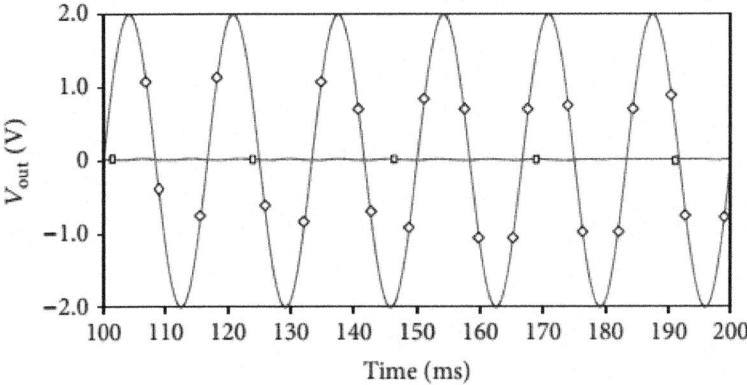

◊ Input waveform
▫ Output waveform

Figure 11: Input and output response at 60 Hz.

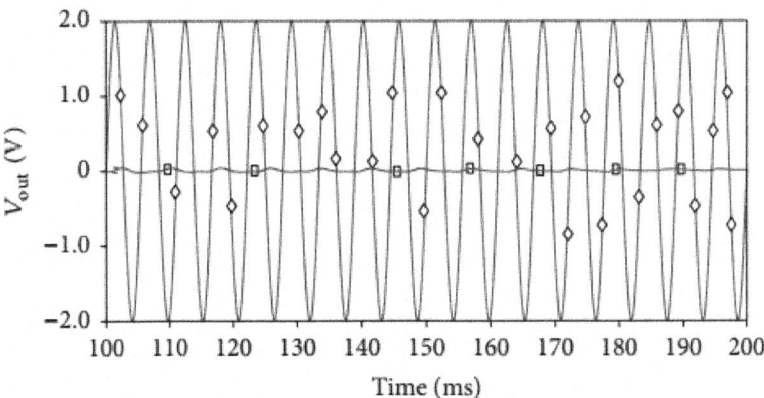

◊ Input waveform
▫ Output waveform

Figure 12: Input and output response at 180 Hz.

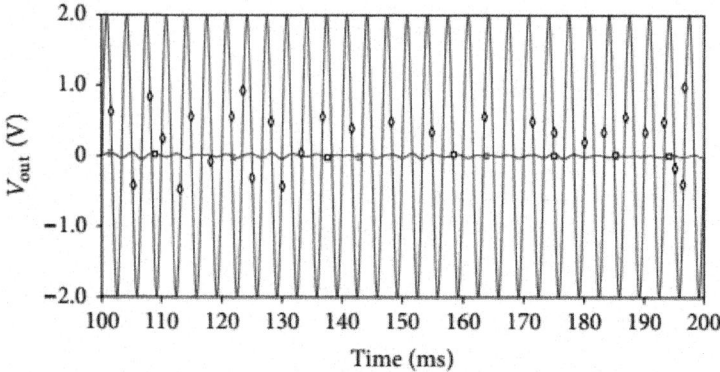

Figure 13: Input and output response at 300 Hz.

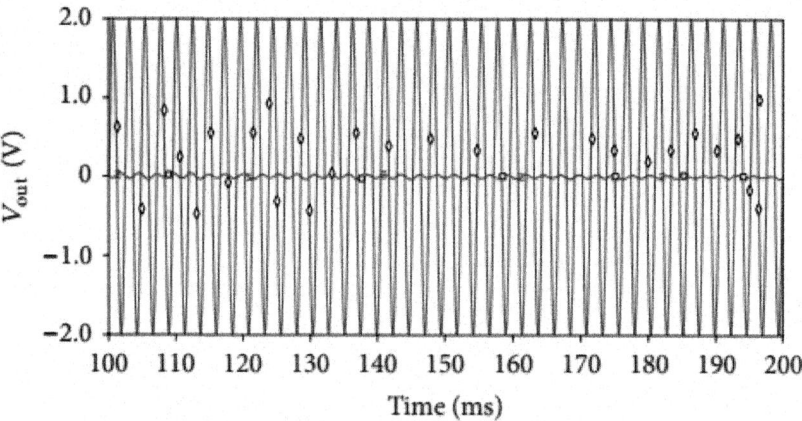

Figure 14: Input and output response at 420 Hz.

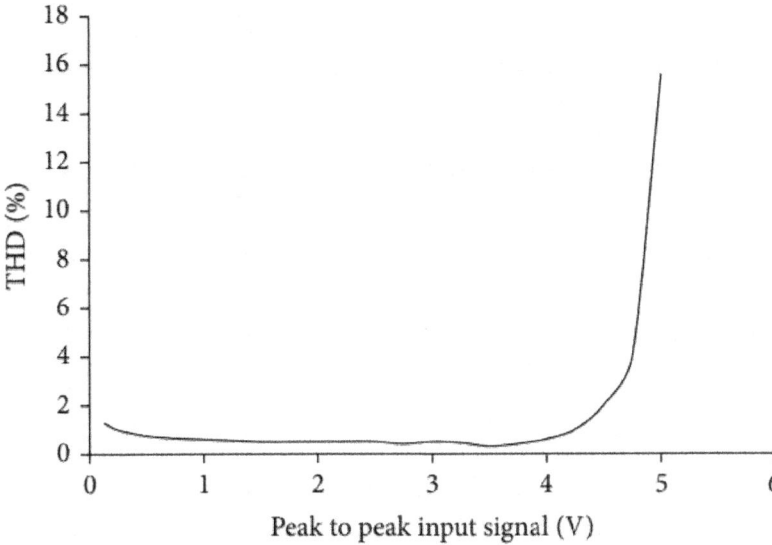

Figure 15: %THD of comb filter.

The noise characteristic of the proposed circuit has been analyzed. The proposed comb filter is designed to reject 60 Hz, 180 Hz, 300 Hz, and 420 Hz signals. The noise of the circuit for an input signal of 100 KHz is obtained as 484 nV/ $\sqrt{Hz}$ at the input and 482 nV/$\sqrt{Hz}$ $\sqrt{Hz}$ at output. However when it is tested with input signal of 60 Hz, the input noise is obtained as 987.7 nV/$\sqrt{Hz}$ and output noise of 0.45 nV/$\sqrt{Hz}$ It shows a significant reduction of noise at stopband. The circuit consumes a power of 21.5 mW. Moreover, 3 dB cutoff frequency of the OTA transconductance is $f_{3\,dB}$ = 65.97 MHz for the biasing current of I_B = 206.3 μA; the same is shown in Figure 9. Therefore the operation of the proposed circuit is limited to 65.97 MHz.

CONCLUSION

A new active comb OTA-C filter is proposed. The workability of the circuit has been tested by using PSPICE in 0.5 μm CMOS Technology. The simulated and theoretical results agreed quite well. The comparison of the proposed work with the reported publications [8, 9] reveals the following features in favour of proposed circuit. (i) It indicates that %THD is very low up to the input signal of 4.5 V (Peak-to-Peak). (ii) Although the number of active components is higher in comparison to [8, 9], the number of passive components is remarkably less. (iii) Parameters of filters such as ω_{on}, Q_{on}, and Δf_n ' can be electronically controlled by varying bias current of OTAs. (iv) Moreover, Q_{on} can be conveniently varied independent of ω_{on} by bias current of OTAs, which is very much useful to achieve high selectivity.

CONFLICT OF INTERESTS

The authors declare that there is no conflict of interests regarding the publication of this paper.

REFERENCES

1. J. C. Huhta and J. G. Webster, "60-Hz interference in electrocardiography," IEEE Transactions on Biomedical Engineering, vol. 20, no. 2, pp. 91–101, 1973.

2. J. Piskorowski, "Power line interference removal from ECG signal using notch filter with non-zero initial conditions," in Proceedings of the IEEE International Symposium on Medical Measurements and Application (MeMeA '12), pp. 1–3, University of Technology, Budapest, Hungary, May 2012.

3. S. M. M. Martens, M. M. Mischi, S. G. Oei, and J. W. M. Bergmans, "An improved adaptive power line interference canceller for electrocardiography," IEEE Transactions on Biomedical Engineering, vol. 53, no. 11, pp. 2220–2231, 2006.

4. G. Duzenlia, Y. Kcili, H. Kuntmanc, and A. Atamanb, "On the design of low-frequency filters using CMOS OTAs operating in the subthreshold region," Microelectronics Journal, vol. 30, no. 1, pp. 45–54, 1999. r

5. T. W. Dawson, K. Caputa, M. A. Stuchly, and R. Kavet, "Pacemaker interference by 60-Hz contact currents," IEEE Transactions on Biomedical Engineering, vol. 49, no. 8, pp. 878–886, 2002.

6. C. Ling, P. Ye, R. Liu, and J. Wang, "A low-pass power notch filter based on an OTA—C structure for electroencephalogram," in Proceedings of the International Symposium on Intelligent Signal Processing and Communications Systems (ISPACS '07), pp. 451–453, Xiamen, China, December 2007.

7. G. Ferri, V. Stornelli, and A. di Simone, "A CCII-based high impedance input stage for biomedical applications," Journal of Circuits, Systems and Computers, vol. 20, no. 8, pp. 1441–1447, 2011

8. C.-D. Tsai, D.-C. Chiou, Y.-D. Lin, H.-L. Chan, and C.-P. Wu, "An active comb filter design for harmonic interference removal," Journal of the

Chinese Institute of Engineers, vol. 21, no. 5, pp. 605–610, 1998. ·

9.   C.-T. Tsai, H.-L. Chan, C.-C. Tseng, and C.-P. Wu, "Harmonic interference elimination by an active comb filter [ECG application]," in Proceedings of 16th Annual International Conference of IEEE Engineering in Medicine and Biology Society, vol. 2, pp. 964–965, Baltimore, Md, USA, November 1994.

10.   Fabre, O. Said, F. Wiest, and C. Boucheron, "High frequency applications based on a new current controlled conveyor," IEEE Transactions on Circuits and Systems I: Fundamental Theory and Applications, vol. 43, no. 2, pp. 82–91, 1996. r

11.   M. T. Abuelma'atti and N. A. Tasadduq, "New current-mode current-controlled filters using the current-controlled conveyor," International Journal of Electronics, vol. 85, no. 4, pp. 483–488, 1998.·

12.   U. Çamam, F. Kaçar, O. Cicekoglu, H. Kuntman, and A. Kuntman, "Novel grounded parallel immittance simulator topologies employing single OTRA," AEU—International Journal of Electronics and Communications, vol. 57, no. 4, pp. 287–290, 2003

13.   K. N. Salama and A. M. Soliman, "Active RC applications of the operational transresistance amplifier,"Frequenz, vol. 54, no. 7-8, pp. 171–176, 2000.

14.   S. Kılınc and U. Çam, "Cascadable all pass and notch filters employing single operational transresistance amplifier," Journal of Computers and Electrical Engineering, vol. 31, no. 6, pp. 391–401, 2005. r

15.   P. Visocchi, J. Taylor, R. Mason, A. Betts, and D. Haigh, "Design and evaluation of a high-precision, fully tunable OTA-C bandpass filter implemented in GaAs MESFET technology," IEEE Journal of Solid-State Circuits, vol. 29, no. 7, pp. 840–843, 1994.

16.   R. L. Geiger and E. Sánchez-Sinencio, "Active filter design using operational transconductance amplifiers: a tutorial," IEEE Circuits and Devices Magazine, vol. 1, no. 2, pp. 20–32, 1985.

17.   S.-H. Yang, K.-H. Kim, Y.-H. Kim, Y. You, and K.-R. Cho, "A novel CMOS operational transconductance amplifier based on a mobility compensation technique," IEEE Transactions on Circuits and Systems II: Express Briefs, vol. 52, no. 1, pp. 37–42, 2005.

Chapter 8

DESIGN OF A 2 GHZ LINEAR-IN-DB VARIABLE-GAIN AMPLIFIER WITH 80-DB GAIN RANGE

Zhengyu Sun and Yuepeng Yan

Institute of Microelectronics of Chinese Academy of Sciences, Beijing 100029, China

ABSTRACT

A broadband linear-in-dB variable-gain amplifier (VGA) circuit is implemented in 0.18 μm SiGe BiCMOS process. The VGA comprises two cascaded variable-gain core, in which a hybrid current-steering current gain cell is inserted in the Cherry-Hooper amplifier to maintain a broad bandwidth while covering a wide gain range. Postlayout simulation results confirm that the proposed circuit achieves a 2 GHz 3-dB bandwidth with wide linear-in-dB gain tuning range from −19 dB up to 61 dB. The amplifier offers a competitive gain bandwidth product of 2805 GHz at the maximum gain for a 110-GHz f_t BiCMOS technology. The amplifier core consumes 31 mW from a 3.3 V supply and occupies active area of 280 μm by 140 μm.

INTRODUCTION

The telecommunication industry continues to drive forward with gigabit-class high-speed data transmission in microwave, millimeter-wave, and optical communication systems. Higher data rate transmission requires wider bandwidth and larger dynamic range for the receiver system. For example in the 60 GHz standards such as IEEE 802.11ad [1], the spectrum is about 2 GHz. And the signal strength at the receiver input can change dramatically which is common in short-range wireless systems. VGA as a key component in the automatic gain control (AGC) loop provides constant signal strength to the baseband processor to maximize the dynamic range of the receiver system and compensates gain variations caused by process, voltage, and temperature (PVT) variations. Research in SiGe and CMOS circuits in the broadband VGA is an active topic in optical, wireline, and millimeter-wave receivers [2–6].

Among various broadband techniques, Cherry-Hooper amplifier [7] has had numerous applications in AGC and limiting amplifiers. Solutions employing Cherry-Hooper amplifier [8] can achieve a 3-dB bandwidth above 2 GHz and satisfy bandwidth requirements such as 60 GHz short-range wireless systems. However, their gain range is not sufficient to meet the target of more than 80 dB if additional margin is needed for the receiver. Moreover, a linear-in-dB control signal for VGA will simplify the system level design for AGC.

In this paper, a VGA circuit is presented for high-speed data communication systems. In Section 2, a new variable gain core is designed by combining a hybrid current-steering cell with the Cherry-Hooper amplifier. Section 3 describes the VGA system which is made up of variable gain core introduced in Section 2 and other circuits. Section 4 shows the postlayout simulation results. Finally, Section 5 draws the conclusions.

VARIABLE GAIN CORE DESIGN

The schematic of the proposed variable gain core is shown in Figure 1. In order to maintain a broad bandwidth while covering a wide gain range, a hybrid current-steering (HCS) current gain cell is inserted in the Cherry-Hooper amplifier. The HCS cell is featured with a constant DC current output and an ability to have same AC characteristics with classical current-steering (CS) cell. The gain is varied through only the HCS cell without degrading the broadband characteristic of Cherry-Hooper amplifier with constant gain.

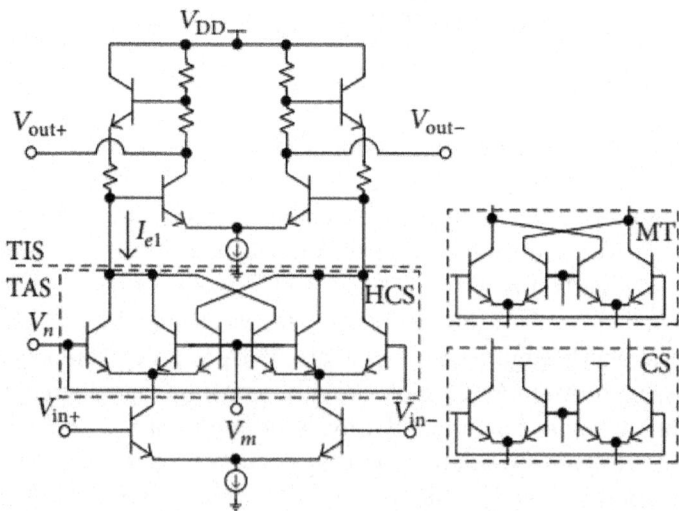

Figure 1: Proposed variable gain core with hybrid current-steering cell.

Broadband Amplifier Topology

If the current gain cell is removed from the gain core shown in Figure 1 the topology is a differential Cherry-Hooper (CH) amplifier with emitter-follower feedback [9]. The Cherry-Hooper amplifier is a cascading combination of a transadmittance stage (TAS) with a transimpedance stage (TIS). The strong mismatch between the two stages results in a large bandwidth [7]. This topology has the advantage of extending the bandwidth by moving the dominant poles from lower pole determined by load capacitance and the output resistance to the much higher poles. The shunt-shunt feedback employs an emitter follower which isolates the bias currents through the TAS and TIS. Both the emitter follower and the resistor in the feedback loop help to raise the gain at high frequency.

Because of the symmetry, the differential Cherry-Hooper amplifier with emitter-follower feedback can be analyzed through single-ended part. Figure 2 is the small signal equivalent circuit for the TAS-TIS combination including major parasitic parameter. Since the TIS is usually followed by transconductor stage or emitter follower, the load of Cherry-Hooper amplifier can be approximately assumed to be a capacitor C_L. Thus, the input impedance of TIS stage can be derived as

$$Z_{\text{in}} = \frac{v_2}{i_1} = \frac{R_f}{1 + g_{m2}R_1}$$

$$\cdot \left((1 + sC_L(R_1 + R_2)) \right.$$

$$\times \left(1 + \frac{C_2R_f + C_L(R_1 + R_2)}{1 + g_{m2}R_1}s \right.$$

$$\left. \left. + \frac{C_2C_LR_f(R_1 + R_2)}{1 + g_{m2}R_1}s^2 \right)^{-1} \right).$$

$$(1)$$

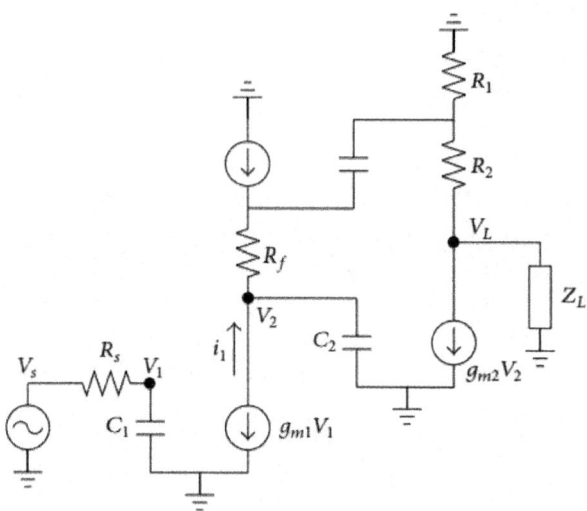

Figure 2: Small signal equivalent circuit for the TAS-TIS combination.

The transimpedance gain of TIS stage can also be derived as

$$Z_t = \frac{v_L}{i_1} = -\frac{g_{m2}}{C_2 C_L \left(s^2 + (w_0/Q) s + w_0^2\right)},$$

(2)

where

$$w_0 = \sqrt{\frac{1 + g_{m2} R_1}{C_2 C_L R_f (R_1 + R_2)}},$$

(3)

$$Q = \sqrt{\frac{C_2 C_L R_f (R_1 + R_2)(1 + g_{m2} R_1)}{\left(C_2 R_f + C_L R_1 + C_L R_2\right)^2}}.$$

(4)

From (2) and (3), there is a trade-off between DC gain and bandwidth through resistors R_f and R_2. Increasing R_2 or R_f will raise the DC gain while reducing bandwidth. From (1), the input impedance of TIS stage is $(1 + gm2R_1)$ times less than R_f and is relatively low impedance compared to the output impedance of TAS stage. Therefore, the overall voltage gain of TAS-TIS combination can be approximately seen as the product of transadmittance gain and transimpedance gain as

$$A_v = \frac{v_L}{v_s} = \frac{i_1}{v_s} \cdot \frac{v_L}{i_1} = G_m Z_t,$$

(5)

with the transadmittance gain as

$$G_m = \frac{i_1}{v_s} = \frac{g_{m1}}{1 + sC_1R_s},$$

(6)

where R_s is the output impedance of preceding stage. Since the input impedance of TAS stage is usually much larger than the R_s, the pole of TAS stage is much larger than of TIS stage. Therefore, the dominant pole of the CherryHooper amplifier is mostly contributed by the poles of TIS stage. The output impedance of TIS stage can also be derived as

$$Z_{out} = \frac{R_1 + R_2}{1 + g_{m2}R_1} \cdot \frac{1 + sC_2R_f}{1 + s\left(C_2R_f / \left(1 + g_{m2}R_1\right)\right)},$$

(7)

which shows relatively low impedance that can drive the following stage.

For broadband amplifier design, the bias currents through the TAS and TIS are constant to maintain the transistors biased at near peak f_t current density. Then the desired gain and bandwidth are determined by the resistors in the feedback and load. If the DC current I_{el} in the TAS changed, the frequency response of the TIS will be changed as well since the bias conditions of transistors in emitter-follower and TIS are changed. Figure 3 shows the transimpedance gain between output voltage and current flowing out of TAS with its 3-dB bandwidth at different bias DC current I_{el}. Decreasing the DC current I_{el}, the bandwidth will decrease while the gain will increase. Such effect is detrimental to the integrity design of the Cherry-Hooper amplifier.

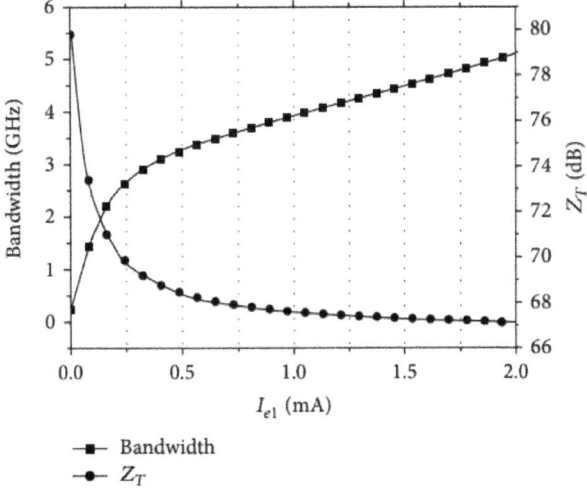

Figure 3: Transimpedance gain and its bandwidth versus bias current . I_{el}

Current Gain Cell

There are two commonly used current gain cells as shown on the right side of Figure 1: currentsteering (CS) cell and multiplier (MT) cell. The currentsteering cell has better dynamic range performance but its DC output current is not constant. The multiplier cell has constant DC output level due to the cross coupling of the quad collectors but its gain polarity will change at certain control voltage leading to a limited tuning range.

In order to maintain the broadband characteristics of the Cherry-Hooper amplifier while covering a wide gain range, a hybrid current-steering current gain cell is employed to change gain while having the least influence to the Cherry-Hooper amplifier. A generalized hybrid current-steering cell is shown in Figure 4. The cell is symmetrical about the Y-axis, thus the DC current output is equal to the DC current input. The output current coming out of the cell is related to the input current as follows:

$$I_{out} = (\alpha_1 + \alpha_2)(I_{in0} + i_{in}) + \alpha_3(I_{in0} - i_{in})$$

$$= I_{in0} + (\alpha_1 + \alpha_2 - \alpha_3)i_{in}$$

$$(8)$$

where α_n is the current proportion of the nth branch dependent on transistor size and bias voltage as follows:

$$\alpha_n = \frac{1}{\sum_{i=1}^{3}(I_{0i}/I_{0n})\exp((V_i - V_n)/V_T)}.$$

$$(9)$$

Thus the current gain is expressed as

$$G_I = \alpha_1 + \alpha_2 - \alpha_3.$$

$$(10)$$

In the proposed VGA, a special case with equal V_2 and V_3 voltage and transistor size arrangement as $Q_1 : Q_2 : Q_3 = 2:1:1$ [10] is employed with current gain derived from (9) and (10) as follows:

$$G_I = \alpha_1 = \frac{1}{1 + \exp((V_2 - V_1)/V_T)}.$$

$$(11)$$

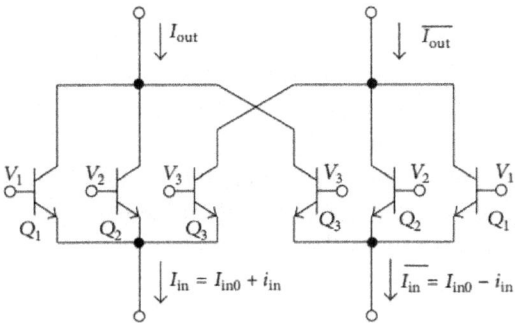

Figure 4: Generalized hybrid current-steering current gain cell.

The current gain expression of the hybrid currentsteering cell is the same as the conventional current-steering cell, which implies that the same linear-dB technology for CS cell can also be used for HCS cell. Figure 5 compares the current gain and DC output current of the three current gain cells: CS, MT, and HCS. The HCS cell has the same wide gaintuning ability as CS cell while having a constant DC output current as MT cell.

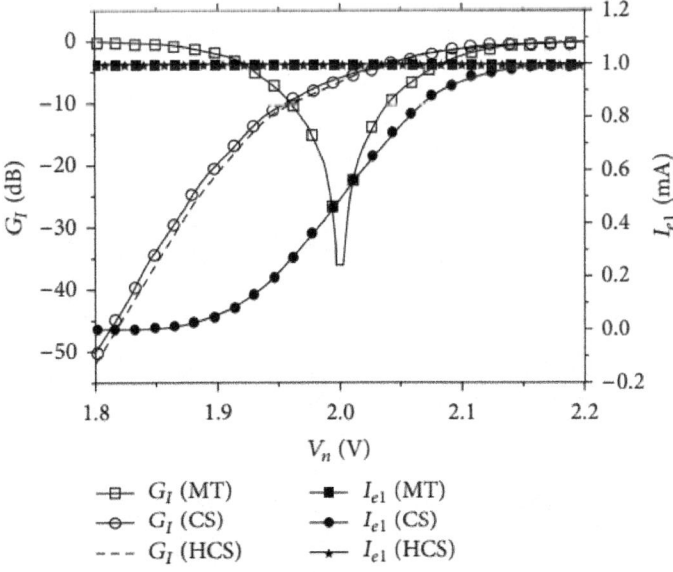

Figure 5: Comparison of current gain and DC current output I_{e1} of MT, CS and HCS cell.

Broadband Variable Gain Core

The strong impedance mismatch between gain stages for broadband technique is maintained with the insertion of hybrid current-steering cell in the middle of the Cherry-Hooper amplifier. The transadmittance stage is composed of the input transconductance amplifier and the hybrid current-steering cell, which has a high input and high output impedance. It is followed by the transimpedance stage which has low input and low output impedance. The bias current of the input transconductance amplifier is reused by the HCS cell and the emitter follower in the feedback path.

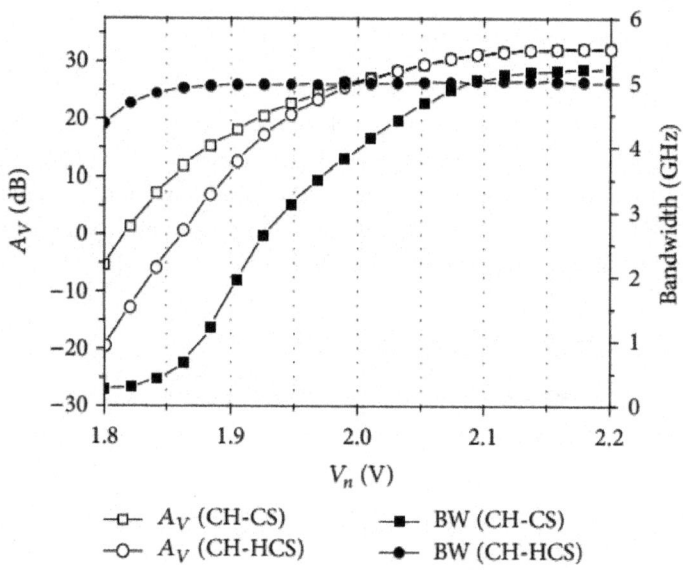

Figure 6: Comparison of voltage gain and its bandwidth of Cherry-Hooper amplifier with CS and HCS cell.

Since the current gain cell operates as common-base amplifier, the bandwidth of it is much wider than that of the Cherry-Hooper amplifier. Thus the dominant poles of Cherry-Hooper amplifier still dominate the 3-dB bandwidth of the variable gain core and can be optimized to strike a balance between characteristics such as gain, bandwidth, gain flatness, and group delay using guidelines [9]. To show the advantage of the proposed structure, another structure with only difference in that its current gain cell was replaced with a current-steering cell was also designed. Figure 6 shows the comparison of the Cherry-Hooper amplifiers inserted with hybrid current-steering (CH-HCS) cell and with current-steering (CH-CS) cell. The 3-dB bandwidth of the CH-CS cell decreases rapidly with decreasing control voltage, while the bandwidth of CH-HCS cell maintains being flat above 4 GHz in spite of control voltage changing. Moreover, the relationship between gain and control voltage of CH-CS cell is deviated from the CH-HCS cell at low gain range due to the transimpedance gain increases as indicated in Figure 3.

In order to further show the advantage of the proposed CH-HCS cell over traditional current-steering type VGA with load resistor R_L (CS-R_L), a CS-R_L cell of the same DC current consumption and peak gain as the CH-HCS cell is designed. The 3-dB bandwidth varying with control voltage is compared in Figure 7(a), which shows that a threefold GBW can be achieved using CH-HCS cell. Figures 7(b) and 7(c) compare the S_{21} and noise figure at the maximum gain, which shows a similar noise performance and wider GBW for the CH-HCS cell over traditional CS-R_L cell.

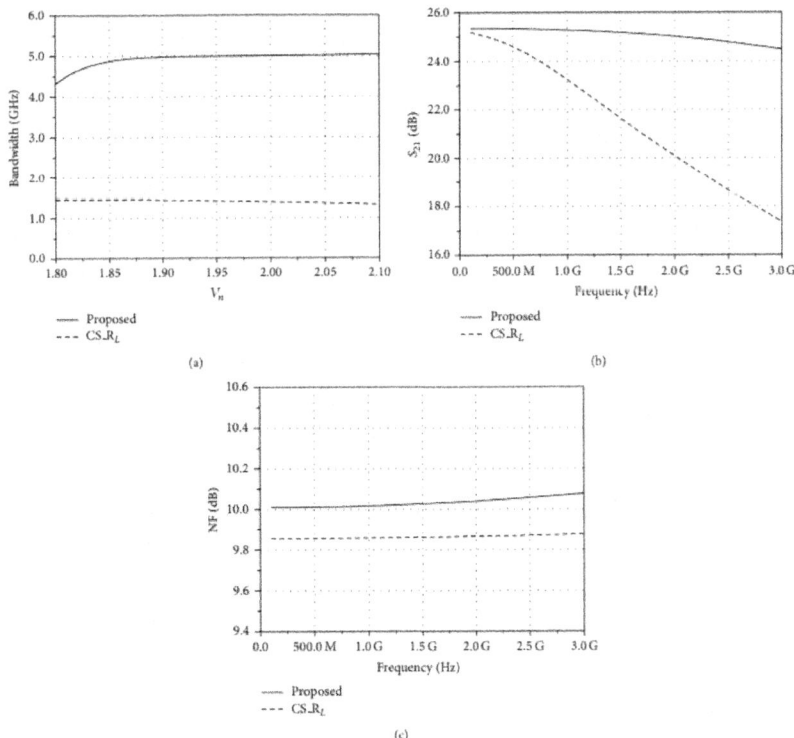

Figure 7: Comparison between proposed CH-HCS cell and CS-R_L cell.

VGA ARCHITECTURE

The proposed VGA architecture is shown in Figure 8. In order to obtain a gain range over 70 dB, two cascaded stages of variable gain core mentioned in Section 2 are employed with each providing about 40-dB gain range. Since the transadmittance and transimpedance stages in one gain core have independent bias currents and allow a dc-coupled output, the first core in cascaded stages behaves as an output buffer to the second core which can eliminate the DC blocking capacitor. For testing purposes, 50 ohm resistors are placed at the differential inputs for input matching and a high-speed ft doubler output buffer is used at the output.

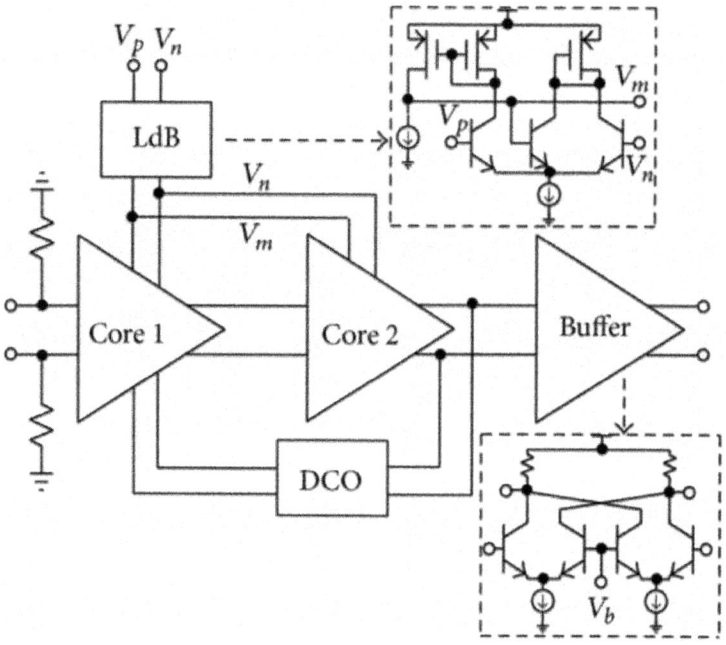

Figure 8: Block diagram of proposed VGA system.

In order to reduce DC offsets due to component mismatch, a differential feedback loop is designed with its schematic shown in Figure 9.

The differential output voltage of the second variable gain core is sensed by a RC LPF at low frequency and stored in the Miller capacitances. Following a high input impedance voltage amplifier using an NFET differential pair, a transconductance amplifier using an NPN differential pair transforms the input voltage into differential current which is injected back to the output of the transadmittance stage of the first gain core to adjust the output DC offset voltage. This DC offset loop is designed to have a low pass frequency response with a cut-off frequency of 800 kHz.

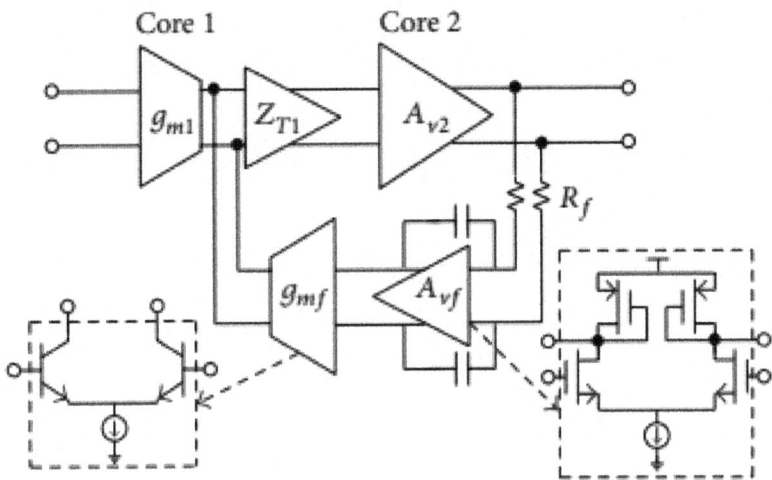

Figure 9: Block diagram of DC offset correction circuit.

As mentioned in Section 2, the hybrid current-steering cell has the same relationship between gain and control voltage with the conventional current-steering cell. Thus a curvature linearization circuit [11] designed for currentsteering cell is employed to the gain cores in two stages to make the gain control linear-in-dB.

Post Layout Simulation Results

The proposed architecture is implemented in HHNEC's 0.18 μm SiGe HBT BiCMOS technology which integrates 0.2 μm, 1.8 V BV_{CEO}, 110 GHz f_t SiGe HBTs, together with 0.18 μm, 1.8 V Si CMOS devices. The layout of the chip is shown in Figure 10. The VGA occupies 280 μm by 140 μm and total chip with testing pads takes an area of 800 μm by 550 μm. Under the 3.3 V power supply, the VGA consumes 9.2 mA excluding the output buffer which draws 8 mA.

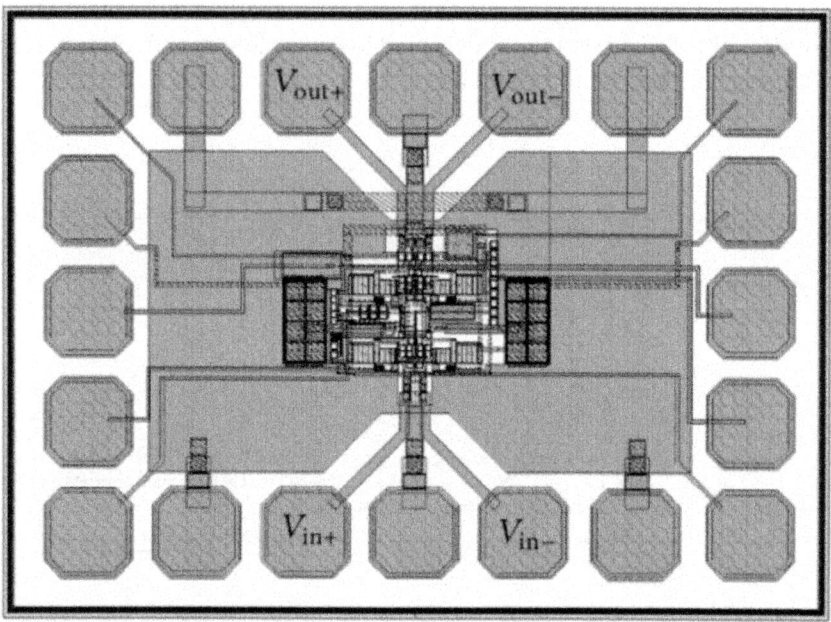

Figure 10: Layout of the VGA.

The frequency response of the proposed VGA is shown in Figure 11 with control voltage V_p ranging from 1.96 V to 2.18 V. At the maximum gain of 61 dB, the simulated 3-dB bandwidth is 2.5 GHz, and at the minimum gain of −19 dB, the bandwidth is 2.0 GHz. Thanks to the broadband character of the Cherry-Hooper amplifier, the gain-bandwidth product at the maximum gain can achieve a quite competitive value of 2805 GHz. Figure 12 shows the voltage gain and linear-indB characteristics. The gain error is 4 dB over 80-dB gain range. The noise figure at maximum gain shown in Figure 13 is smaller than 15 dB over the entire band.

The simulated input P1 dB is −10 dBm at −19 dB minimum gain and −65 dBm at 61 dB maximum gain as shown in Figure 14.

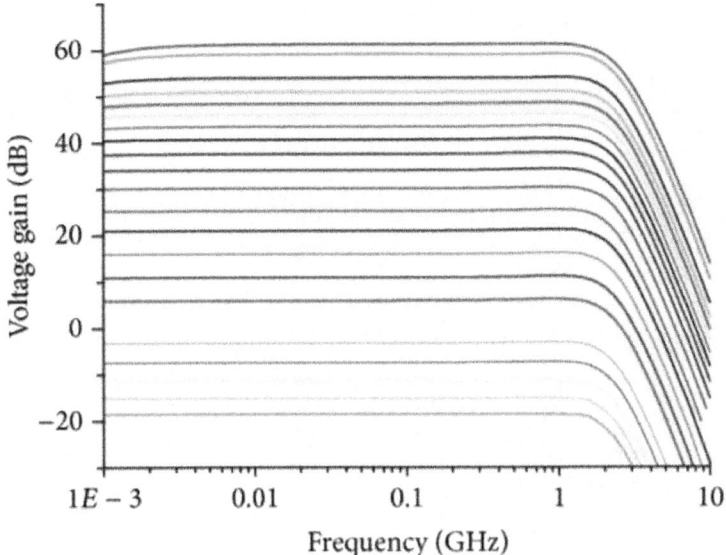

Figure 11: Frequency response versus various control voltage $V_{p^{\cdot}}$.

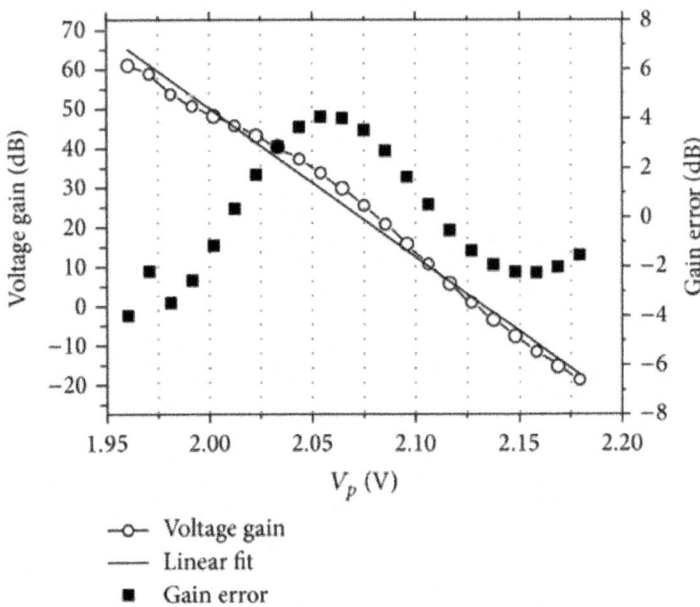

Figure 12: Voltage gain and gain linearity versus control voltage $V_{p^{\cdot}}$.

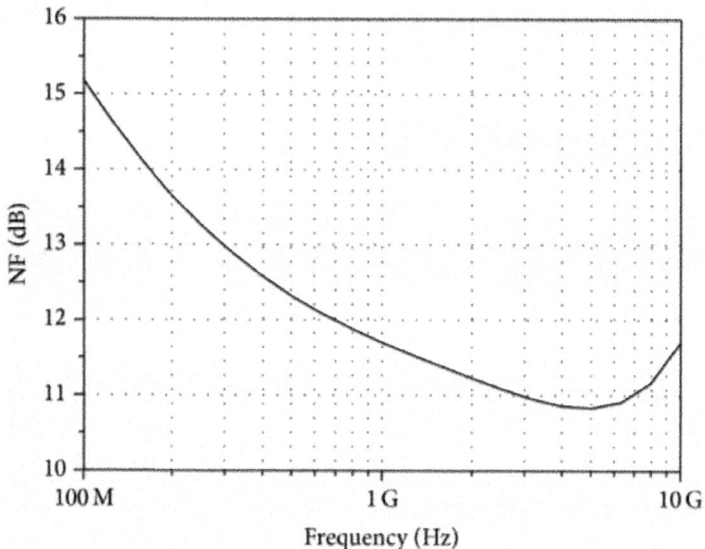

Figure 13: Noise figure at maximum gain.

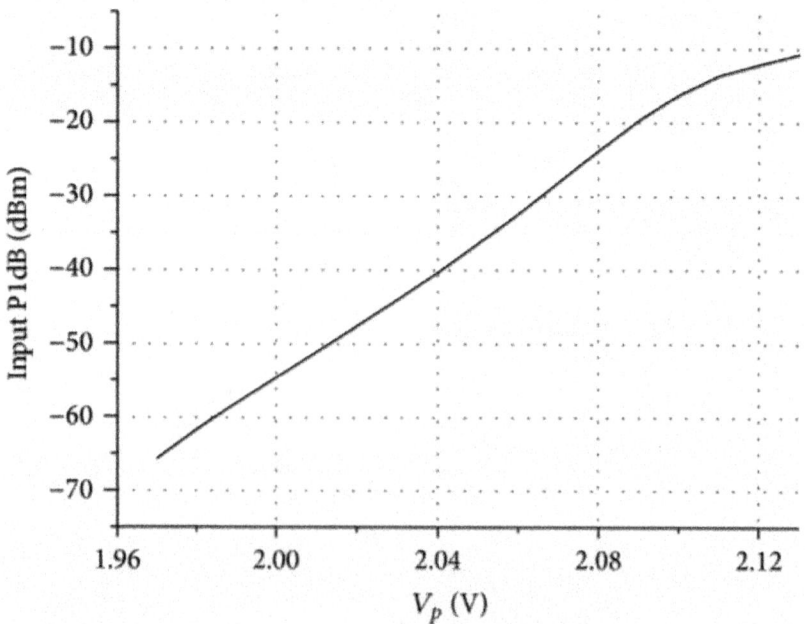

Figure 14: Input P1 dB versus control voltage $V_{P'}$.

Table 1 compares the performance of the proposed VGA with other works and shows a competitive gain bandwidth product with the largest gain tuning range among other works.

Table 1: VGA performance comparison.

	Gain range (dB)	Bandwidth (GHz)	GBP (GHz)	Power (mW)	Technology	Linear-in-dB
Wang et al. 2012 [8]	−10~50	2.2	700	2.5	CMOS 90 nm	No
Kumar et al. 2010 [2]	−10~10	4	12	9	SiGe 0.18 um	No
Manstretta and Dauphinee 2007 [4]	−30~30	1	32	250	SiGe 0.18 um	No
Jianhong et al. 2007 [5]	−17~16	0.4~0.8	6	22	CMOS 0.18 um	No
Chang et al. 2009 [6]	−10~17	0.9	7	40	SiGe 0.35 um	Yes
This work	−19~61	2	2805	31	SiGe 0.18 um	Yes

CONCLUSIONS

The proposed VGA comprises two cascaded variable gain core, in which a hybrid current-steering current gain cell is inserted in the Cherry-Hooper amplifier to maintain a broad bandwidth while covering a wide gain range. The introduced hybrid current-steering cell can provide current transfer function equal to the conventional current-steering cell while maintaining a constant DC output current. Postlayout simulation results confirm that the proposed circuit achieves a 2 GHz 3-dB bandwidth with wide linear-in-dB gain tuning range from −19 dB up to 61 dB. The amplifier offers a competitive gain bandwidth product of 2805 GHz at the maximum gain for a 110-GHz f_t BiCMOS technology. The amplifier core consumes 31 mW from a 3.3 V supply and occupies active area of 280 μm by 140 μm.

CONFLICT OF INTERESTS

The authors declare that they have no conflict of interests regarding the publication of this paper.

REFERENCES

1. Agilent, "Wireless LAN at 60 GHz—IEEE 802.11ad explained," Application Note, 2012.

2. T. B. Kumar, K. Ma, K. S. Yeo et al., "A DC to 4 GHz fully differential wideband digitally controlled variable gain amplifier," in Proceedings of the Asia-Pacific Microwave Conference (APMC '10), pp. 2295–2298, Yokohama, Japan, December 2010.

3.   C. Liu, Y. P. Yan, W. L. Goh, Y. Z. Xiong, L. J. Zhang, and M. Madihian, "A 5-Gb/s automatic gain control amplifier with temperature compensation," IEEE Journal of Solid-State Circuits, vol. 47, no. 6, pp. 1323–1333, 2012.

4.   D. Manstretta and L. Dauphinee, "A highly linear broadband variable gain LNA for TV applications," in Proceedings of the IEEE Custom Integrated Circuits Conference (CICC '07), pp. 531–534, San Jose, Calif, USA, September 2007.

5.   X. Jianhong, I. Mehr, and J. Silva-Martinez, "A high dynamic range CMOS variable gain amplifier for mobile DTV tuner," IEEE Journal of Solid-State Circuits, vol. 42, no. 2, pp. 292–301, 2007.

6.   C.-H. Chang, C.-F. Lee, and Y.-L. Tsou, "A wide-band low noise amplifier using cross-coupled technique with linear in dB gain-control," in Proceedings of the IEEE International Conference on Ultra-Wideband (ICUWB '09), pp. 251–254, September 2009.

7.   E. M. Cherry and D. E. Hooper, "The design of wide-band transistor feedback amplifier," Proceedings of the Institution of Electrical Engineers, vol. 110, no. 2, pp. 375–389, 1963.

8.   Y. Wang, B. Afshar, L. Ye, V. C. Gaudet, and A. M. Niknejad, "Design of a low power, inductorless wideband variable-gain amplifier for high-speed receiver systems," IEEE Transactions on Circuits and Systems I: Regular Papers, vol. 59, no. 4, pp. 696–707, 2012.

9.   C. D. Holdenried, J. W. Haslett, and M. W. Lynch, "Analysis and design of HBT Cherry-Hooper amplifiers with emitter-follower feedback for optical communications," IEEE Journal of Solid-State Circuits, vol. 39, no. 11, pp. 1959–1967, 2004.

10.   H. Ishihara, "Variable gain amplifier circuit," US Patent 6177839, 2001.

11.   S. Aggarwal, A. Khosrowbeygi, and A. Daanen, "A single-stage variable-gain amplifier with 70-dB dynamic range for CDMA2000 transmit applications," IEEE Journal of Solid-State Circuits, vol. 38, no. 6, pp. 911–917, 2003.

Chapter 9

THE ANALYSIS OF THE PERFORMANCE OF MULTI-BEAMFORMING IN MEMORY NONLINEAR POWER AMPLIFIER

Huiyong Li[1], Xun Li[1] and, Chen Wei[1]

School of Electronic Engineering, University of Electronic Science and Technology of China, Chengdu 611731, China

ABSTRACT

With the increasingly diverse and complex requirements of radar systems and communication systems, the application of multifunction-phased array radar has become a trend, and the digital multi-beamforming technology plays a crucial role in it. In practice, power amplifier (PA) is an essential component in radar systems and communication systems. Unfortunately, it is always nonlinear to provide a high output power. With the purpose of a high output power and efficiency, it is necessary to study the influence of PA nonlinear characteristics on the digital multi-beamforming. In this paper, a form of the multi-beamforming signal and a nonlinear model with memory for PA are given. The output signal *via* the PA model has been analyzed subsequently. As the result of analysis, it can be found that the output signal is divided into the original signal and the interferential signal. The power ratio of original signal to interference signal can reflect the influence of PA nonlinear characteristics on the digital multi-beamforming. Finally, according to the ratio, the results of computer simulation show that the memory effect plays a key role for the small power signal, while the nonlinearity plays an important role for the large power signal.

INTRODUCTION

In recent years, with the development of the military radar technology and the increasing demand for information processing of communications, the application of integrated electronic information system with multifunction-phased array radar has become a trend, and the digital multi-beamforming technology is crucial to the implementation of this system. At present, in contrast to the mature receiving digital multi-beamforming technology, the transmitted digital multi-beamforming technology is still under development. Transmitted multi-beamforming has many advantages. For example, when the radar array intends to search and track objects, the working mode of the common beam is time division, which means that only one job, tracking or searching, can be done at the same time. If the transmitted simultaneous multi-beamforming is used, the two work can be carried out at the same time by using two beams, which are added and synthesized in the digital side and can be transmitted simultaneously. However, because the transmitted signal is the sum of signals, the major bottleneck of the realization of the transmitted simultaneous multi-beamforming is that the transmitted signal envelope is not constant. Therefore, it has a higher requirement on linearity of the transmitter power amplifier.

In order to design an optimal PA, it is necessary to analyze the effect of the PA nonlinearities on transmitted multi-beamforming firstly. The nonlinear distortion of PA has always been a hot research area. A memory PA model is proposed in [1] and the performance of behavioral models is analyzed in [2]. H Ku has analyzed behavioral modeling of nonlinear RF power amplifiers with memory effects in [3]. The major consequence of memory effects has been introduced by W Bosch in [4]. The papers [5, 6] present that memory effects of PA in the wideband system are obvious. However, here the definition of wideband is the high ratio of the transmitted signal bandwidth to the device bandwidth, rather than the signal bandwidth to the carrier frequency. For a transmitted narrowband signal, the memory effects of the PA are evident when the signal bandwidth matches the device bandwidth [7]. The constant envelope signal in the nonlinear PA is studied in [8]. Kohls [9] has used a Bessel series fit to the measured amplifier transfer functions with a computer model to predict third-order intermodulation product beam patterns. In [10], Hemmi describes the nonlinear PA response of an active linear array by a third-order polynomial and develops equations for the beam-pointing angle of the harmonics and intermodulation products. An active phased array multibeam antenna model, including the nonlinear Shimbo model of the amplifiers, has been developed and validated experimentally in [11], which is useful for solid state power amplifiers (SSPAs). In [12], only the AM-AM varieties are considered to affect the beamforming, regardless of the AM-PM varieties and memory effects.

In this paper, the principle of transmitted multi-beamforming is presented in section 2. Then, section 3.1 presents a model based on the Hammerstein model to describe the AM-AM varieties, AM-PM varieties and memory effects. The output signal model from PA is analyzed in section 3.2. The simulation results are shown in section 4. Finally, section 5 concludes the study. Assumed that each channel and array elements are ideal and each PA is identical. A narrowband signal is considered and the attention will be focused on the effects of the amplitude and phase nonlinearities.

THE PRINCIPLE OF TRANSMITTING MULTI-BEAM-FORMING

As shown in Figure 1, consider an L-element array with elements uniformly spaced on the line of distance equal to d. The transmitted signals are $s_1(n)$, ..., $s_N(n)$ and the weight vectors are $\mathbf{w}_1$, ..., $\mathbf{w}_N$.

$$\mathbf{w}_i = \left[1, e^{j\varphi_i}, \cdots, e^{j(L-1)\varphi_i}\right]^{\mathrm{T}}, i = 1, \cdots, N$$
$$\varphi_i = \frac{2\pi d \sin\theta_i}{\lambda}$$

$$(1)$$

where φ_i is the space phase of the i th signal, λ is the wavelength and θ_i is the transmitted angle of the i th signal.

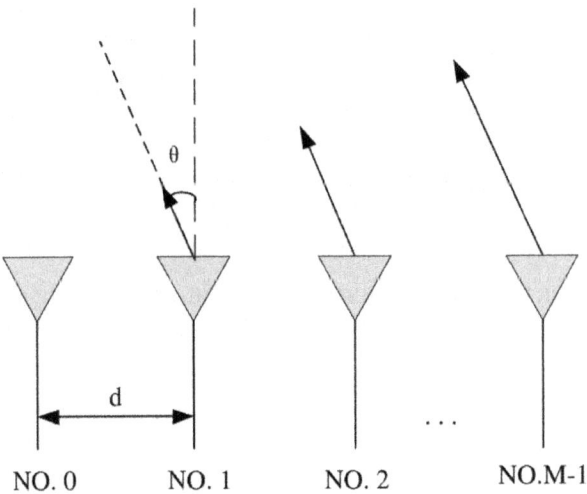

Figure 1: The model of transmitting array.

The space-power-function of transmitted multi-beam is given by

$$f(\theta) = \left| \sum_{i=1}^{N} \mathbf{w}_i^H \mathbf{a}(\theta) s_i(n) \right|^2$$

$$(2)$$

where $a(\theta) = [1, e^{j\varphi}, \cdots, e^{j(L-1)\varphi}]^T$ is the steering vector of the scanning beam. When the transmitted signals are orthogonal and the powers of them are equal to each other, the multi-beam pattern can be written as

$$E(\theta) = \frac{f(\theta)}{P_s} = \sum_{i=1}^{N} |\mathbf{w}_i^H \mathbf{a}(\theta)|^2$$

$$(3)$$

where P_s is the power of transmitted signal.

THE ANALYSIS OF TRANSMITTED MULTI-BEAMFORM-ING BASED ON NONLINEAR MEMORY POWER AMPLIFIER

The model of PA

A behavior model for PA can be divided into two types: the band-pass PA model for the RF signal processing and the baseband PA model for the envelope information processing. In practical engineering, the input signal of PA is a real signal with radio frequency. Therefore, the band-pass PA model can be very accurate to analyze a variety of components in the nonlinear devices, including harmonics, intermodulation, etc. However, in fact, the processed harmonics will not be transmitted. As the result, it is not convenient to analyze the signal of the band-pass PA model. As we all know, the useful information is carried only by the signal envelope. Moreover, nonlinear characteristics reflected by the behavioral model are only related to the signal amplitude, rather than the frequency. Therefore, a baseband PA model can be used for the signal analysis.

In this paper, the Hammerstein model (Figure 2), a modification of Volterra, has been used, which is given by:

$$y(n) = \sum_{k=1}^{P} a_k x^k(n)$$
$$z(n) = \sum_{i=0}^{M-1} h_i y(n-i)$$

$$(4)$$

where $x(n)$, $y(n)$, and $z(n)$ denote the input signal of the memoryless nonlinear PA model, memory linear PA and Hammerstein, respectively. M is the memory depth and P is the order of nonlinear system.

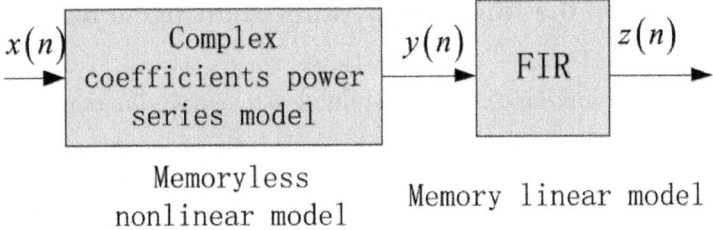

Figure 2: The model of memory nonlinear PA.

The memory linear model uses an $M-$order FIR filter as description. In order to reflect its AM-AM and AM-PM characteristics simultaneously, the memoryless nonlinear model employs a complex coefficients power series model. Assume that $a_k = a_{Ik} + ja_{Qk}$, $x(n) = re^j$. So,

$$
\begin{aligned}
y(n) &= \sum_{k=1}^{P} a_k x(n)|x(n)| k-1 \\
&= re^{j\theta} \sum_{k=1}^{P} a_k r^{k-1} \\
&= \left[\sum_{k=1}^{P} a_{Ik} r^k + j \sum_{k=1}^{P} a_{Qk} r^k \right] e^{j\theta} \\
&= \sqrt{\left(\sum_{k=1}^{P} a_{Ik} r^k \right)^2 + \left(\sum_{k=1}^{P} a_{Qk} r^k \right)^2} e^{j\left(\theta + \arctan\left(\sum_{k=1}^{P} a_{Qk} r^k / \sum_{k=1}^{P} a_{Ik} r^k \right) \right)}
\end{aligned}
$$

$$(5)$$

The functions of AM-AM and AM-PM can be given by:

$$
\begin{aligned}
A(r) &= \sqrt{\left(\sum_{k=1}^{P} a_{Ik} r^k \right)^2 + \left(\sum_{k=1}^{P} a_{Qk} r^k \right)^2} \\
\Phi(r) &= \arctan\left(\sum_{k=1}^{P} a_{Qk} r^k / \sum_{k=1}^{P} a_{Ik} r^k \right)
\end{aligned}
$$

$$(6)$$

From the above conclusion, power series with complex coefficients can describe the characteristics of AM-AM and AM-PM. If the PA has no memory, the output of the PA is determined only by the odd-component of the input signal. The even-component of the input signal will not affect the output of the PA, which can be expressed as:

$$
y(n) = \sum_{k=0}^{(P-1)/2} a_{2k+1} |x(n)|^{2k} x(n)
$$

$$(7)$$

The Analysis of the Transmitted Multi-Beam Signal Based on the Taylor model

After frequency mixing of two transmitted real signals, the output signals can be represented as:

$$s_1(t) = A_1(t) \cos(\omega_0 t + \varphi_1(t))$$
$$s_2(t) = A_2(t) \cos(\omega_0 t + \varphi_2(t))$$

$$(8)$$

The signal input into the nonlinear PA is given by:

$$s_{in}(t) = A_{w_1} A_1(t) \cos(\omega_0 t + \varphi_1(t) - \phi_{w1})$$
$$+ A_{w_2} A_2(t) \cos(\omega_0 t + \varphi_2(t) - \phi_{w2})$$

$$(9)$$

where A_{w_1}, A_{w_2} and ϕ_{w_1}, ϕ_{w_2} are the amplitudes and phases of the two weight vectors w_1 and w_2, respectively. $A_1(t)$, $A_2(t)$ and $\varphi_1(t)$, $\varphi_2(t)$ are the amplitudes and phases of the two signals, respectively. ω_0 is the carrier frequency.

Even order harmonic component, produced by PA, can be removed by filters. Therefore, the odd-order harmonic component, which cannot be removed by filters, has a great effect on input signals. According to the power series model, after the signal passes the nonlinear PA, the output signal is given by:

$$s_{out}(t) = a_1 s_{in}(t) + a_3 s_{in}(t)^3$$

$$(10)$$

where a_1, a_3 is a real number.

Expanding the cubic term and removing the non-fundamental frequency component, we can get:

$$a_3 s_{in}(t)^3 \approx a_3 \left[\left(\frac{3}{4} A_{s_1}(t)^3 + \frac{3}{2} A_{s_1}(t) A_{s_2}(t)^2 \right) \cos(\omega_0 t + \varphi_1(t) - \phi_{w_1}) \right.$$
$$+ \left(\frac{3}{4} A_{s_2}(t)^3 + \frac{3}{2} A_{s_2}(t) A_{s_1}(t)^2 \right) \cos(\omega_0 t + \varphi_2(t) - \phi_{w_2})$$
$$+ \left(\frac{3}{4} A_{s_2}(t) A_{s_1}(t)^2 \right) \cos(w_0 t + 2(\varphi_1(t) - \phi_{w_1}) - \varphi_2(t) - \phi_{w_2})$$
$$\left. + \left(\frac{3}{4} A_{s_1}(t) A_{s_2}(t)^2 \right) \cos(w_0 t + 2(\varphi_2(t) - \phi_{w_2}) - \varphi_1(t) - \phi_{w_1}) \right]$$

$$(11)$$

where $A_{s_1}(t) = A_{w_1} A_1(t), A_{s_2}(t) = A_{w_2} A_2(t).$

The real signal is a double-sideband signal. To analyze conveniently, we study one side of the spectrum, and the analytic signal of output is given by:

$$s_{\text{out}}(t) = a_1 \left(A_{s_1}(t)e^{j\left(\omega_0 t + \varphi_1(t) - \phi_{w_1}\right)} + A_{s_2}(t)e^{j\left(\omega_0 t + \varphi_2(t) - \phi_{w_2}\right)} \right)$$
$$+ a_3 \left[\left(\frac{3}{4}A_{s_1}(t)^3 + \frac{3}{2}A_{s_1}(t)A_{s_2}(t)^2 \right)e^{j\left(\omega_0 t + \varphi_1(t) - \phi_{w_1}\right)} \right.$$
$$+ \left(\frac{3}{4}A_{s_2}(t)^3 + \frac{3}{2}A_{s_2}(t)A_{s_1}(t)^2 \right)e^{j\left(\omega_0 t + \varphi_2(t) - \phi_{w_2}\right)}$$
$$+ \left(\frac{3}{4}A_{s_2}(t)A_{s_1}(t)^2 \right)e^{j\left(w_0 t + 2\left(\varphi_1(t) - \phi_{w_1}\right) - \varphi_2(t) - \phi_{w_2}\right)}$$
$$\left. + \left(\frac{3}{4}A_{s_1}(t)A_{s_2}(t)^2 \right)e^{j\left(w_0 t + 2\left(\varphi_2(t) - \phi_{w_2}\right) - \varphi_1(t) - \phi_{w_1}\right)} \right]$$

$$(12)$$

According to expression in (12), whether the nonlinear term is the same as the original signal is related to the input signal. Under amplitude modulation, the desired signal can be represented as:

$$s_{\text{desire}}(t) = a_1 \left(A_{s_1}(t)e^{j\left(\omega_0 t + \varphi_1(t) - \phi_{w_1}\right)} + A_{s_2}(t)e^{j\left(\omega_0 t + \varphi_2(t) - \phi_{w_2}\right)} \right)$$

$$(13)$$

All items of a_3 in expression (12), which are different from previous signals, are interference signals.

$$s_{\text{jam}}(t) = a_3 \left[\left(\frac{3}{4}A_{s_1}(t)^3 + \frac{3}{2}A_{s_1}(t)A_{s_2}(t)^2 \right)e^{j\left(\omega_0 t + \varphi_1(t) - \phi_{w_1}\right)} \right.$$
$$+ \left(\frac{3}{4}A_{s_2}(t)^3 + \frac{3}{2}A_{s_2}(t)A_{s_1}(t)^2 \right)e^{j\left(\omega_0 t + \varphi_2(t) - \phi_{w_2}\right)}$$
$$+ \left(\frac{3}{4}A_{s_2}(t)A_{s_1}(t)^2 \right)e^{j\left(\omega_0 t + 2\left(\varphi_1(t) - \phi_{w_1}\right) - \varphi_2(t) - \phi_{w_2}\right)}$$
$$\left. + \left(\frac{3}{4}A_{s_1}(t)A_{s_2}(t)^2 \right)e^{j\left(\omega_0 t + 2\left(\varphi_2(t) - \phi_{w_2}\right) - \varphi_1(t) - \phi_{w_1}\right)} \right]$$

$$(14)$$

When it is not an amplitude modulation, the cubic term contains a part of desired signal component, and the desired signal can be represented as:

$$s_{\text{desire}}(t) = \left[a_1 A_{s_1} + a_3 \left(\frac{3}{4}A_{s_1}^3 + \frac{3}{2}A_{s_1}A_{s_2}^2 \right) \right]e^{j\left(\omega_0 t + \varphi_1(t) - \phi_{w_1}\right)}$$
$$+ \left[a_1 A_{s_2} + a_3 \left(\frac{3}{4}A_{s_2}^3 + \frac{3}{2}A_{s_2}A_{s_1}^2 \right) \right]e^{j\left(\omega_0 t + \varphi_2(t) - \phi_{w_2}\right)}$$

$$(15)$$

The interference signal can be represented as:

$$s_{jam}(t) = a_3 \left[\left(\frac{3}{4} A_{s_2} A_{s_1}{}^2 \right) e^{j\left(\omega_0 t + 2\left(\varphi_1(t) - \phi_{w_1}\right) - \varphi_2(t) - \phi_{w_2}\right)} \right.$$
$$\left. + \left(\frac{3}{4} A_{s_1} A_{s_2}{}^2 \right) e^{j\left(\omega_0 t + 2\left(\varphi_2(t) - \phi_{w_2}\right) - \varphi_1(t) - \phi_{w_1}\right)} \right]$$

$$(16)$$

According to expression in (14) and (16), on the condition of amplitude modulation, the power of interference signal mainly consists of the signal which has the same weight vector with the desired signal, and the power of interference signal has the same pointing direction with the desired signal. However, on the condition of non-amplitude modulation, it is the weight vectors $w_1^{H2} w_2$ and $w_2^{H2} w_1$ that determining the pointing direction of interference signal power.

The Analysis of the Transmitted Multi-Beam Signal Based on The Hammerstein Model

A complex envelope of the input signal which is the sum of two narrow-band signals can be expressed by:

$$x(n) = A_{w_1} A_1(n) e^{j\varphi_1(n) - \phi_{w_1}} + A_{w_2} A_2(n) e^{j\varphi_2(n) - \phi_{w_2}}$$
$$= r_1(n) e^{j\varphi_1(n) - \phi_{w_1}} + r_2(n) e^{j\varphi_2(n) - \phi_{w_2}}$$

$$(17)$$

where $Aw1 A_{w_1}$, $Aw2 A_{w_2}$, $\varphi w1 \phi_{w_1}$, and $\varphi w2 \phi_{w_2}$ are the amplitudes and phases of the two weight vectors, and $A_1(n)$, $A_2(n)$, $\varphi_1(n)$, and $\varphi_2(n)$ are the amplitudes and phases of the two signals. and $r_2(n) = A_{w_2} A_2(n)$.

After the signal passes the memoryless nonlinear model, the output signal is given by

$$y(n) = a_1 x(n) + a_3 x(n) |x(n)|^2$$
$$= (a_1 + a_3 r^2(n)) x(n)$$

$$(18)$$

where $r^2(n) = r_1(n)^2 + r_2(n)^2 + 2r_1(n) r_2(n) \cos[\varphi_1(n) - \phi_{w_1} - \varphi_2(n) + \phi_{w_2}]$.

If the input signals have constant envelopes, that is $r_1(n) = r_1$ and $r_2(n) = r_2$ then:

$$y(n) = \left[a_1 r_1 + a_3 r_1 (r_1^2 + r_2^2) \right] e^{j\varphi_1(n) - \phi_{w_1}}$$
$$+ 2a_3 r_1{}^2 r_2 \cos\left[\varphi_1(n) - \phi_{w_1} - \varphi_2(n) + \phi_{w_2}\right] e^{j\varphi_1(n) - \phi_{w_1}}$$
$$+ \left[a_1 r_2 + a_3 r_2 (r_1^2 + r_2^2) \right] e^{j\varphi_2(n) - \phi_{w_2}}$$
$$+ 2a_3 r_2{}^2 r_1 \cos\left[\varphi_1(n) - \phi_{w_1} - \varphi_2(n) + \phi_{w_2}\right] e^{j\varphi_2(n) - \phi_{w_2}}$$

$$(19)$$

We notice that the first and third coefficients are time-independent complexes, so they can be regarded as original signals. However, the second and fourth coefficients are time-dependent complexes, which means that the information carried by the signal envelope has changed. Hence, they should be considered as interferential signals. The effect of nonlinearity on the transmitted signal can be measured by the power ratio of the original signal to interference.

The output of memoryless PA model is seen as the input of linear FIR filter. Assume that the memory depth is 2. Then, the output signal of memory PA model can be represented as:

$$z(n) = h_0 y(n) + h_1 y(n-1) \qquad (20)$$

According to the concept of phase modulation and frequency modulation, the phase difference of $\varphi(n-1)$ and $\varphi(n)$ is decided by the modulation signal. So, $\varphi(n-1)$ and $\varphi(n)$ are different information. Finally, Equation 10 also can be described as:

$$
\begin{aligned}
z(n) = &\, h_0 \left[a_1 r_1 + a_3 r_1 (r_1^2 + r_2^2) \right] e^{j\varphi_1(n) - \phi_{w_1}} \\
&+ h_0 \left[a_1 r_2 + a_3 r_2 (r_1^2 + r_2^2) \right] e^{j\varphi_2(n) - \phi_{w_2}} \\
&+ h_1 \left[a_1 r_1 + a_3 r_1 (r_1^2 + r_2^2) \right] e^{j\varphi_1(n-1) - \phi_{w_1}} \\
&+ h_1 \left[a_1 r_2 + a_3 r_2 (r_1^2 + r_2^2) \right] e^{j\varphi_2(n-1) - \phi_{w_2}} \\
&+ 2 h_0 a_3 r_1{}^2 r_2 \cos\left[\varphi_1(n) - \phi_{w_1} - \varphi_2(n) + \phi_{w_2} \right] e^{j\varphi_1(n) - \phi_{w_1}} \\
&+ 2 h_1 a_3 r_1{}^2 r_2 \cos\left[\varphi_1(n-1) - \phi_{w_1} - \varphi_2(n-1) + \phi_{w_2} \right] e^{j\varphi_1(n-1) - \phi_{w_2}} \\
&+ 2 h_0 a_3 r_2{}^2 r_1 \cos\left[\varphi_1(n) - \phi_{w_1} - \varphi_2(n) + \phi_{w_2} \right] e^{j\varphi_2(n) - \phi_{w_2}} \\
&+ 2 h_1 a_3 r_2{}^2 r_1 \cos\left[\varphi_1(n-1) - \phi_{w_1} - \varphi_2(n-1) + \phi_{w_2} \right] e^{j\varphi_2(n-1) - \phi_{w_2}}
\end{aligned}
\qquad (21)
$$

According to the expression in (21), it can be seen that the first and second signals are still original signals. However, the third and fourth signals have different phases from the original signals, and the rest of the signals have become new signals which the amplitudes vary with time. That is to say, except for the first and second signals, the others are interferential signals.

Simulation

Consider a 16-element array with elements uniformly spaced on the line of distance equal to half wavelength. Two transmitted beams are assumed to be transmitted at angle $30°$ and $-20°$. Several null points are set at angle $-10°$, $10°$, and $50°$. One thousand sampling points are chosen. The transmitted signals are two linear frequency modulation signals with $s1(t)=As1ejKt2+n1(t)$ $s_1(t) = A_{s_1} e^{jKt^2} + n_1(t)$ and $s_2 = A_{s_2} e^{j2\pi Bt + jKt^2} + n_2(t)$, where $K = B/T$. $B = 10$ M is the bandwidth of the signal and T is pulse period, $T = NT_s = N/f_s$, $N = 1{,}000$ is the sampling numbers and $f_s = 3.01/2 \times 10^9$ Hz is the sampling frequency. $n_1(t)$ and $n_2(t)$ are system thermal noises. The transmitted signal before it is input to

PA has an SNR = 90 dB. Although the amplitudes, As1A_{s_1} and As2A_{s_2}, are the same, the two signals which have different carrier frequencies are orthogonal. The first coefficient of PA is $a_1 = 1$. The first tap coefficient of FIR is $h_0 = 1$ and the memory depth is 2. The beam pattern affected by the nonlinear PA is produced by a beamforming technique based on orthogonal projection algorithm.

a_1/a_3 is set to measure the memoryless nonlinearity of PA and h_0/h_1 is used to measure the memory effect of PA. The effect of nonlinearity PA on the multi-beamforming at −20° can be shown by changing the two parameters.

Figure 3 and Figure 4 show AM-AM and AM-PM variation curves versus different PA coefficients. According to these figures, it can be found that the linearity of $a_3 = -0.01 \times (2 + j)$ is less than $a_3 = -0.01 \times (1 + j)$.

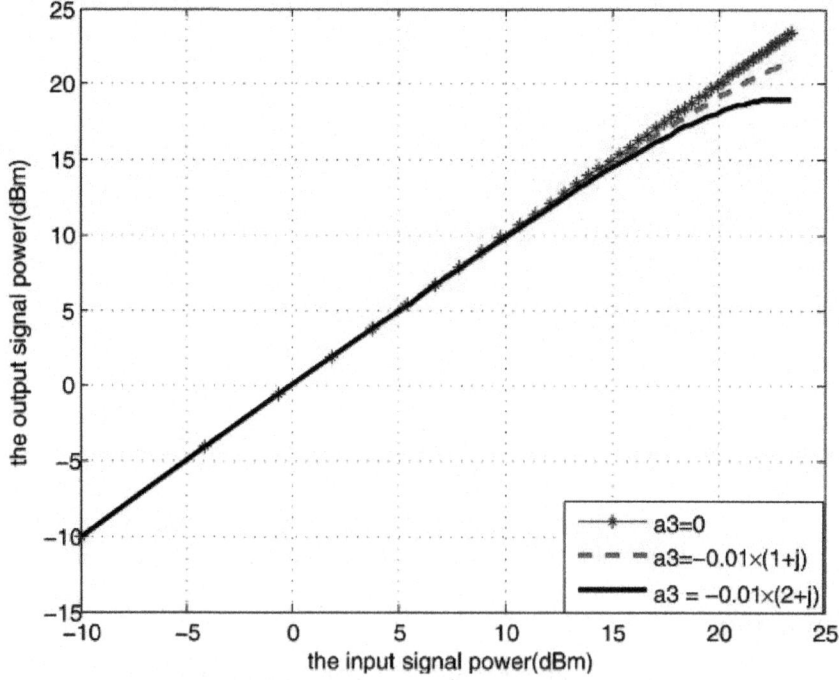

Figure 3: AM-AM conversion.

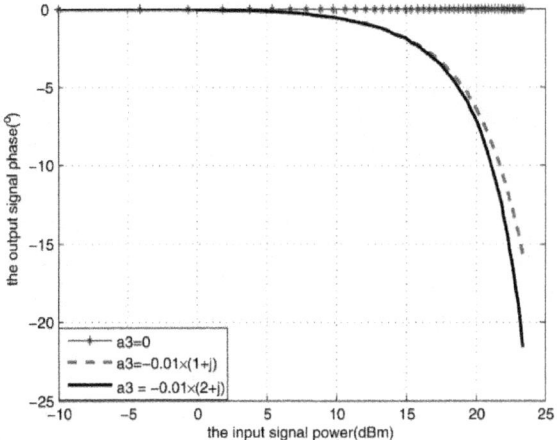

Figure 4: AM-PM conversion.

The power ratio of the original signal to the interference can show the effect of the nonlinearity of PA on the input signal. Figure 5 shows that when $h_1 = 10^{-2}$ and the input signal power is below 7 dBm, the power ratio keeps 40 dB with different nonlinearities. It can be explained that when the input signal power is small, the memory effect plays a dominant role. On the other hand, when the input signal power is 20 dBm, owing to the nonlinearity, the ratio of $a_3 = -0.01 \times (2+j)$ is more than $a_3 = -0.01 \times (1+j)$.

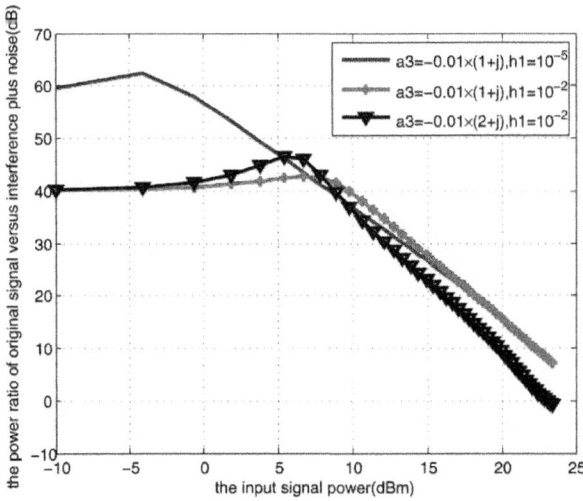

Figure 5: The power ratio of original signal versus interference plus noise of input signal.

Equation 9 shows that the equation value is significant only if a_3 and h_1a_1 have the same dimension, which means we can ignore the effect of h_1a_3 on beamforming. In addition, the nonlinearity represents gain compression characteristics. So, a_3 is negative and h_1 is positive. Under this premise, the total power of interference can be considered as a positive, when the memory effect plays a major role. The power of interference becomes weak as the memory effect is reduced. On the contrary, when nonlinearity plays a key role, the total power of interference is a negative. The power of interference increases as the memory effect is reduced. This is why the power ratio has an upward trend.

To measure the memoryless nonlinearity of PA, we assumed that $a_3 = -0.01 \times (x+j)$, $h_1 = 10^{-5}$ and $A_1 = 19.3$ dBm, where x is an independent variable. Figure 6 shows that when the input signal power A_1 is 19.3 dBm, the power ratio of the different h_1 is almost identical. However, when A_1 is 9.8 dBm, the power ratio of the different h_1 is different. The same conclusion can be found in Figure 6.

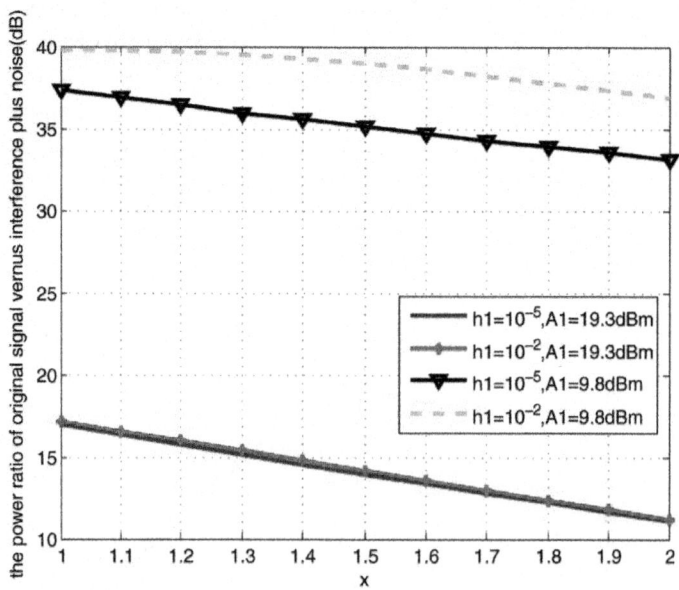

Figure 6: The power ratio of original signal versus interference plus noise with x .

$a_3 = -0.01 \times (1+j)$, $A_1 = 9.8$ dBm, and $A_1 = 19.3$ dBm are assumed to measure the memory effect of PA. From Figure 7, it can be seen that when the input signal power A_1 is 9.8 dBm, h_1 which is more than 10^{-4} will affect the ratio. However, it hardly affects the ratio when h_1 is less than 10^{-5}. On the other hand, when the power of input signal A_1 is 19.3 dBm, it is a turning point that h_1 is

10^{-2}. So, it is concluded that the low power of input signal is sensitive to the memory effect. The reason for the upward trend in Figure 7 has already been explained in Figure 5.

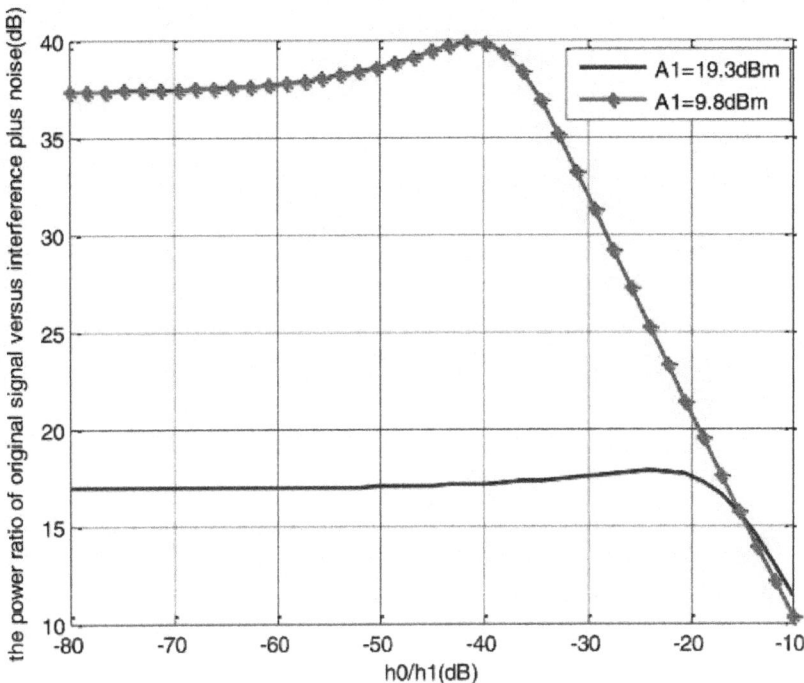

Figure 7: The power ratio of original signal versus interference plus noise with memory effect.

CONCLUSION

In this paper, starting from the signal form, the output signal of PA is obtained *via* the establishment of the PA model. Then the influence of the memory effect and nonlinearity of PA is presented by analyzing the composition of the final output signal. Lastly, computer simulations verify that the memory effect plays a key role, in the presence of the small power signal, while the nonlinearity plays an important role in the presence of the large power signal. A detailed theoretical basis and reference to linearize PA is provided in this paper for other researchers. When the transmitted signals are small power signals, the memory effect should gain attention in linearizing PA. However, if large power signals are transmitted, the importance of the nonlinearity of PA should be attached.

ACKNOWLEDGEMENTS

This research was supported by Applied Basic Research Programs of Sichuan Province (No. 2013JY0004), National Natural Science Foundation of China (No. 61371184) and the Fundamental Research Funds for the Central Universities (No. ZYGX2012J018).

REFERENCES

1.   Ding L, Zhou GT, Morgan DR, Ma Z, Kenney JS, Kim J, Giardina CR: A robust digital baseband predistorter constructed using memory polynomials. *IEEE Trans. Commun.* 2004, 52(1):159-165. 10.1109/ TCOMM.2003.822188

2.   Isaksson M, Wisell D, Ronnow D: A comparative analysis of behavioral models for RF power amplifiers. *IEEE Trans. Microw. Theory Tech.* 2006, 54(1):348-359.

3.   Ku H, Kenney JS: Behavioral modeling of nonlinear RF power amplifiers considering memory effects. *IEEE Trans. Microw. Theory Tech.* 2003, 51(12):2495-2504. 10.1109/TMTT.2003.820155

4.   Bosch W, Gatti G: Measurement and simulation of memory effects in predistortion linearizers. *IEEE Trans. Microw. Theory Tech.* 1989, 37: 1885-1890. 10.1109/22.44098

5.   Vuolevi JHK, Rahkonen T, Manninen JPA: Measurement technique for characterizing memory effects in RF power amplifiers. *IEEE Trans. Microw. Theory Tech* 2001, 49: 1383-1388. 10.1109/22.939917

6.   Kim J, Konstantinou K: Digital predistortion of wideband signals based on power amplifier model with memory. *Electron. Lett.* 2001, 37(23):1417-1418. 10.1049/el:20010940

7.   Ku H, Mckinley M, Kenney JS: Quantifying memory effects in RF power amplifiers. *IEEE Trans. Microw. Theory Tech.* 2002, 50(12):2843-2849. 10.1109/TMTT.2002.805196

8.   Wu Y, Liu Y: The analysis of the effect of high power amplifier and bandwidth to the constant envelop guidance signal. *J Telemetry. Tracking. Command.* 2011, 32(3):14-20.

9.   Kohls EC, Ekelman EP, Zaghloul AI, Assal FT: Intermodulation and bit-error ratio performance of a Ku-band multibeam high-power phased array. In *Proceedings of the IEEE International Symposium on Antennas and Propagation. Volume 3*. Piscataway: IEEE; 1995:1404-1408.

10. Hemmi C: Pattern characteristics of harmonic and intermodulation products in broadband active transmit arrays. *IEEE Trans. Antennas Propag.* 2002, 50(6):858-865. 10.1109/TAP.2002.1017668

11. Maalouf KJ, Lier E: Theoretical and experimental study of interference in multibeam active phased array transmit antenna for satellite communications. *IEEE Trans. Antennas Propag.* 2004, 52(2):587-592. 10.1109/TAP.2004.823900

12. Real EC, Charette DP: Non-linear amplifier effects in transmit beamforming arrays. In *Proceedings of the IEEE International Conference on Acoustics Speech Signal Processing. Volume 5.* Piscataway: IEEE; 1995:3635-3638.

Chapter 10

POWER AMPLIFIER LINEARIZATION TECHNIQUE WITH IQ IMBALANCE AND CROSSTALK COMPENSATION FOR BROADBAND MIMO-OFDM TRANSMITTERS

Fernando Gregorio[1], Juan Cousseau[1], Stefan Werner[2,] Taneli Riihonen[2] and Risto Wichman[2]

[1]Conicet Department of Electrical and Computer Engineering, Universidad Nacional del Sur, Av. Alem 1253, Bahía Blanca 8000, Argentina

[2] Aalto University School of Electrical Engineering P.O. Box 13000, FI-00076 Aalto, Finland

ABSTRACT

The design of predistortion techniques for broadband multiple input multiple output-OFDM (MIMO-OFDM) systems raises several implementation challenges. First, the large bandwidth of the OFDM signal requires the introduction of memory effects in the PD model. In addition, it is usual to consider an imbalanced in-phase and quadrature (IQ) modulator to translate the predistorted baseband signal to RF. Furthermore, the coupling effects, which occur when the MIMO paths are implemented in the same reduced size chipset, cannot be avoided in MIMO transceivers structures. This study proposes a MIMO-PD system that linearizes the power amplifier response and compensates nonlinear crosstalk and IQ imbalance effects for each branch of the multiantenna system. Efficient recursive algorithms are presented to estimate the complete MIMO-PD coefficients. The algorithms avoid the high computational complexity in previous solutions based on least squares estimation. The performance of the proposed MIMO-PD structure is validated by simulations using a two-transmitter antenna MIMO system. Error vector magnitude and adjacent channel power ratio are evaluated showing significant improvement compared with conventional MIMO-PD systems.

INTRODUCTION

Emerging broadband communication systems require high spectral efficiency and robustness against multipath channels. For this reason, OFDM has been adopted in the majority of modern wireless communication standards. Furthermore, multiantenna transceivers represent one of the most prominent techniques to enhance system capacity. Mobile WiMAX, LTE, Ultra Wide Band, and WLAN (IEEE 802.11n) allow the use of MIMO-OFDM (multiple input multiple output-OFDM) in their specifications. However, several factors should be considered to obtain the advantages promised by MIMO techniques. The high dynamic range of OFDM signals imposes the use of linear amplifiers (class A and class AB). The requirement of linear amplifiers, with a poor power duty, creates a problem accentuated by the use of multiple antennas. The high-data transmission rates, reached with the combination of OFDM and MIMO contrast with the loss of portability of the product, because of their elevated power consumption. Therefore, there is a trade-off between the high data rate obtained by employing MIMO techniques and the high power consumption of the OFDM system. Furthermore, when considering low-cost components, there are also several imperfections/impairments that degrade the system performance and need to be taken into account in the design of a compensation system.

Despite several advantages, OFDM is sensitive to distortions introduced at the RF front-end. Inexpensive OFDM transceivers employing direct conversion architectures (zero intermediate frequency) are seriously affected by front-end distortions, e.g., in-phase and quadrature (IQ) baseband imbalance and phase noise. In addition, OFDM transceivers are also intrinsically sensitive to power amplifier (PA) nonlinear distortion. Nonlinear PA creates spectral regrowth (out-of-band distortion) and in-band distortion that degrades the system's bit error rate (BER). The trade-off between power efficiency and linearity motivates the development of novel signal-processing techniques to reduce the nonlinear distortion.

Nonlinear distortion can be compensated either at the transmitter side or the receiver side. For the former case, the signal to be transmitted is modified before the PA, and among the well-known methods for this purpose are predistortion and PAPR reduction techniques. SISO predistortion methods aim to model the inverse of the PA nonlinear response. The predistorter (PD) is placed before the PA such that the cascade PD-PA produces a linearly amplified signal. The most simple PD structure is memoryless, where the current output depends only on the current input. In that case, the PD is described by a static nonlinear function often implemented with polynomial models [1]. In broadband OFDM implementations, however, memory effects that appear in the PA response need to be considered in the PD design. Volterra, Wiener, Wiener-Hammerstein, and memory polynomial (MP) models are generally employed for these cases of PD design [2, 3].

Receiver-side compensation can be justified for uplink transmission moving the processing task to the base station where higher computational complexity is allowed. In this form, mobile terminals are kept simple and power efficient [4]. It is worthwhile to mention that receiver-side compensation techniques need to deal with the estimation problems associated to the channel (i.e., memory effects and time-varying characteristics).

Imperfections in the IQ modulator represent another important issue in the design of a baseband PD. The predistorted baseband signal is up-converted to RF using an IQ modulator. Phase and amplitude imbalances of the modulator affect the estimation of the PD coefficients. This problem motivates the joint compensation of PA nonlinear response and the distortion introduced in the up-conversion process. The latter distortion is due to phase and amplitude imbalances of the local oscillator (LO) and mismatch in the cascade of digital-analog converters (DAC)-low-pass filters in the I and Q branches. The effects of IQ imbalance on predistortion techniques have been presented in [5–8]. Joint PD and IQ compensation techniques have been previously considered in several studies for the SISO case [9–11].

On the other hand, the implementation of PD techniques in MIMO systems introduces specific problems mostly related to the crosstalk between the different branches. In multi-antenna systems, the crosstalk between the different MIMO channels affects severely the system performance. The crosstalk can be (a) after the transmitter antennas and (b) at the transmitter RF front-end (before PA). For case (a), the coupling process can be modeled as a linear crosstalk (it occurs after the transmitter PA), and is usually mitigated at the receiver side (in the channel equalization process) [12, 13]. In spite of this, the use of MIMO transceiver structures implemented on reduced size leads to the coupling case (b) between the branches at the front-end that cannot be avoided. This kind of crosstalk is amplified by a nonlinear device (amplifier) and is denoted as nonlinear crosstalk. The *nonlinear crosstalk* and the PA nonlinear response should be jointly compensated by a MIMO predisorter to get a reliable system performance.

A MIMO-PD for broadband systems including coupling effects has been reported in [13]. In the cited article, it is shown that the use of multiple SISO PDs assuming N-independent paths (no crosstalk) gives poor results in terms of in-band distortion and adjacent channel power ratio (ACPR). On the other hand, the proposed PD coefficient estimation technique has high computational complexity, and IQ imbalance effects, and tracking the changes of PA characteristics in time due to temperature or bias point variations have not been considered in [13].

In this study we propose a MIMO-PD based on a parallel MP model that renders the following properties:

(a) *Compensation of nonlinear crosstalk and IQ imbalance effects*: We propose a modified (parallel) MP [14] structure to linearize each broadband PA response and to compensate the (frequency-dependent) crosstalk and IQ imbalance effects.

(b) *Reduced complexity*: The proposed MIMO-PD, based on the indirect learning scheme [1], uses efficient recursive least squares (RLS) and stochastic gradient (SG) algorithms (when compared with [13], for example) to estimate the MIMO-PD coefficients.

(c) *Robustness against measurement noise*: Even when indirect-learning architectures are affected by measurement noise at the PA output (creating a bias in the estimated PD coefficients) [15], the performance of the MIMO-PD proposed is validated for practical scenarios (SNR > 30 dB), showing good results.

(d) *Tracking capability*: Contrary to what is required in [5], the estimation of PD coefficients can be performed at the system initialization without a special training sequence. The proposed recursive implementation allows for updating the PD coefficients on-line to track the changes in the PA parameters.

The organization of the study is as follows. The MIMO-OFDM system model (including nonlinear PA, crosstalk effects and IQ imbalance imperfections) is described in Section 2. In Section 3, the proposed PD structure is introduced, and the specific efficient estimation algorithms are derived. Also, an analysis of complexity and implementation issues is included in this section. Section 4 presents simulation results to validate the performance of the new MIMO-PD scheme. In this section, the figures of merit employed to evaluate the proposed MIMO-PD are the error vector magnitude (EVM) and the ACPR. Finally, Section 5 concludes the article.

To simplify the notation, we develop the MIMO-PD for $M = 2$ transmit antennas. However, the proposed technique is not restricted to this case and can easily be generalized to $M > 2$ transmit antennas.

Throughout this article, we employ the following abbreviations. MIMO-PD denotes MIMO predistorter, and CPD represents conventional predistorter. MIMO-PD SG, MIMO-PD RLS, and MIMO-PD LS are employed to define the MIMO predistorters coefficients of which were obtained using stochastic gradient, recursive least squares, and least squares algorithms, respectively. The acronyms MP and MLP denote memory and memoryless polynomials.

SYSTEM MODEL

The transmitter front-end considered uses direct-conversion architecture [16]. This architecture presents several advantages when compared with the conventional super-heterodyne structure: small number of parts, low-mixing product spurs, few filters, and low current consumption [17].

Let $\{X_\ell(k)\}_{k=0}^{N-1} \in \mathcal{C}$ be the modulated data symbols associated with carrier k to be transmitted by antenna $l = 1, 2,..., M$. The time-domain OFDM symbols $\{x_\ell(n)\}_{n=0}^{N-1}$ are obtained via the inverse discrete Fourier transform:

$$x_\ell(n) = \frac{1}{\sqrt{N}} \sum_{k=0}^{N-1} X_\ell(k) e^{j\frac{2\pi}{N}kn}, \quad n = 0, 1, ..., N - 1.$$

(1)

The OFDM signal $x \ell (n)$ at the transmitter is separated into real and imaginary (IQ) digital baseband components, $x i \ell (n)$ and $x q \ell (n)$. The IQ components are filtered by the I and Q branches equivalent low-pass filters, $h i \ell (n)$ and $h q \ell (n)$, and converted to continuous-time baseband signals, $\tilde{x}_{i\ell}(t)$ and $\tilde{x}_{q\ell}(t)$.

The low-pass filters $h i \ell (n)$ and $h q \ell (n)$, which model the cascade of DAC and the analog low-pass filters, are represented as FIR filters of lengths $L i$ and $L q$, respectively [18]. Generically, the impulse responses $h i \ell (n)$ and $h q \ell (n)$ are different, creating an IQ frequency-dependent mismatch.

The IQ components at the low-pass filters output (continuous-time baseband signals) are directly modulated to RF, $x_{rfl}(t)$, using two LO signals ideally in quadrature. However, in "real-life" implementations, LO signals present phase and amplitude imbalances in the I and Q branches. Amplitude- and phase imbalance parameters of the IQ-modulator associated to the branch l are denoted as β_l and ϑ_l, respectively [19]. Finally, the RF signal $x_{rfl}(t)$ is amplified and transmitted through the channel, $y_l(t)$. A block diagram of the transmitter front-end of the l branch is illustrated in Figure 1.

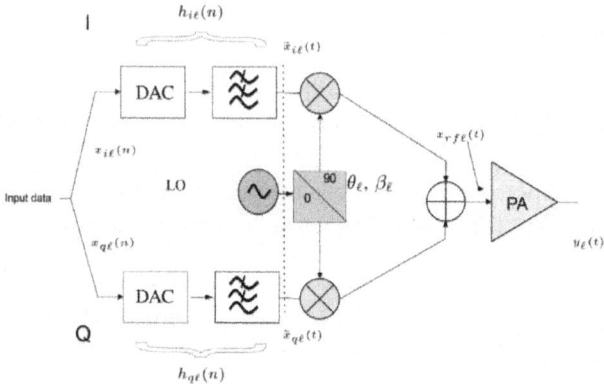

Figure 1: Ofdm ℓ branch transmitter block diagram.

Following the model described in [18], the equivalent discrete-time baseband signal after IQ modulator for each branch of the MIMO transmitter can be represented by

$$s_\ell(n) = g_{1\ell}(n) \otimes x_\ell(n) + g_{2\ell}(n) \otimes x_\ell^*(n) \tag{2}$$

where $\otimes$ denotes convolution, $g_{1\ell}(n)$ and $g_{2\ell}(n)$ are *equivalent* filters with impulse response given by

$$g_{1\ell}(n) = h_{i\ell}(n) + \beta_\ell h_{q\ell}(n)e^{j\vartheta_\ell}$$
$$g_{2\ell}(n) = h_{i\ell}(n) - \beta_\ell h_{q\ell}(n)e^{j\vartheta_\ell}. \tag{3}$$

Besides IQ imbalance, direct conversion transceivers suffer from DC offset because of LO leakage [20], the mixing of the LO signal with itself and noise from the mixers, filters, and DAC converters. The output of the IQ modulator including the DC offset term can be written as

$$u_\ell(n) = g_{1\ell}(n) \otimes x_\ell(n) + g_{2\ell}(n) \otimes x_\ell^*(n) + \varepsilon_\ell \tag{4}$$

where ε_{l} is the DC offset term due to imperfections at the up-converter. A block diagram of the described two-antenna MIMO-OFDM transmitter front-end (equivalent baseband model) is illustrated in Figure 2.

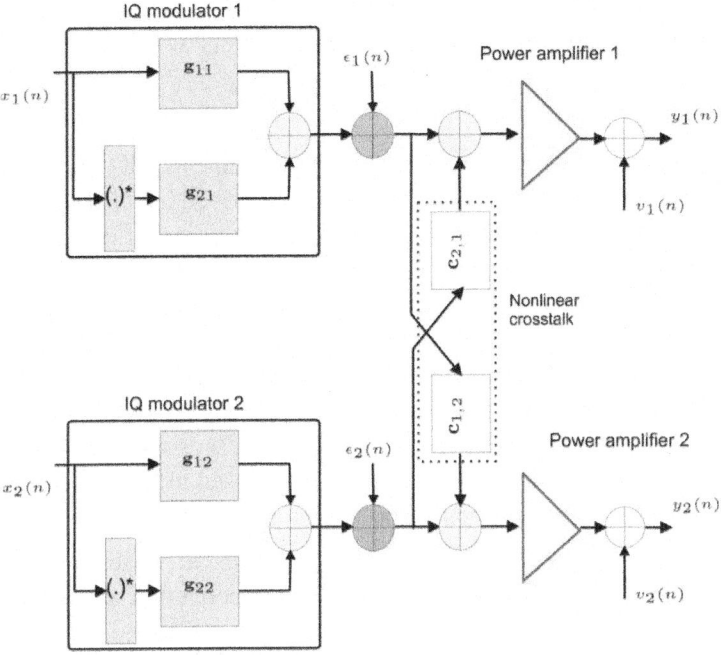

Figure 2: 2 × 2 MIMO transceiver front-end baseband equivalent model.

The IQ imbalance model given by (2) is composed by two branches and motivates the parallel structure of our MIMO-PD as presented in the next section.

MIMO transceiver RF front-end requires a careful design to isolate the different branches. Nevertheless, when considering reduced-size implementation (chipset), the coupling between the MIMO branches cannot be fully eliminated. To model this kind of crosstalk which is assumed frequency dependent, we consider the output of the PAs, written as

$$y\ell(n) = p\ell \left[u_\ell(n) + \sum_{m=1, m \neq \ell}^{M} c_{m\ell}(n) \otimes u_m(n) \right] + v_\ell(n)$$

(5)

where $p\ell[\cdot]$ is the PA response of each branch, $u_m(n)$ is the output of the IQ modulator of the m path of the transceiver, and $c_{m\ell}(n)$ is the filter representing the crosstalk with impulse response, $c_{m\ell} = [c_{m\ell}(n), c_{m\ell}(n-1), \ldots c_{m\ell}(n - L_{c_m} + 1)]^T$ and modeling the coupling between path m to path ℓ. The measurement noise at the output of each PA is denoted by $v \ell (n)$. Equation $\underline{5}$ allows inferring that to obtain a distortion-free signal $x \ell (n)$, the PD should be able to invert the PA response $p \ell [\cdot]$, remove the undesired coupled signal, and mitigate the effects of the IQ imbalance.

MIMO Predistorter

Owing to the effects of crosstalk and IQ imbalance, the MIMO transmitter to be linearized follows a characteristic that can be described by a parallel nonlinear model. We consider, for the derivation, the linearization of one MIMO path. The PD coefficients are estimated using an indirect learning structure [1]. In this methodology, the MIMO-PD parameters are estimated and copied to the predistorter avoiding the inverse model estimation required by direct learning techniques. However, despite several advantages, the indirect learning structure is affected by measurement noise at the PA output [9, 15]. Measurement noise creates a bias in the estimated model, which increases with the model order. The effects of the measurement noise on the proposed technique are discussed and evaluated following a specific application in Section 4.

The proposed identification structure requires a feedback path where the RF signal at the output of the PA is down-converted and translated to baseband. The components of the down-converter, filters, DAC, and mixer need to be carefully designed in order to minimize its harmful effects over the performance of the identification technique. In this approach of this study, an ideal feedback path is considered. It is assumed that the demodulation is implemented digitally minimizing the demodulation errors. A feedback path without IQ demodulator imbalance and nonlinear effects was also considered in previous publications [8–10]. In [5, 11], errors in the feedback loop and techniques to remove its harmful effects are addressed. However, only frequency-independent imbalances are considered.

A. MIMO predistorter structure

Even when other alternatives are possible, the proposed MIMO predistorter is based on the MP model [14]. That model has been employed in predistortion techniques showing a very good performance [2]. The main characteristics of the MP model, which we exploit regarding real-time applications are its modularity and simplicity. Furthermore, alternative modeling of the static part of the MP can also be considered. Orthogonal polynomials alleviate the ill-conditioned problems associated with the conventional polynomial models [21]. Generalized MP proposed by Morgan [14] should also be an interesting option with improved stability at a reasonable increase of the implementation complexity, but its use is not discussed here.

To include all the impairments, i.e., the nonlinear distortion and memory effects due to the PA, crosstalk coupling due to the MIMO structure, and the IQ imbalance distortion, we propose for each PA (of the M-antenna MIMO system) a $(2 + 2(M - 1)) \times 1$ MISO PD. Each branch of the MISO PD is formed by a MP [14]. There are two branches to model the own PA nonlinear distortion (associated to the IQ components of the IQ imbalance characterization) and $2(M - 1)$ to model the crosstalk associated to the other PAs. The proposed MISO PD structure, for the case M = 2, is depicted in Figure 3. Each block $\mathcal{P}_{\ell,i}$ denotes the MP associated to the branch i of antenna l. Based on the $M = 2$ case, the (4×1) PD output associated to antenna l can be written as.

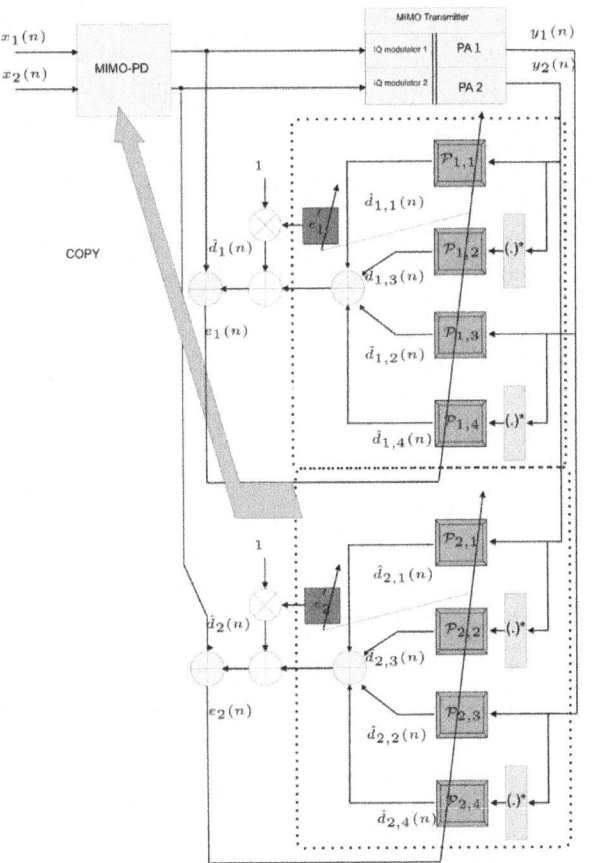

Figure 3: Indirect learning structure for identifying the MIMO-PD coefficients.

Blocks $P_{1,1}$ to $P_{1,4}$ represent the memory polynomials associated to branch 1, and $P_{2,1}$ to $P_{2,4}$ represent the memory polynomials associated to branch 2.

$$\hat{d}_\ell(n) = \sum_{p=0}^{P_{\ell 1}-1} \sum_{k=0}^{M_{\ell 1}} \theta_{pk}^{(\ell,1)}(n)\psi_{1_p}^*(n-k) + \sum_{p=0}^{P_{\ell 2}-1} \sum_{k=0}^{M_{\ell 2}} \theta_{pk}^{(\ell,2)}(n)\psi_{1_p}(n-k)$$

$$+ \sum_{p=0}^{P_{\ell 3}-1} \sum_{k=0}^{M_{\ell 3}} \theta_{pk}^{(\ell,3)}(n)\psi_{2_p}^*(n-k) + \sum_{p=0}^{P_{\ell 4}-1} \sum_{k=0}^{M_{\ell 4}} \theta_{pk}^{(\ell,4)}(n)\psi_{2_p}(n-k) + \varepsilon'_\ell$$

$$= \sum_{i=1}^{4} \hat{d}_{\ell,i}(n) + \varepsilon'_\ell$$

(6)

where $\theta_{pk}^{(\ell,1)}$ and $\theta_{pk}^{(\ell,2)}$ denote the MP coefficients associated to the input signal and its conjugate, respectively; $\theta_{pk}^{(\ell,3)}$ and $\theta_{pk}^{(\ell,4)}$ are the coefficients associated to the crosstalk signal and its conjugate. The basis function of the corresponding MPs are defined by $\psi_{1_p}(n) = \gamma_1(n)|\gamma_1(n)|^{2p}$ and $\psi_{2_p}(n) = \gamma_2(n)|\gamma_2(n)|^{2p}$. $P_{\ell i}$ and $M_{\ell i}$ are the polynomial order and memory depth of the branch i, respectively. The coefficient ε'_ℓ represents the DC offsets that arise from the IQ modulators associated to the branches 1 and 2 of the transceiver.

It is straightforward to extend (6) to the more general case of M-antennas MIMO-PD by including, instead of the terms $\hat{d}_{\ell,3}(n)$ and $\hat{d}_{\ell,4}(n)$ the corresponding $2(M-1)$ terms characterizing the nonlinear crosstalk from the other branches.

According to the two-antenna PD to simplify the notation, we define the coefficient vector: $\boldsymbol{\theta}_\ell(n) = [\boldsymbol{\theta}^{(\ell,1)^T}(n)\boldsymbol{\theta}^{(\ell,2)^T}(n)\boldsymbol{\theta}^{(\ell,3)^T}(n)\boldsymbol{\theta}^{(\ell,4)^T}(n)]$, where

$$\boldsymbol{\theta}^{(\ell,i)}(n) = [\theta_{10}^{(\ell,i)}(n) \cdots \theta_{P_{\ell i}0}^{(\ell,i)}(n) \cdots \theta_{1M_{\ell i}}^{(\ell,i)}(n) \cdots \theta_{P_{\ell i}M_{\ell i}}^{(\ell,i)}(n)]^T$$

(7)

with $i = 1,...,4$. Then, by defining

$$\psi^{(\ell,1)}(n) = [\psi_{1_0}(n) \cdots \psi_{1_{P_{\ell 1}}}(n)]^T$$
$$\psi^{(\ell,2)}(n) = [\psi_{1_0}^*(n) \cdots \psi_{1_{P_{\ell 3}}}^*(n)]^T$$
$$\psi^{(\ell,3)}(n) = [\psi_{2_0}(n) \cdots \psi_{2_{P_{\ell 2}}}(n)]^T$$
$$\psi^{(\ell,4)}(n) = [\psi_{2_0}^*(n) \cdots \psi_{2_{P_{\ell 4}}}^*(n)]^T$$

(8)

with $i = 1,... 4$, the basis function vector can be written as

$$\phi_\ell(n) = [\psi^{(\ell,1)^T}(n) \cdots \psi^{(\ell,1)^T}(n-M_{\ell 1})\psi^{(\ell,2)^T}(n) \cdots \psi^{(\ell,2)^T}(n-M_{\ell 2})$$
$$\psi^{(\ell,3)^T}(n) \cdots \psi^{(\ell,3)^T}(n-M_{\ell 3})\psi^{(\ell,4)^T}(n) \cdots \psi^{(\ell,4)^T}(n-M_{\ell 4})]^T.$$

(9)

The PD output of the branch l can be written as

$$\hat{d}_\ell(n) = \phi_\ell^H(n)\theta_\ell(n) + \varepsilon'_\ell. \qquad (10)$$

To account for the DC offset from the LO an extra coefficient, ε'_ℓ, is added to the coefficient vector. Using an augmented coefficient vector, the PD output signal can be expressed as

$$\hat{d}_\ell(n) = \phi_\ell^H(n)\bar{\theta}_\ell(n) \qquad (11)$$

where

$$\bar{\phi}_\ell(n) = [1, \phi_\ell]^T$$
$$\bar{\theta}_\ell(n) = [\varepsilon, \theta_\ell]^T \qquad (12)$$

where the augmented coefficient vector has dimensions $1 \times (P_{\ell 1}M_{\ell 1} + P_{\ell 2}M_{\ell 2} + P_{\ell 3}M_{\ell 3} + P_{\ell 4}M_{\ell 4} + 1)$.

B. MIMO predistorter identification schemes

To estimate the MP coefficients, $\theta_l(n)$ and to track the time-varying characteristics of the PA, adaptive estimation algorithms are considered. We propose two different algorithms: RLS and stochastic gradient algorithms. At the initialization, the PD is bypassed, and the PD coefficients are obtained by minimizing the error signal given by

$$e_\ell(n) = x_\ell(n) - \hat{d}_\ell(n) = x_\ell(n) - \bar{\phi}_\ell^H(n)\bar{\theta}_\ell(n). \qquad (13)$$

Using the instantaneous squared error $|e_l(n)|^2$ as an objective function, a stochastic gradient algorithm that updates $\bar{\theta}_\ell(n)$ is given by

$$\bar{\theta}_\ell(n+1) = \bar{\theta}_\ell(n) - \mu\nabla_{\bar{\theta}_\ell}[e_\ell(n)] = \bar{\theta}_\ell(n) + \delta(n)\bar{\phi}_\ell(n)e_\ell^*(n) \qquad (14)$$

where $\delta(n) = \dfrac{\mu}{\bar{\phi}H_\ell^H(n)\bar{\phi}_\ell(n)}$ and μ is a step size controlling the convergence speed and algorithm stability.

In the case of the recursive least squares algorithm, the deterministic objective function to be minimized is given by

$$\Gamma_\ell(n) = \sum_{i=0}^{n} \lambda^{n-i}|e_\ell(i)|^2 = \sum_{i=0}^{n} \lambda^{n-i}|x_\ell(i) - \bar{\phi}_\ell^H(i)\bar{\theta}_\ell(n)|^2 \qquad (15)$$

where λ is the forgetting factor. The update process for the PD coefficients at each time instant n can be summarized as follows:

$$\hat{d}_\ell(n) = \bar{\phi}_\ell^H(n)\bar{\theta}_\ell(n-1)$$

$$e_\ell(n) = x_\ell(n) - \hat{d}_\ell(n)$$

$$k(n) = \frac{\lambda^{-1}P(n-1)\bar{\phi}_\ell(n)}{1 + \lambda^{-1}\bar{\phi}_\ell^H(n)P(n-1)\bar{\phi}_\ell(n)}$$

$$\bar{\theta}_\ell(n) = \bar{\theta}_\ell(n-1) + k(n)e_\ell^*(n)$$

$$P(n) = \lambda^{-1}\left(P(n-1) - k(n)\bar{\phi}_\ell^H(n)P(n-1)\right)$$

(16)

where $k(n)$ is the gain vector, and $P(n)$ is the inverse correlation matrix.

In the RLS algorithm, the coefficient vector $\boldsymbol{\theta}\ell$ is initialized by $\boldsymbol{\theta}\ell\,(0) = [0, 1,... 0]$, and the inverse correlation matrix is initialized to $P(0) = \alpha\,\boldsymbol{I}$ where $\boldsymbol{I}$ is a $(P_{\ell1}M_{\ell1} +$ $\begin{matrix} P_{\ell2}M_{\ell2} + P_{\ell3}M_{\ell3} + P_{\ell4}M_{\ell4} + 1) \times (P_{\ell1}M_{\ell1} + P_{\ell2}M_{\ell2} + \\ P_{\ell3}M_{\ell3} + P_{\ell4}M_{\ell4} + 1) \end{matrix}$ identity matrix, and α is a large constant.

For comparison, we discuss an extension of the least squares (LS) algorithm of [13] that also includes IQ imbalance distortion at each MIMO-PD branch, in addition to nonlinear crosstalk. Also to maintain simple notation, we discuss the two-antenna predistorter. The estimated input signal for this case can be expressed as

$$\tilde{x} = \boldsymbol{\Psi}^H\tilde{\boldsymbol{\theta}} \qquad (17)$$

where

$$x_1 = [x_1(0) \quad x_1(1) \quad \cdots \quad x_1(N-1)]^T \quad x_2 = [x_2(0) \quad x_2(1) \quad \cdots \quad x_2(N-1)]^T$$
$$\boldsymbol{\Psi}_1 = [\bar{\phi}_1(0) \quad \bar{\phi}_1(1) \quad \cdots \quad \bar{\phi}_1(N-1)] \quad \boldsymbol{\Psi}_2 = [\bar{\phi}_2(0) \quad \bar{\phi}_2(1) \quad \cdots \quad \bar{\phi}_2(N-1)]$$

18)

Then $\tilde{x} = [x_1^T x_2^T]^T$ is a $(2N \times 1)$ vector representing the N samples of the desired PD output of each branch, $\boldsymbol{\Psi} = [\boldsymbol{\Psi}_1 \ \boldsymbol{\Psi}_2]$ is an $(L_1 + L_2 + 1) \times 2N$ matrix formed by the basis function defined by (9) and (12) (augmented basis function that includes an unitary term), and $\tilde{\theta} = [\bar{\theta}_1^T\bar{\theta}_2^T]^T$ is an $(L_1 + L_2 + 1) \times 1$ vector formed by the MIMO-PD coefficients (including DC compensation coefficient). The coefficients vector size is defined as $L_\ell = \sum_{i=1}^4 M_{\ell,i}P_{\ell,i} + 1$ with $\ell = 1,2$. The LS solution for (17) is given by [22]

$$\tilde{\theta} = \left(\boldsymbol{\Psi}\boldsymbol{\Psi}^H\right)^{-1}\boldsymbol{\Psi}\tilde{x} \qquad (19)$$

In this estimator, the measurement noise affects the data matrix, $\boldsymbol{\Psi}$ while in the ordinary LS solution, the measurement noise lies in the observation vector, $\tilde{x}$. In this case, the estimator defined by Equation (19) is called Data Least Squares [23].

The performance and characteristics of this extension of the MIMO-PD of [13] are studied and compared with our proposal in Section 4.

IMPLEMENTATION ASPECTS OF THE MIMO PREDISTORTER

The implementation of predistortion techniques involves two steps: PD coefficient estimation and predistortion using the estimated coefficients.

A. Predistorter coefficients estimation

The dimensions of MP models, i.e., memory depth and polynomial order, need to be carefully chosen. Using a large polynomial order allows for coping with strong nonlinear responses which leads an improvement in the linearization capabilities of the MIMO-PD. On the other hand, overmodeling could deteriorate the numerical stability of the PD identification algorithms [15] and could make the estimation algorithms more sensitive to the measurement noise [24]. Large polynomial order also decreases the interval of successful compensation (that reduces the PA dynamic range) [25]. We study the trade-off between implementation complexity, linearization capabilities, robustness against noise, and algorithm stability by simulations in this section.

1)   *Estimation algorithms*: The complexity of the estimation technique is directly related to the size of the coefficient vector. For example, the two-antenna PD proposed is formed by two independent PD blocks composed of four branches, each formed by a MP. The length of the coefficient vector of each block is given by the sum of the coefficients length of each branch: $L_\ell = \left(\sum_{i=1}^{4} P_{\ell i} M_{\ell i} \right) + 1$. In case of LS implementation, its complexity is proportional to L_ℓ^3, $O[L_\ell^3]$. For the RLS algorithm, the complexity is reduced to L_ℓ^2, $O[L_\ell^2]$. On the other hand, the complexity of stochastic gradient algorithms is proportional to L_ℓ, $O[L_\ell]$. However, if the PA characteristic is highly nonlinear, it results in a poorly conditioned covariance matrix. This ill-conditioned covariance is reflected in slow convergence, when this kind of stochastic gradient algorithms is employed.

In order to reduce the implementation complexity of the RLS algorithm, the widely linear-RLS (WL-RLS) [26] algorithm can be evaluated in a future research. It is an interesting approach to reduce the implementation complexity of the RLS version of the MIMO-PD, obtaining similar convergence speed and robustness. This algorithm has an implementation complexity proportional to $O[2(L_\ell/2)^2]$, which is computationally more economical than the conventional RLS.

Note that the identification algorithms are not executed for every sample. It is done periodically depending on the variation of the PA parameters due to thermal effects (which usually vary slowly with the time). The PD parameter identification step is not a big consumer of computational resources because it is carried out only periodically. However, it should be kept in mind that when using LS algorithm, a large portion of memory is required to store the block of samples. In addition to the large complexity associated to the inversion of a huge matrix, as required by the LS implementation, memory requirements are another point that motivates the use of an RLS algorithm.

B. MIMO-PD implementation

To estimate the MP coefficients, $\theta_\ell(n)$ and to track the time-varying characteristics of the PA, adaptive estimation algorithms are considered. We propose two different algorithms: RLS and stochastic gradient algorithms. At the initialization, the PD is bypassed, and the PD coefficients are obtained by minimizing the error signal given by $\psi_{\ell,p}^*(n) = x_\ell^*(n)|x_\ell(n)|^{2p}.$ For this reason, its implementation does not require extra multiplications (only a conjugation operation).

The implementation of each MLP (considering the input signal $x_l(n)$ and its conjugate) requires $\max\{P_{li}\}$ complex-valued products to implement the basis function, $P_{li} + P_{li+1}$ complex-valued products to weight the basis function with the polynomial coefficients, P_{li+1} conjugations, and $P_{li} + P_{li+1}$ additions. An extra addition is required to compensate the DC offset.

All PD block of the structure proposed ($l = 1,..., M$) employs the same basis functions, and so the complete PD requires the implementation of only one basis function (the one which have the largest polynomial order). Each branch i of the path l composes the MIMO-PD, is formed by a MP implemented with M_{li} MLPs. For this reason, to build a MP branch, the operations required to implement a MLP needs to be executed M_{li} times with $i = 1,..., 2 + 2(M - 1)$ and $l = 1,..., M$. The block diagram of a 2×2 MIMO-PD is depicted in Figure 4a. The structure of one branch of the MIMO-PD is illustrated in Figure 4b. In Figure 4c, the implementation of a MLP is detailed.

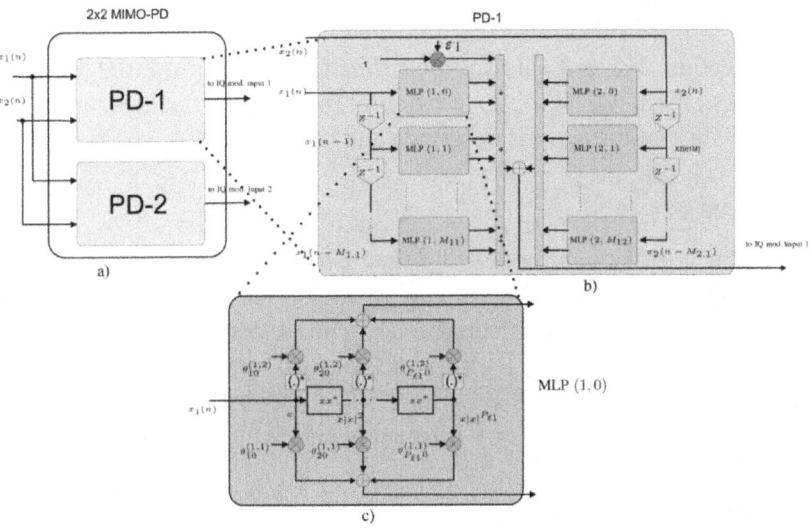

Figure 4: The proposed MIMO-PD structure. (a) 2×2 MIMO-PD implementation, **(b)** one branch of the MIMO-PD composed by $2(M_{1,1} + M_{1,2})$ memoryless polynomials, and **(c)** memoryless polynomial implementation of two branches (conjugate and non-conjugate).

Table 1 summarizes the operations required to implement the proposed MIMO-PD. For comparison, we also include the required operations to implement a conventional PD, CPD-1 assuming no coupling and an ideal IQ modulator, and a conventional PD, CPD-2 assuming nonlinear crosstalk compensation but not IQ imbalance reduction [13].

Table 1: PD implementation complexity for the proposed MIMO-PD and conventional PDs, CPD-1 and CPD-2

Operations	Proposed MIMO-PD	CPD-1	CPD-2
Products	$\sum_{\substack{k=1 \\ odd}}^{2M} (\max\{M_{\ell j}\})(\max\{P_{\ell j}\})\rvert_{\ell=1,M}^{j=k:k+1} +$ $\sum_{\ell=1}^{M}\sum_{j=1}^{2+2(M-1)} M_{\ell j}P_{\ell j o}\,(240)$	$2\sum_{\ell=1}^{M}M_{\ell 1}P_{\ell 1}\,(96)$	$\sum_{\ell=1}^{M}M_{\ell 1}P_{\ell 1}+\sum_{\ell=1}^{M}\sum_{j=1}^{1+(M-1)}M_{\ell j}P_{\ell j}\,(144)$
Additions	$\sum_{\ell=1}^{M}\sum_{j=1}^{2+2(M-1)}M_{\ell j}P_{\ell j}+1\,(193)$	$\sum_{\ell=1}^{M}M_{\ell 1}P_{\ell 1}\,(48)$	$\sum_{\ell=1}^{M}\sum_{j=1}^{1+(M-1)}M_{\ell j}P_{\ell j}\,(96)$
Conjugations	$\sum_{\substack{k=1 \\ odd}}^{2M} (\max\{M_{\ell j}\})(\max\{P_{\ell j}\})\rvert_{\ell=1,M}^{j=k:k+1}\,(48)$	-	-

The numbers in brackets are calculated for a 2×2 MIMO-PD implemented using identical memory depth and polynomial order, $M_{l,i} = 6$ and $P_{l,i} = P = 4$ with $i = 1,..., 4$ and $l = 1, 2$.

PERFORMANCE EVALUATION

In this section, we evaluate the performance of the proposed linearization techniques. First, we discuss the figures of merit evaluated and then the complete simulation setup is described and discussed.

A. Figures of merit

The nonlinear effects introduced by the distortions considered (the nonlinear PA, IQ imbalances, and crosstalk), create in-band and out-of-band distortions. Two figures of merit are considered to evaluate the performance of the proposed linearization technique: the EVM (which quantifies the in-band distortion and is directly related to the BER), and the ACPR (which is a measure of the effects of the out-of-band distortion on adjacent channels).

EVM

In most of the standards, EVM is adopted to quantify the amount of in-band distortion that occurs at the transmitter side. EVM is expressed as the difference, $e_l(k)$, between the original constellation points, $X_l(k)$, and the recovered signal affected by system imperfections at the k th subcarrier of the l th PA. The EVM for an OFDM system formed by N subcarriers is given by the average of EVM over active subcarriers:

$$\rho\ell = \frac{1}{N} \sum_{k=0}^{N-1} \rho\ell(k) = \frac{1}{N} \sum_{k=0}^{N-1} \sqrt{\frac{E[|e_\ell(k)|^2]}{E[|X_\ell(k)|^2]}} \qquad (20)$$

where E[·] denotes the expectation operator, $E[|X_\ell(k)|^2]$ is the desired signal energy per symbol, $\rho \ell(k)$ is the EVM at subcarrier k, and $e\ell(k)$ is the error signal defined as $e_\ell(k) = X_\ell(k) - Y_\ell(k)$, where $Y_\ell(k)$ is the received frequency-domain signal after down-conversion. In order to get the constellation after the PA and evaluate the degradation in terms of the EVM, the output signal is down-converted and demodulated via FFT. Since we are interested in the degradation of the transmitted signal, we assume that down-conversion and FFT demodulation processes, performed to demodulate the transmitted baseband signal, are carried by employing an ideal receiver.

ACPR

The out-of-band distortion is directly related to the PA operation point. The out-of-band emission increases when the PA is driven into its nonlinear operation region. This is also the region that allows high power efficiency. The ACPR is employed to characterize the spectral regrowth and is defined as

$$ACPR = 10 \log_{10} \left(\frac{\int_f Y(f) df}{\int_{fmain} Y(f) df} \right)$$

(21)

where $Y(f)$ is the power spectral distribution at the output of the linearized PA. f_{ad} and f_{main} define the frequency bands of the adjacent and the main channels, respectively.

For example, mobile WiMAX standard defines an spectral mask that should be fulfilled. The PA requires to be operated as close as possible to the maximum efficiency point that met the ACPR and spectrum mask requirements. If these requirements cannot be met, then the PA operation point must be moved, where the mask and the ACPR are fulfilled reducing the PA efficiency. The adjacent channel frequency varies with the system application. In this study the ACPR is evaluated in an adjacent band (frequency offset) shifted 7 MHz from the main band.

B. Simulations

The performance is validated with a two-antenna MIMO-OFDM transmitter, 16-QAM modulation on $N = 1,024$ subcarriers, and a bandwidth of 20 MHz. An oversampling factor of $R = 8$ with an interpolation filter based on root-raised cosine pulse shape with a roll-off factor of 20% has been employed. The coupling between the twobranch transmitter (after PA) is assumed to be frequency dependent and modeled by a FIR filter with impulse response $c_{21} = \rho_2[1, 0.2]$ (coupling from branch 2 to branch 1) and $c_{12} = \rho_1[1, 0.15]$ (coupling from branch 1 to branch 2) where ρ is a coupling factor employed to define the crosstalk level.

The amplitude and phase imbalance are $\beta_1 = 5\%$ and $\vartheta_1 = 5°$; and $\beta_2 = 3\%$ and $\vartheta_1 = 3°$, respectively. The frequency-dependent imbalance are modeled by FIR filters with the following impulse response: $h_{i_1} = [1, 0.15]$ and $h_{q_1} = [1, 0.12]$ for the branch 1, and $h_{i_2} = [1, 0.1]$ and $h_{q_2} = [1, 0.12]$. The DC offset are $|\varepsilon_1| = 0.1$ and $|\varepsilon_2| = 0.07$ with signal power normalized to 1.

The proposed MIMO-PD is evaluated for two different PA models:

Power amplifier 1 (PA-1): Class A amplifier. The PA is modeled by a Wiener model where static nonlinearity corresponds to a solid-state power amplifier (SSPA) which is modeled by the Saleh model [27], i.e.,

$$g[x(n)] = \frac{|x(n)|}{\left[1 + \left(\frac{|x(n)|}{A_s} \right)^{2p} \right]^{1/p}} \exp(j \angle x(n))$$

(22)

where the parameter $p = 1.2$ adjusts the smoothness of the transition from the linear region to the saturation region, and A_s is the amplifier input saturation. The PA operation point is set 2 dB from the 1-dB compression point. PA memory effects are modeled with a FIR filter with coefficients $\mathbf{h}_p = [1, 0.25, 0.1]$.

Power amplifier 2 (PA-2): Class AB amplifier. The PA is modeled by a Wiener model taken from [28]. The static nonlinearity is modeled with a polynomial and can be expressed as

$$\hat{y}(n) = \sum_{p=0^2} b_{2p+1} x(n) |x(n)|^{2p}$$

(23)

where the complex-valued coefficients are $b_1 = 14.97 + 0.0519j$, $b_3 = -23.0954 + 4.968j$, and $b_5 = 21.3936 + 0.4305j$. The linear filter is given by

$$A(z) = \frac{1 + 0.3z^{-2}}{1 - 0.2z^{-1}}$$

(24)

We assume identical polynomial order and memory depth for each PD branch, i.e., $M\ell_{,i} = 6$ and $P\ell_{,i} = P = 4$ for $i = 1,..., 4$ and $\ell = 1, 2$. However, owing to the difference between the power of the useful signal and the power of the conjugate component, different polynomial order and memory depth can be considered in each branch to optimize the implementation complexity [9].

For PD coefficient estimation, we evaluate the RLS, the stochastic gradient, and the LS algorithms. For comparison, we also evaluate the performance of a conventional PD (CPD-1) that compensates PA memory effects but neglects IQ imbalance and crosstalk, i.e., it is assumed that each branch of the MIMO transceiver is decoupled. We also implemented a conventional predistorter, denoted CPD-2, that compensates PA effects and crosstalk effects neglecting IQ imbalances [13].

Learning curves

Figure 5a, b shows the learning curves of the MIMO-PD for the stochastic gradient and RLS algorithms using the SSPA and class AB models, respectively. We see that for the PA-1, the RLS algorithm only requires five OFDM symbols to reach the MSE steady state. The stochastic gradient algorithm, which has a reduced implementation complexity, requires ten OFDM symbols to reach the convergence. When conventional PDs (CPD-1, CPD-2) are used, the MSE floor is increased, reflecting this performance in EVM and ACPR. For the class AB PA, which presents a moderate nonlinearity at low and high amplitudes, the stochastic gradient algorithm is seriously affected by a poorly conditioned covariance matrix and cannot reach the convergence even for a large training sequence (30 OFDM symbols). These results lead us to conclude that the RLS algorithm is the best option to the estimation of PD coefficients.

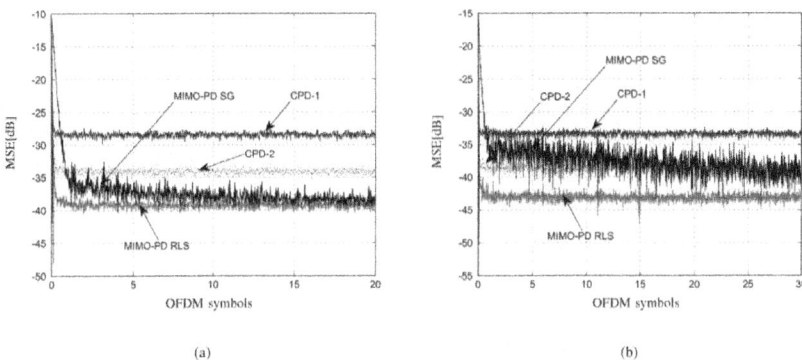

Figure 5: Learning curves of MIMO-PD for SG and RLS algorithms. Curves for conventional PDs are included for comparison. Crosstalk ρ = -20 dB and SNR = 40 dB. **(a)** PA-1 and **(b)** PA-2.

Linearization capabilities

Figure 6 shows the AM-AM and AM-PM curves before and after linearization for PA-1. Identical results are observed for our MIMO-PD estimated using stochastic gradient, RLS and LS algorithms. These curves also show that conventional PDs are unable to linearize the MIMO transmitter. Residual memory effects are observed in AM-AM and AM-PM figures. These results can also be observed in the constellation map at the linearized PA output depicted in Figure 7, where the impact of the crosstalk and IQ imbalance can be observed. Figures 6 and 7had been obtained including a DC offset, $|\varepsilon_1| = |\varepsilon_2|$ = 0.1.

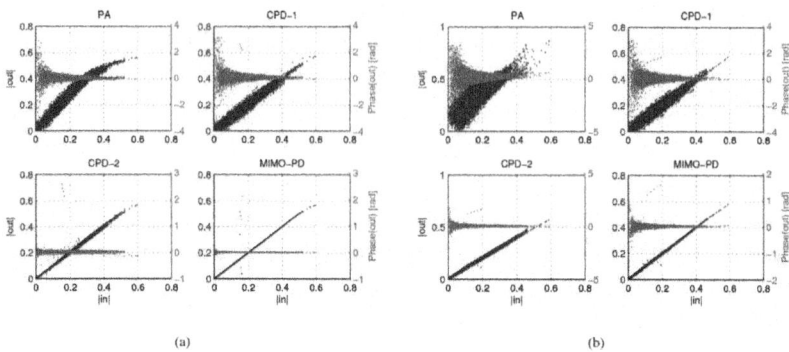

(a)   (b)

Figure 6: AM/AM and AM/PM curves at the output of the PA without linearization, with conventional predistorters and with our MIMO-PD with a SNR = 40 dB. Crosstalk ρ = -20 dB, IQ imbalances of β_1 = 5% and $\vartheta 1$ = 5°, β_2 = 3% and ϑ_2 = 3°, and DC offset, $|\varepsilon_1|$ = 0.1 and $|\varepsilon_2|$ = 0.07 for **(a)** PA-1, **(b)** PA-2.

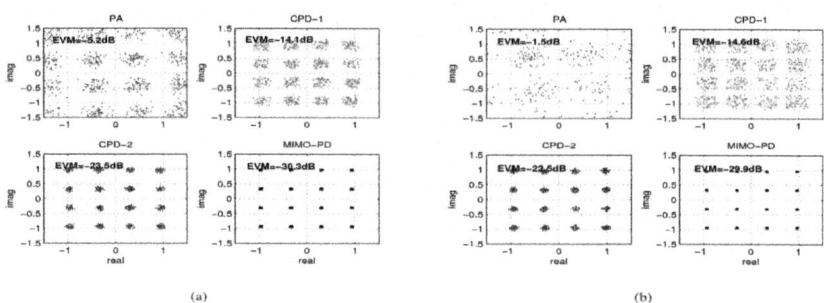

(a)   (b)

Figure 7: Constellation map at the output of the PA without linearization, with conventional predistorters and with our MIMO-PD with a SNR = 40 dB. Crosstalk ρ = -20 dB, IQ imbalance of β_1 = 5% and ϑ_1 = 5°, β_2 = 3% and ϑ_2 = 3°, and DC offset, $|\varepsilon_1|$ = 0.1 and $|\varepsilon_2|$ = 0.07 for **(a)** PA-1, **(b)** PA-2.

Spectral regrowth curves are shown in Figure 8 for a measurement noise of 40 dB and crosstalk coupling ρ = -20 dB, using a class A PA (PA-1). In an identical scenario, the in-band distortion was evaluated showing EVM values around -30 dB, when using LS, RLS, and SG identification algorithms, and for a practical SNR = 40 dB. Conventional PDs present a poor performance when the in-band distortion is evaluated, giving unsuitable levels of EVM. ACPR and EVM results are summarized in Table 2 with crosstalk coupling of ρ = -20 dB, ρ = -30 dB and without coupling. *Effects of polynomial order.*

The effects of the PD polynomial order over the MIMO-PD performance were also evaluated. MIMO-PDs with orders 3, 4, 5 (only odd-orders) and a full odd and even order ($P = 3$) are considered. ACPR curves, depicted in Figure 9, indicate that when the polynomial order is increased, the performance of the MIMO-PD is diminished for moderate SNR. For infinite SNR (without measurement noise), the increment of polynomial order has no noticeable effect in the system performance. The performance of the different implementations in terms of EVM and ACPR is summarized in Table 3. These results demonstrate that overmodeling degrades the PD performance or at least increases the implementation complexity without any extra advantages in terms of performance.

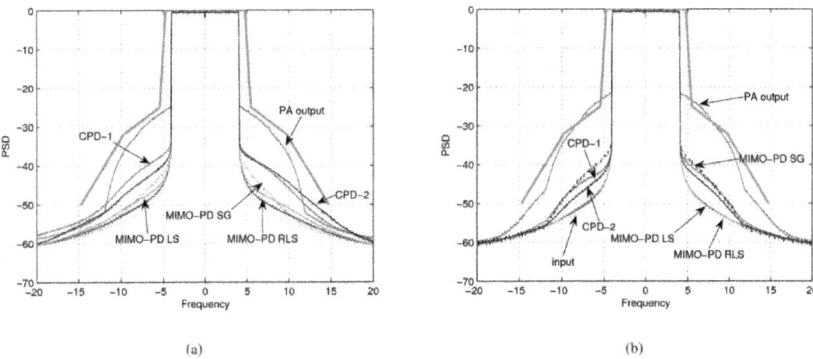

(a)   (b)

Figure 8: Spectral regrowth with and without predistortion including crosstalk between the transmitters $\rho = -20$ dB (spectral mask defined for mobile Wimax is plotted). Measurement noise SNR = 40 dB, IQ imbalance of $\beta_1 = 5\%$ and $\vartheta_1 = 5°$, $\beta_2 = 3\%$ and $\vartheta_2 = 3°$, and DC offset, $|\varepsilon_1| = 0.1$ and $|\varepsilon_2| = 0.07$ for **(a)** PA-1, **(b)** PA-2.

Table 2: PD performance for several coupling levels ρ with SNR = 40 dB and IQ imbalance of $\beta_1 = 5\%$ and $\vartheta_1 = 5°$, $\beta_2 = 3\%$ and $\vartheta_2 = 3°$ and (PA-2)

Coupling level ρ	LS P = 4		RLS P = 4		SG P = 4		CPD-1 P = 4		CPD-2 P = 4	
	ACPR (dB)	EVM (dB)	ACPR (dB)	EVM (dB)	ACPR (dB)	EVM (dB)	ACPR (dB)	EVM (dB)	ACPR (dB)	EVM (dB)
-20 dB	-50.5	-30.0	-50.1	-30.0	-45.8	-28.9	-40.0	-14.1	-40.5	-21.2
-30 dB	-50.6	-30.2	-50.2	-30.0	-46.2	-29.1	41.7	-18.5	-41.9	-21.6
Without coupling	-50.7	-30.3	-50.2	-30.3	-47.0	-29.1	-42.1	-22.5	-42.1	-22.6

The DC offset are $|\varepsilon_1| = 0.1$ and $= |\varepsilon_2| = 0.7$.

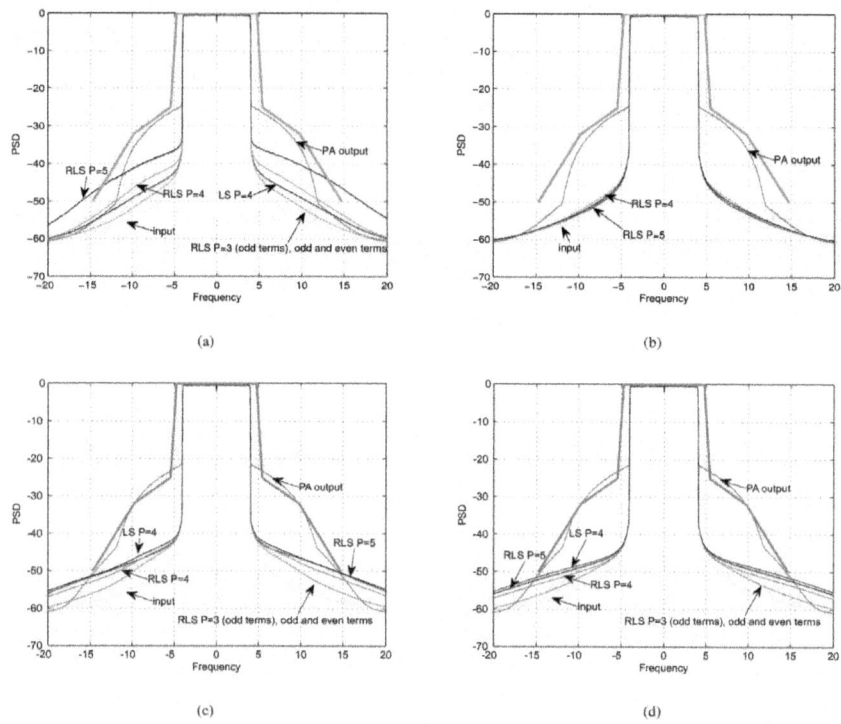

Figure 9: Spectral regrowth versus polynomial order with crosstalk between the transmitters ρ = - 20 dB (spectral mask defined for mobile Wimax is plotted). The DC offset is $|\varepsilon_1| = |\varepsilon_2| = 0$. **(a)** PA-1 with SNR = 40 dB, **(b)** PA-1 with SNR = ∞ (without measurement noise), **(c)** PA-2 with SNR = 40 dB, **(d)** PA-2 with SNR = ∞ (without measurement noise).

Table 3: MIMO-PD performance versus polynomial order for coupling levels ρ = -20 dB with SNR = 30 dB and SNR = 40 dB, and IQ imbalance of $\beta_1 = \beta_2 = 5\%$ and $\vartheta_1 = \vartheta_2 = 5°$ (PA-1)

SNR	LS P = 4		RLS P = 3		RLS P = 4		RLS P = 5		RLS P = 3 (even and odd)	
	ACPR (dB)	EVM (dB)	ACPR (dB)	EVM (dB)	ACPR (dB)	EVM (dB)	ACPR (dB)	EVM (dB)	ACPR (dB)	EVM (dB)
30 dB	- 41.0	- 28.2	- 44.8	- 28.2	- 37.8	- 26.5	- 30.6	- 26.3	- 44.8	- 27.7
40 dB	- 50.2	- 31.0	- 49.3	- 29.5	- 49.8	- 29.6	- 49.9	- 29.6	- 49.8	- 29.4

The DC offset is $|\varepsilon_1| = |\varepsilon_2| = 0.1$.

Effects of measurement noise

We evaluate the MIMO-PD performance against measurement noise for both PAs. ACPR and EVM curves versus measurement noise are illustrated in Figure 10a,b, respectively, using PA-1. Figure 10c,d, and 10b shows ACPR and EVM results for PA-2, respectively. These curves indicate that reasonable values of ACPR and EVM are obtained for SNR larger than 35 dB. This SNR level can be easily obtained in a practical scenario. These results show that PD using the largest polynomial order give the worst performance in terms of EVM and ACPR in the low SNR region. At high SNR, MIMO-PDs using different polynomial orders reach similar results.

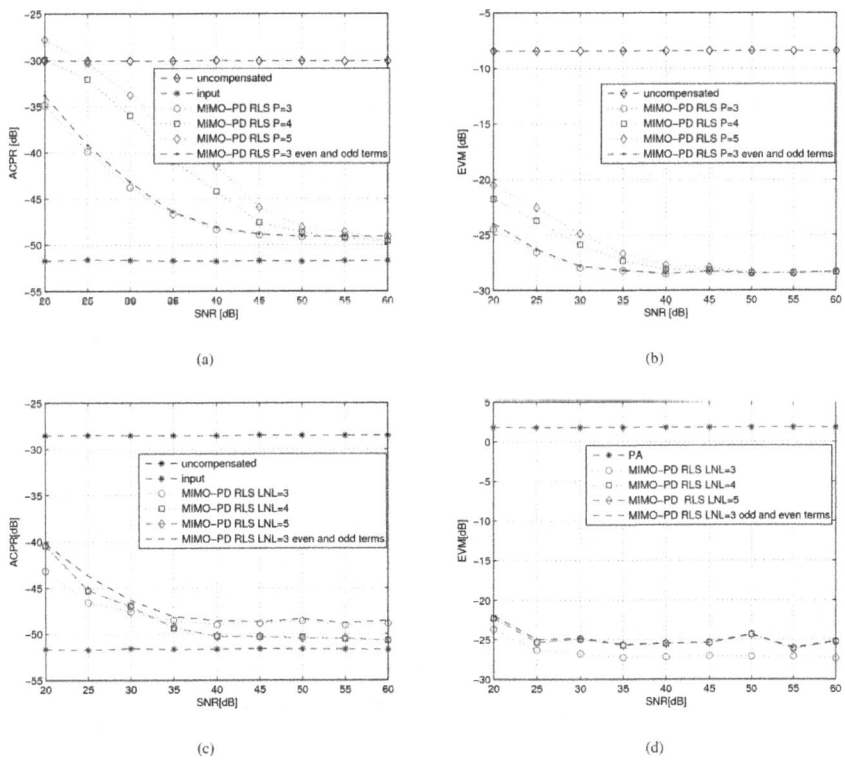

(a)

(b)

(c)

(d)

Figure 10: Measurement noise effects over MIMO-PD performance using RLS algorithm. The crosstalk between the transmitters is ρ = -20 dB and the DC offset is $|\varepsilon_1|$ = $|\varepsilon_2|$ = 0.1. (**a**) ACPR (PA-1), (**b**) EVM (PA-1), (**c**) ACPR (PA-2), and (**d**) EVM (PA-2).

CONCLUSIONS AND FUTURE WORK

We have presented a MIMO-PD that combines PA response linearization, IQ imbalance, and crosstalk compensation. Our PD shows an improved performance compared with conventional isolated PD structures. EVM values around -30 dB, and ACPR nearer to 50 dB are obtained with the proposed MIMO-PD in a scenario that includes impairments at the upconversion block modulator and crosstalk between the different branches. Conventional PDs are unable to operate in this scenario, giving EVM and ACPR values that fail to comply with the specifications of the majority of the wireless standards. The proposed technique shows a moderate implementation complexity and also includes tracking capabilities to follow PA parameter variations. Simulation results show that the MIMO-PD works appropriately in a realistic measurement noise scenario.

Through a future study, the WL-RLS algorithm can be evaluated. The WL-RLS approach is computational more economical than the conventional RLS obtaining similar convergence speed and robustness. Reduced complexity techniques and robustness are interesting issues which need to be addressed in future research.

ACKNOWLEDGEMENTS

This work was partially supported by the Academy of Finland, Smart Radios (SMARAD) Center of Excellence, Agencia Nacional de Promoción Científica y Tecnológica PICT 2008-00104 and PICT 2008-0182, and Universidad Nacional del Sur, Argentina, Project 24/K044.

REFERENCES

1.   Eun C, Powers EJ: A new Volterra predistorter based on the indirect learning architecture. IEEE Trans Signal Process1997,45(1):223-227. 10.1109/78.552219

2.   Ding L, Zhou GT, Morgan DR, Ma Z, Kenney JS, Kim J, Giardina CR: A robust digital baseband predistorter constructed using memory polynomials. IEEE Trans Commun 2004,52(1):159-165. 10.1109/TCOMM.2003.822188

3.   Gilabert P, Montoro G, Bertran E: On the Wiener and Hammerstein models for power amplifier predistortion. Proceedings of the Asia-Pacific Microwave Conference, APMC 2005 2005.

4.   Gregorio F, Werner S, Cousseau J, Laakso T: Receiver cancellation technique for nonlinear power amplifier distortion in SDMA-OFDM

systems. IEEE Trans Veh Technol 2007,56(5 Part I):2499-2516.

5.   Cavers J: The effect of quadrature modulator and demodulator errors on adaptive digital predistorters for amplifier linearization.IEEE Trans Veh Technol 1997,46(2):456-466. 10.1109/25.580784

6.   Cavers J, Liao M: Adaptive compensation for imbalance and offset losses in direct conversion transceivers. IEEE Trans Veh Technol 1993,42(4):581-588. 10.1109/25.260752

7.   Cavers J: New methods for adaptation of quadrature modulators and demodulators in amplifier linearization circuits. IEEE Trans Veh Technol 1997,46(3):707-716. 10.1109/25.618196

8.   Ding L, Ma Z, Morgan D, Zierdt M, Tong Zhou G: Compensation of frequency-dependent gain/phase imbalance in predistortion linearization systems. IEEE Trans Circuits Syst I 2008,55(1):390-397.

9.   Anttila L, Handel P, Valkama M: Joint mitigation of power amplifier and IQ modulator impairments in broadband direct-conversion transmitters. IEEE Trans Microw Theory Tech 2010,58(4):730-739.

10. Zareian H, Vakili VT: New adaptive method for IQ imbalance compensation of quadrature modulators in pre-distortion systems. EURASIP J Adv Signal Process 2009, 10. Article ID 181285

11. Kim Y-D, Jeong E-R, Lee YH: Adaptive compensation for power amplifier nonlinearity in the presence of quadrature modulation/demodulation errors. IEEE Trans Signal Process 2007,55(9):4717-4721.

12. Palaskas Y, Ravi A, Pellerano S, Carlton B, Elmala M, Bishop R, Banerjee G, Nicholls R, Ling S, Dinur N, Taylor S, Soumyanath K: A 5-GHz 108-Mb/s 2×2 MIMO transceiver RFIC with fully integrated 20.5-dBm power amplifiers in 90-nm CMOS. IEEE J Solid-State Circuits2006,41(12):2746-2756.

13. Bassam S, Helaoui M, Ghannouchi F: Crossover digital predistorter for the compensation of crosstalk and nonlinearity in MIMO transmitters, IEEE Trans Microw. Theory Tech 2009,57(5):1119-1128.

14. Morgan D, Ma Z, Kim J, Zierdt M, Pastalan J: A generalized memory polynomial model for digital predistortion of RF power amplifiers. IEEE Trans Signal Process 2006,54(10):3852-3860.

15. Morgan D, Ma Z, Ding L: Reducing measurement noise effects in digital predistortion of RF power amplifiers. Proceedings of the IEEE International Conference on Communications, ICC 2003, 4: 2436-2439.

16. Mak P-I, Seng-Pan U, Martins R: Transceiver architecture selection: review, state-of-the-art survey and case study. IEEE Circuits Syst

Mag 2007,7(2):6-25.

17. Masse C, Luu Q: A 2.4 GHz WiMAX direct conversion transmitter, Analog Devices. AN-826 Application note 2007, 1-16.

18. Tandur D, Moonen M: Joint compensation of OFDM frequency-selective transmitter and receiver IQ imbalance. EURASIP J Wireless Commun Networking 2007, 10. Article ID 68563

19. Liu CL: Impacts of IQ imbalance on QPSK-OFDM-QAM detection. IEEE Trans Consum Electron 1998,44(3):984-989. 10.1109/30.713223

20. Li M, Hoover L, Gard KG, Steer MB: Behavioral modeling and impact analysis of physical impairments in quadrature modulators.IET, Trans Microwaves Antennas Propag 2010,4(12):2144-2154. 10.1049/iet-map.2009.0278

21. Raich R, Qian H, Zhou GT: Orthogonal polynomials for power amplifier modeling and predistorter design. IEEE Trans Veh Technol2004,53(5):1468-1479. 10.1109/TVT.2004.832415

22. Golub G, Loan CFV: Matrix Computations. The Johns Hopkins University Press, Baltimore; 1993.

23. DeGroat RD, Dowling EM: The data least squares problem and channel equalization. IEEE Trans Signal Process 1993,41(1):407-411. 10.1109/TSP.1993.193165

24. Messaoudi N, Fares M-C, Boumaiza S, Wood J: Complexity reduced odd-order memory polynomial pre-distorter for 400-watt multi-carrier Doherty amplifier linearization. 2008 IEEE MTT-S International Microwave Symposium Digest 2008, 419-422.

25. Tsimbinos J: Identification and compensation of nonlinear distortion. Ph.D thesis, Institute for Telecommunications Research, School of Electronic Engineering, University of South Australia 1995.

26. Douglas S: Widely-linear recursive least-squares algorithm for adaptive beamforming. IEEE International Conference on Acoustics, Speech and Signal Processing 2009, 2041-2044.

27. Saleh AAM: Frequency-independent and frequency-dependent nonlinear models of TWT amplifiers. IEEE Trans Commun1981,29(11):1715-1720. 10.1109/TCOM.1981.1094911

28. Ding L: Digital predistortion of power amplifiers for wireless applications. Ph.D thesis, School of Electrical and Computer Engineering, Georgia Institute of Technology 2004.

Chapter 11

RADIO FREQUENCY SOLID STATE AMPLIFIERS

J. Jacob
ESRF, Grenoble, France

ABSTRACT

Solid state amplifiers are being increasingly used instead of electronic vacuum tubes to feed accelerating cavities with radio frequency power in the 100 kW range. Power is obtained from the combination of hundreds of transistor amplifier modules. This paper summarizes a one hour lecture on solid state amplifiers for accelerator applications.

INTRODUCTION

The aim of this lecture was to introduce some important developments made in the generation of high radio frequency (RF) power by combining the power from hundreds of transistor amplifier modules. Such RF solid state amplifiers (SSA) were developed and implemented at a large scale at SOLEIL to feed the booster and storage ring cavities [1, 2]. At ELBE FEL four 1.3 GHz–10 kW klystrons have been replaced with four pairs of 10 kW SSAs from Bruker Corporation (now Sigmaphi Electronics, Haguenau/France), thereby doubling the available power [3]. The company Cryolectra GmbH (Wuppertal/Germany) delivers SSA solutions at various frequencies and power levels, such as a 72 MHz–150 kW SSA for a medical cyclotron [4]. Following a transfer of technology from SOLEIL, the company ELTA (Blagnac/France), a subsidiary of the French group AREVA, has delivered seven 352.2 MHz–150 kW RF SSAs to ESRF [5, 6]. These SSAs are used as an example to illustrate this technology. A compact SSA making use of a cavity combiner with fully planar RF amplifier modules that are suited for mass production is under development at ESRF: the aim is to reduce the required space and fabrication costs

Typical Radio Frequency System Layout

The typical layout of an RF transmitter powering an accelerating cavity in a particle accelerator is shown in Fig. 1. The low-level RF signal generated by the master source is distributed to all RF stations and other equipment such as, for example, the beam diagnostics systems. It is modulated in amplitude and phase by the low-level RF (LLRF) system, pre-amplified and amplified to high power by means of an RF power amplifier, which is the subject of this paper.

RF power amplifiers are often protected against reverse power by an isolator built with an RF power circulator that transmits the incident power to the accelerating cavity and deviates the reverse power into a high power load. The RF transmitter generally includes various auxiliaries, a power supply, a modulator, and fast interlock protection systems.

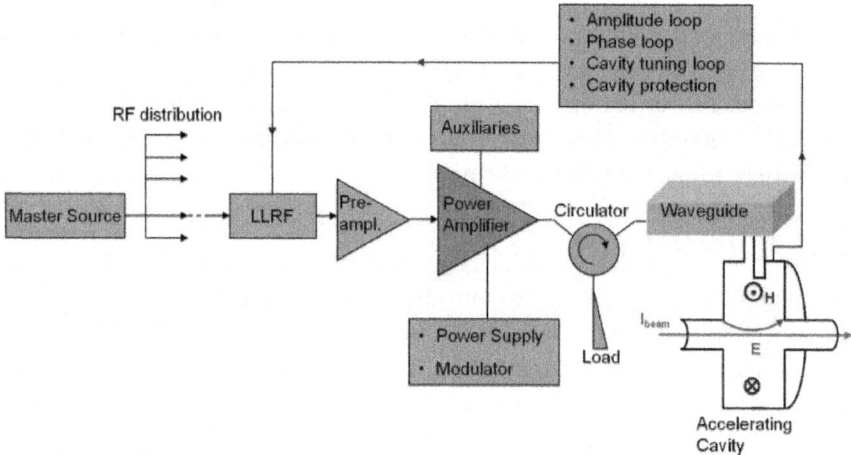

Figure 1: RF transmitter for an accelerating cavity.

Example: the ESRF Transmitters

The 1.3 MW 352.2 MHz super klystron from Thales Electron Devices (TED, Vélizy/France) in Fig. 2(a) was initially developed for the CERN/LEP ring. Similar klystrons are or were manufactured by other tube manufacturers. Four transmitters using this type of klystron were implemented at ESRF in the 1990s: one to deliver 600 kW in 10 Hz pulses to the booster cavities, and three others to power the storage ring cavities in a continuous wave (CW). As shown in Fig. 2(b), the power from these klystrons needs to be split to feed two to four cavities with several hundred kilowatts each.

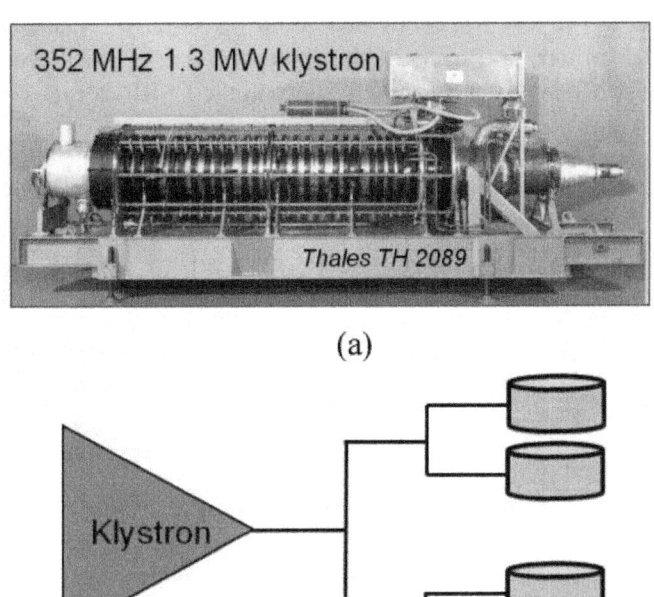

(a)

(b)

Figure 2: (a) 1.3 MW klystron from Thales Electron Devices (TED, Vélizy/France) used at ESRF. DC to RF efficiency η_{max} = 62%, gain G_{max} = 42 dB. (b) Power splitting to provide several hundreds of kilowatts of RF to a number of ESRF cavities.

These klystron tubes have a high gain and a good efficiency. They are fed from 100 kV, 20 A DC high voltage power supplies with a sophisticated crowbar protection. In modern systems, fast insulatedgate bipolar transistor (IGBT) switched power supplies are used. Klystrons require many auxiliary power supplies for the modulation anode, the filament, and the focusing coils. Due to the high voltage, the electrons generate a substantial level of high energy X-rays when hitting the collector, which necessitates a sophisticated lead shielding hutch. Out of the original three manufacturers, only one still produces the ESRF klystron shown in Fig. 2(a), and in the early 2000s there remained only a few customers. Seeing the potential risk of obsolescence of these klystrons, it was necessary to become prepared for an alternative RF power source in order to safeguard the operation of the facility. Following SOLEIL's great success with the development and operation of 180 kW SSAs [1] it was decided to implement an initial series of seven SSAs in the frame of the ESRF upgrade phase I over the years 2009 to 2015. Figure 3 shows one of the seven new 150 kW SSAs at ESRF. Four SSAs have replaced the booster klystron transmitter, and each of the three remaining SSAs feeds one new Higher Order Mode (HOM) damped cavity in the ESRF storage ring [5, 6].

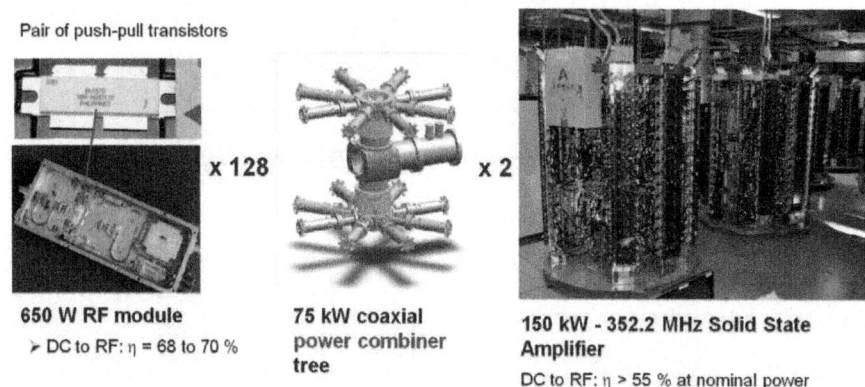

Pair of push-pull transistors

x 128   x 2

650 W RF module
➤ DC to RF: η = 68 to 70 %

75 kW coaxial power combiner tree

150 kW - 352.2 MHz Solid State Amplifier
DC to RF: η > 55 % at nominal power

Figure 3: One of the 150 kW–352.2 MHz solid state amplifiers obtained by combining the power of 256 RF transistor amplifier modules each delivering up to 650 W (seven such SSAs are in operation at ESRF, provided by ELTA/AREVA following a transfer of technology from SOLEIL). DC to RF power conversion efficiencies are denoted η

An SSA combines the power from as many transistor modules as needed to provide the required nominal output power for one cavity with a given safety margin: 256 modules in the example shown in Fig. 3. The architecture and design of SSAs will be described in detail in Section 2 and a flavour of their performance will be given.

Radio Frequency Power Sources for Accelerating Cavities – a Brief History

This section concludes with a brief history and overview of the RF power sources used to feed accelerating cavities [7]. The graph in Fig. 4 shows the technologies used in various accelerator applications as function of the operating frequency and the required unitary power level.

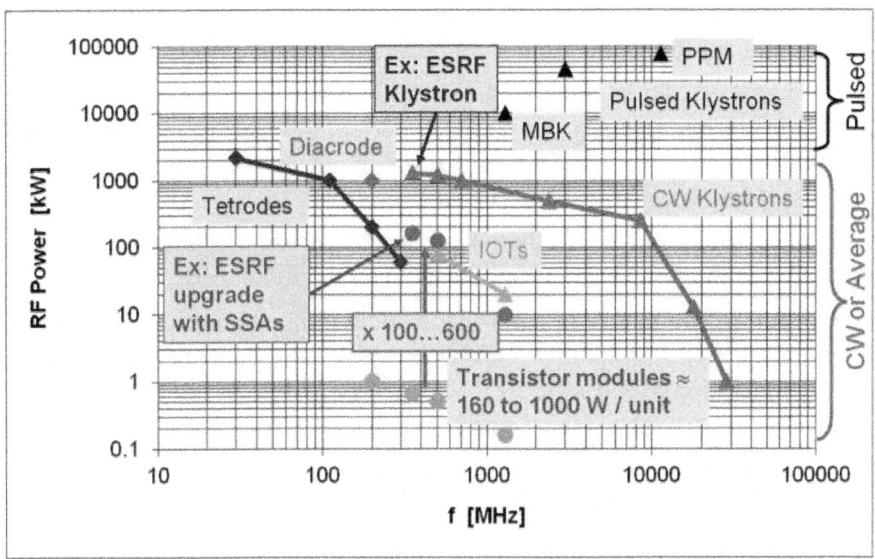

Figure 4: Typical RF sources used in accelerators

Electronic vacuum tubes like triodes and tetrodes have existed since the early twentieth century. Due to their finite electron drift time they are basically limited to frequencies below 1 GHz. Tetrodes are still in use today for broadcast and accelerator applications to generate from kilowatts at 1 GHz up to hundreds of kilowatts below 100 MHz. Note that a small 3.5–5 GHz triode delivering 2 kW pulses exists for radar applications. With its optimized geometry the diacrode developed by Thales can deliver 1 MW at 200 MHz. In the 1940s and 1950s tubes were developed for high frequency applications that benefit from the electron drift time. They are still widely in use, providing high RF power at high frequencies:

- klystrons: 0.3 GHz–10 GHz, delivering from 10 kW to 1.3 MW in CW and 45 MW in pulsed mode, with applications in TV transmitters, accelerators, and radar;

- IOTs (Inductive Output Tubes or klystrodes): a high efficiency mixture of a klystron and a triode, providing typically 90 kW at 500 MHz to 20 kW at 1.3 GHz, initially developed for SDI (Strategic Defence Initiative) in 1986, with current applications in TV and recently in several accelerators;

- travelling wave tubes (TWT): 0.3 GHz to 50 GHZ, broadband, highly efficient, with applications in satellite and aviation transponders;

- magnetrons: 1 GHz to 10 GHz, narrow band, mostly oscillators, highly efficient for applications in radar and microwave ovens;
- gyrotron oscillators: high power millimetre waves, 30 GHz to 150 GHz, delivering typically 0.5–1 MW pulses of several seconds duration, still subject to much R&D, used for plasma heating for fusion and military applications.

In the 1950s and 1960s the invention and spread of transistor technology also opened the way for many applications in the field of RF:

- bipolar, MOSFET, etc. transistors: delivering several tens of watts up to frequencies of about 1.5 GHz, recently up to 1 kW per unit with sixth generation MOSFETs;
- RF SSAs are increasingly used in broadcast applications, in particular in pulsed mode for digital modulation: 10 kW–20 kW obtained by combining several modules;
- SOLEIL: from 2000 to 2007 SOLEIL pioneered the development of high power 352 MHz MOSFET SSAs for accelerators: 40 kW for their booster, then 2×180 kW for their storage ring;
- ESRF: recent commissioning of seven 150 kW SSAs, delivered by ELTA/AREVA following a technology transfer from SOLEIL;
- increasing numbers of accelerator labs use, develop, or consider solid state technology for RF amplification, e.g. 1.3 GHz/10 kW SSAs from Bruker at ELBE/Rossendorf, 500 MHz SSAs for LNLS and Sesame designed by SOLEIL, etc.

RF Solid State Amplifier

In Fig. 5 is shown an RF SSA with its main components. The RF power from the drive amplifier system is split (2) and distributed to n amplifier modules (1) the outputs of which are recombined (3) and coupled to the RF waveguide feeder. A power converter system (4) supplies the amplifier modules with DC power. Components (1) to (4) are described in detail below.

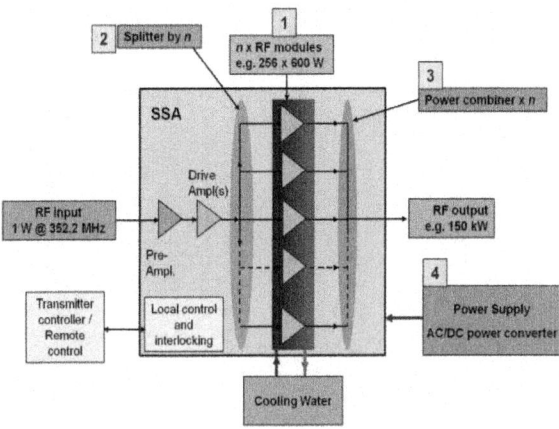

Figure 5: Main components of an RF SSA. The example shown is an ESRF 352.2 MHz–150 kW amplifier.

RF Amplifier Module

Transistor

Thanks to the latest developments in high power transistors such as the laterally diffused metal oxide semiconductor field effect transistor (LDMOS FET) and, in particular, the latest sixth generation devices with 50 V drain polarization, it is now possible to generate several 100 W up to 1 kW in CW with a single RF module, for frequencies up to about 1 GHz. Such transistors are, for instance, manufactured by NXP (Eindhoven/Netherlands) and Freescale semiconductor Inc. (Austin Texas/USA).

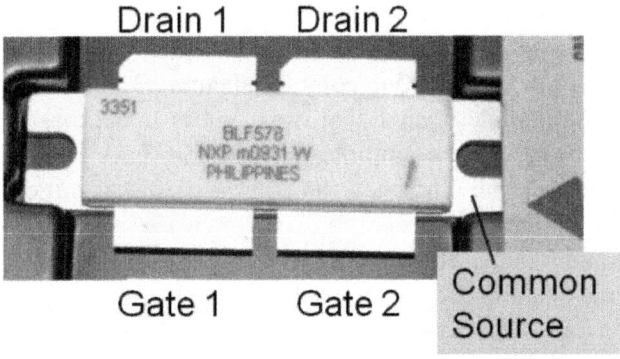

Figure 6: Pair of push-pull transistors.

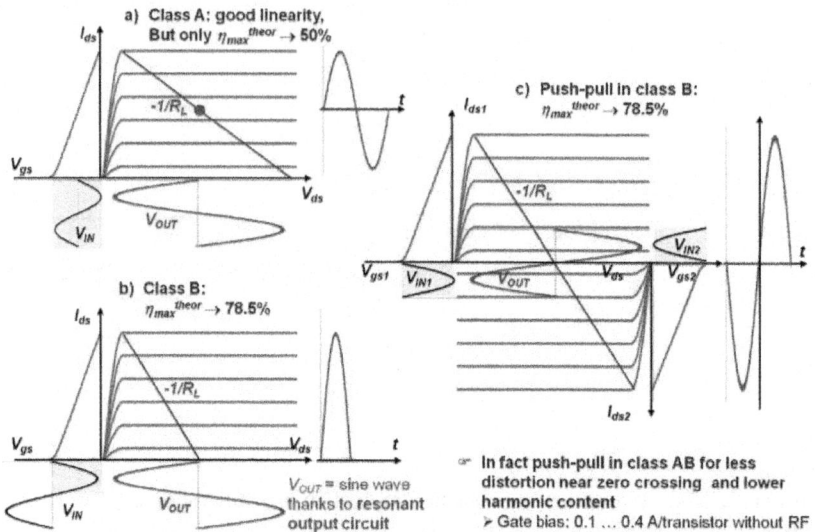

Figure. 7: Schematic of MOSFET operation modes. (a) Class A; (b) class B; (c) class B with two transistors in pushpull. In (a, b) drain-source current I_{ds} is given as a function of gate-source voltage V_{gs} on the left sides and as a function of the drain-source voltage V_{ds} on the right sides of the figures; In (c) drain-source currents I_{ds1} and I_{ds2} of both push-pull transistors are given as function of gate source voltages V_{gs1} and V_{gs2} on the left and right sides of the graphs, respectively, and as a function of the respective drain-source voltages V_{ds1} and V_{ds2} with V_{gs1} and Vgs2 as a parameters in the centre of the graph. The red lines are the working lines of the load resistance RL.

The ELTA/SOLEIL SSA implemented at ESRF uses the BLF 578 transistor from NXP in Fig. 6, which in fact contains two push-pull transistors with a common source, which are operated in class AB. As explained schematically in Fig. 7, this allows optimizing the DC to RF conversion efficiency while keeping the distortion and thereby the harmonic content low: the odd characteristic $I_{ds}(-V_{gs}) = -I_{ds}(V_{gs})$ minimizes the second harmonic H2 and higher even harmonics. As shown in Fig. 7, the working point for an amplifier in class A gives a good linearity as the complete input sine wave Vin is folded on the working line $-1/RL$. But in class A the amplifier always consumes power, even without RF power. At best, the efficiency can approach 50% at full output power. In class B, the amplifier only consumes power when it is driven by RF. This provides a much better efficiency, approaching 78.5% for an ideal transistor. However, only one half of a sine wave is amplified and the signal must be filtered by a narrowband resonator in order to recover the RF signal: it is still polluted by a high harmonic content. Operating two transistors in push-pull in class B allows high efficiency to be maintained, and strongly reducing at least the even harmonics. In fact, the RF modules of the high power SSAs described in this paper are operated in push-pull in class AB with a slight gate bias to further linearize the amplifier characteristics and thereby minimize the harmonic distortions.

RF circuit

As shown in Fig. 8, the RF modules are fed with the unbalanced (unsymmetrical) RF signal from an incoming coaxial cable, which is transformed by a balun circuit into two RF signals that are 180° out of phase and drive the pair of transistors. The matching circuit, together with the bias circuit, determine the class AB working point to reach about 1 dB gain compression and a maximum efficiency of about 68–70% at a nominal output RF power of 650 W for the module shown in Fig. 8(c). Each RF module is protected by a circulator with a 1200 W load against reverse power, and can therefore withstand operating with full reflection. The small amount of power leaking backwards through the circulator slightly affects the gain; however, the impact is negligible when operating the device under normal conditions with up to 30% power reflection. The seven ESRF SSAs have been designed such that no power circulator is needed at an output of 150 kW after combining all of the RF modules.

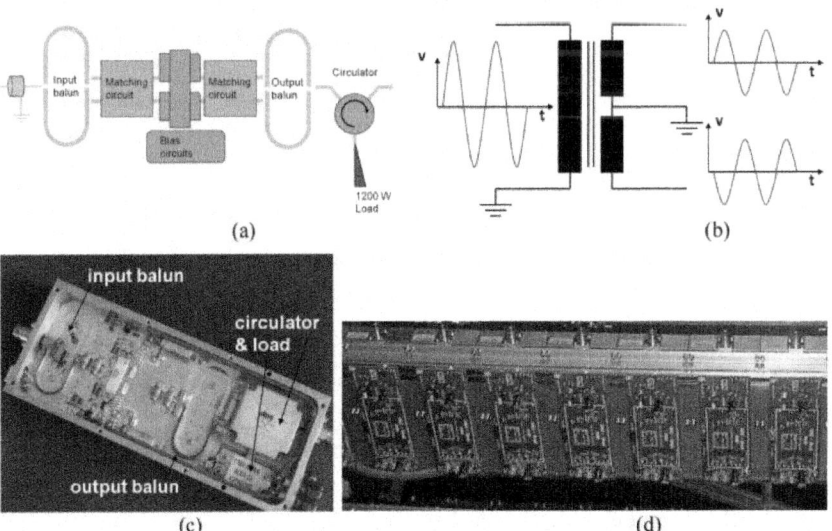

(a) (b)

(c) (d)

Figure 8: RF circuit. (a) Schematic of the RF circuit comprising input and output balun transformers, matching circuits, a DC bias circuit, and a circulator with a 1200 W load to protect each transistor individually against excessive reverse power. (b) Input balun for the transformation of the unbalanced signal from the coaxial input into a balanced push-pull signal that drives the pair of push-pull transistors. (c) A 650 W ELTA/SOLEIL RF module as installed at ESRF. (d) Water-cooled plate supporting 16 RF modules on the rear side and 32 of the 280 V_{DC}/50 VDC converter units (two per RF module) on the front side to feed each module with up to 1.2 kW DC power.

Each RF module is integrated in an individual shielded case. Sensors monitor the temperatures of the transistor socket and the load. The drain current of each pair of transistors is also monitored. These parameters allow the identification of damaged modules during operation. Note that the input and output balun transformers in Fig. 8(c) are built with hand-soldered coaxial lines. For the SSA under development at ESRF the fully planar 700 W RF module in Fig. 9 has been designed, and a series of these has been produced to equip a 75 kW prototype SSA. Hand soldering has been almost completely avoided by implementing a planar balun concept from Motorola, by replacing the RF drain chokes with planar quarter-wave transmission lines and minimizing the number of components, all of them surface-mounted devices (SMD) and prone for automated manufacturing. All of these measures allow cost-effective series production of the RF modules.

Figure 9: Development of a fully planar 700 W RF module at ESRF suited for cost-effective mass production. (a) Planar balun patented by Motorola; (b) fully planar ESRF module, efficiency 66% and gain 19.5 dB at 700 W output power.

The parameters given in the caption for Fig. 9(b) are averages measured over the first series of 66 RF modules fabricated with automated SMD techniques. With a gain dispersion of 19.8 ± 0.6 dB and a phase dispersion of ±6°, a good combining efficiency is expected. Many factors contribute to these dispersions, such as the transistor dispersion and other components' parameters, as well as geometrical tolerances. The latter, however, are minimized by the automated pick-and-place technique. Apart from the gate bias, no other parameter has been tweaked. There is still room for improvement in the achievable efficiency, which is still lower by 3% as compared to the ELTA/SOLEIL RF module. R&D is ongoing in the frame of a collaboration between ESRF and Uppsala University for the optimization of the circuit board. Note that, as shown in Fig. 5, a few RF amplifier modules are required to amplify the low-level input RF power before splitting it again and distributing it to the inputs of all of the RF amplifier modules. In the ELTA/SOLEIL design, one pre-amplifier feeds four RF modules, each of which then feeds 64 RF power modules.

Power Splitters

Power splitters are key components, designed to distribute the RF power from the drive amplifiers to the individual RF modules with minimum amplitude and phase dispersion. At 352 MHz a matched splitter distributing the drive power without reflections can be built as shown in Fig. 10. Each quarterwavelength line transforms the 50 Ω at its output to $Z^2/(50\ \Omega) = n \times 50\ \Omega$ as seen from its input. The parallel connection of these n lines is thus matched to the 50 Ω of the incoming line.

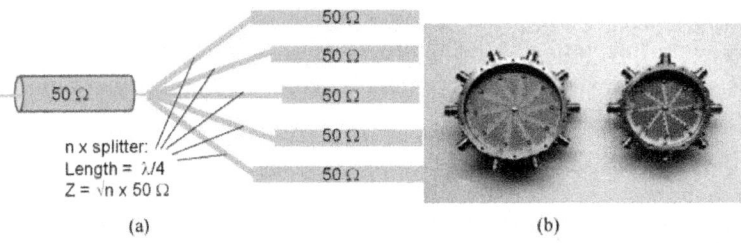

Figure 10: Matched RF stripline splitters for n outputs using quarter-wavelength ($\lambda/4$) transformers. (a) Schematic representation; (b) SOLEIL 10 and 8 line splitters (lids removed) [1, 2].

The Wilkinson splitter in Fig. 11 works in the same manner as the simple quarter-wavelength splitter shown in Fig 10. The additional 50 Ω resistors are not seen by the incoming split signals as they have the same amplitude and phase (common mode). However, differential signals are absorbed by these loads, thereby decoupling the outgoing arms from each other. With a Wilkinson splitter, the connected amplifier inputs are therefore decoupled. Such Wilkinson splitters are implemented in the SSA in development at ESRF.

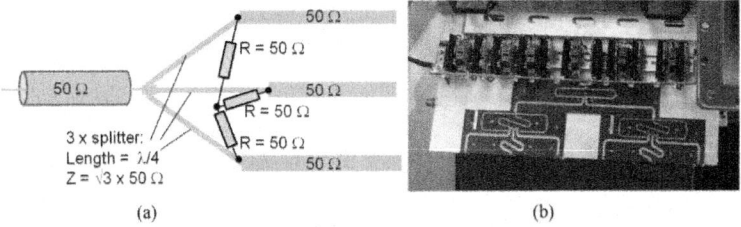

Figure 11: Wilkinson splitters: resistors absorb differential signals without perturbing the common mode, thereby decoupling the connected outputs from each other. (a) Schematic representation; (b) implementation on the SSA under development at ESRF.

Power Combiner

SSAs are often built with several stages of power combiners that use quarter-wavelength transformers like the splitters shown in Fig. 10, except that they are operated in reverse and that the striplines are replaced with transmission lines that are adapted to higher power levels, mostly larger coaxial lines. As for the power splitters, the quarter-wavelength transformers guarantee that the combiner tree is matched in impedance to the connected high power output, provided that all of the inputs are at 50 Ω, which is the case thanks to the circulator/load system at the output of each RF amplifier module. However, as the combiners are non-directional, each individual amplifier does not at all see a matched impedance.

General Considerations on Power Combiners

Before entering into the technology, let us first address some general characteristics of such power combiners. We consider a power combiner with n inputs in a single stage. One can easily derive the Sparameter matrix, which has the following general form:

$$\begin{bmatrix} b_1 \\ b_2 \\ b_3 \\ \cdots \\ b_n \\ b_{n+1} \end{bmatrix} = \begin{bmatrix} (1-n)/n & 1/n & 1/n & \cdots & 1/n & 1/\sqrt{n} \\ 1/n & (1-n)/n & 1/n & \cdots & 1/n & 1/\sqrt{n} \\ 1/n & 1/n & (1-n)/n & \cdots & 1/n & 1/\sqrt{n} \\ \cdots & \cdots & \cdots & \cdots & \cdots & \cdots \\ 1/n & 1/n & 1/n & \cdots & (1-n)/n & 1/\sqrt{n} \\ 1/\sqrt{n} & 1/\sqrt{n} & 1/\sqrt{n} & \cdots & 1/\sqrt{n} & 0 \end{bmatrix} \times \begin{bmatrix} a_1 \\ a_2 \\ a_3 \\ \cdots \\ a_n \\ a_{n+1} \end{bmatrix} ,$$

(1)

where

$a_1 \dots a_n$ are the incident wave amplitudes at the input arms (coming from the amplifier modules);

$b_1 \dots b_n$ are the reverse wave amplitudes at the input arms (coming back to the amplifier modules);

b_{n+1} is the wave amplitude leaving the output arm of the combiner;

a_{n+1} is the reverse wave amplitude incident on the output arm (e.g. the reflection from the cavity).

   Note that wave amplitudes a_i, bi are measured in $\sqrt{W}$ and are generally complex, their arguments representing the respective phase of the wave at port i:

$|a_i|^2$ is the power of the wave incident on port i, where $a_i = |a_i|\, e^{j\varnothing_i}$, and $\varnothing_i$ is the phase of the incident wave i;

$|b_i|^2$ is the power of the wave coming from port i, where $b_i = |b_i|\, e^{j\varnothing'_i}$, and $\varnothing'_i$ is the phase of the reverse wave i.

The different terms in the S-matrix are also generally complex, their arguments depending on the exact geometry of the combiner. However, pertaining to the necessary symmetry of a single stage power combiner, all of the S-parameters $(1 - n)/n$ have the same arguments, as do all of $1/n$ and all of $1/\sqrt{n}$. The phases of the $(1 - n)/n$ and the $1/n$ are $180°$ apart. But the conclusions given below hold even if we neglect the additional phase terms.

- Each connected RF module sees individually a strong mismatch with a reflection coefficient $(1 - n)/n$ that approaches -1 for a large n, corresponding to nearly full reflection.

- The individual inputs for the amplifier modules are coupled to each other with a coefficient $1/n$ showing that this kind of combiner has no directivity at all.

- However, when all of the input waves are equal in amplitude and phase, say with an amplitude a_0, their signals interfere perfectly well in the following way.

For any $i,j \leq n$,

$$a_i = a_0 \Rightarrow b_j = a_j (1 - n)/n + \Sigma\, [a_{i\neq k,\leq n}\,(1/n)] = [(1 - n)/n + (n - 1) \times 1/n]\, a_0 = 0 , \qquad (2)$$

meaning that any of the RF amplifier modules operates in matched condition without any reflection. The strong reflection of its own signal is exactly compensated by the sum of the signals coupled from all other amplifier modules that have the opposite phase,

$$b_{n+1} = \Sigma\, [a_{i,\leq n}\,(1/\sqrt{n})] = [n/\sqrt{n}]\, a_0 \Rightarrow |b_{n+1}|^2 = n\, |a_0|^2 , \qquad (3)$$

meaning that, neglecting imperfections, the power from all n inputs combines to 100% in the output arm of the combiner, as expected.

So, the power adds up correctly at the output of a combiner only by constructive interference, i.e. if and only if the signals at their inputs have the same amplitude and phase. Any differential mode power is distributed as reverse power to the individual input arms, i.e. to the output stages of the RF amplifier modules, where it is absorbed by the circulator loads. So, good control of a flat amplitude and phase distribution are crucial for obtaining a good combining efficiency.

- If just one RF amplifier on arm i is switched off, it will receive the reverse power coupled from all of the other arms without compensation by its own reflection:

$$b_i = \Sigma \, [a_{j \neq i, \leq n} \, (1/n)] = [(n-1)/n] \, a_0 \,, \tag{4}$$

which approaches a_0 for a large n. The consequences are as follows.

i.   When one RF module is not providing power to the combiner, it receives from all the other modules a reverse power that is equivalent to the power of one module.

ii)   When one RF module is not providing any power to the combiner, the output power is lacking the equivalent power from two modules: the power from the missing module plus the power absorbed in the circulator load of the missing module.

iii)   When the SSA output is working on a mismatch with a reflection coefficient $r = |r| \, e^{j\varnothing}$, the reverse wave of amplitude $a_{n+1} = r \, b_{n+1}$ couples back to each module, including the unpowered RF module, as much as

$$a_{n+1}/\sqrt{n} = r \, b_{n+1}/\sqrt{n} \approx r \, (n \times a_0/\sqrt{n})/\sqrt{n} = r \, a_0 \,. \tag{5}$$

In total, an unpowered amplifier module on an arm i will see reverse power corresponding to the following reverse wave amplitude:

$$b_i = [(n-1)/n + r] \, a_0 \approx [1 + |r| \, e^{j\varnothing}] \, a_0 \Rightarrow |b_i|^2 \approx |1 + |r| \, e^{j\varnothing}|^2 \, |a_0|^2 \,, \tag{6}$$

with the worst case

$$\varnothing = 0 \Rightarrow |b_i|^2 \approx (1+|r|)^2 \, |a_0|^2 \tag{7}$$

For full reflection $|r| = 1$, a non-powered RF module can therefore see reverse power as high as four times the nominal power of one module.

For the ELTA SSA, ESRF had specified a maximum reflection of one third of the incident power. Given the nominal module power of 630 W, this means that in the worst case as much power as

$$P_{\text{reverse}}^{\text{max}} = |b_i|^2 \approx (1 + 1/\sqrt{3})^2 \times 630 \text{ W} = 1570 \text{ W} \tag{8}$$

can come back to the circulator and then into the load of an unpowered RF module. Such values were indeed measured on the real SSA shown in Fig. 3. The way to avoid overloading the 1200 W circulator loads of the RF modules is described in Section 2.3.2.

Coaxial Combiner

In most of the existing high power SSAs the power from the individual RF amplifier modules is combined by means of several stages of cascaded coaxial combiners. The SOLEIL SSAs and the 150 kW SSAs in operation at ESRF are literally built around coaxial combiner trees that become amplifier 'towers' in the final assembly. As shown in Fig. 3, the power from two such 75 kW towers is further combined to feed up to 150 kW into an outgoing WR2300 waveguide. The size of the coaxial combiners depends on the power level as, for example, shown in Fig. 3:

- first stage: 630 W × 8 by means of EIA2 1 5/8″ coaxial combiners yielding 5 kW;
- second stage: 5 kW × 8 by means of EIA 6 1/8″ coaxial combiners yielding 40 kW;
- third stage: 40 kW × 2 by means of EIA 6 1/8″ coaxial combiners yielding 80 kW;
- fourth stage: 80 kW × 2 by means of an EIA 9 3/16″ coaxial combiner yielding 160 kW (maximum power, the nominal value being 150 kW).

Dimensioning of the Circulator Loads and fine Tuning of the Coaxial Combiner

Initially, the circulator loads of the 150 kW SSAs for ESRF were dimensioned so as to withstand full module power up to 650 W maximum in full reflection at any phase. This was met by selecting a corresponding circulator and an 800 W load. The SSAs were also tested successfully at full power with up to 30% power reflection at any phase. The SSAs had furthermore been designed with sufficient power reserve to benefit from the intrinsic redundancy of many combined RF amplifier modules, and full output power could easily be fed into a matched load with up to six missing RF modules! However, when testing the first SSA at full power with 30% reflection and some of the modules switched off, the output circuits and the loads of the unpowered RF modules were damaged, while the phase of the mismatch was varied. In fact, operating conditions as described in Eqs. (6–8) lead to an overloading of the circulator loads of the non-powered RF modules. As much as 1500–1700 W was measured depending on the location in the combiner, confirming experimentally the 1570 W predicted in Eq. (8).

Fortunately, the coaxial combiner in the 150 kW ESRF SSA has multiple stages. Its S-parameter matrix does not have the exact symmetry of the single-stage combiner in Eq. (1). When all n of the RF modules feed the combiner with equal amplitudes and phases, their power still combines perfectly into the output port n + 1. Depending on the length of the output arms of the combiners, for instance in the first stage as shown in Fig. 12, the reverse power coming to an RF module from its seven immediate neighbouring RF modules interferes more or less constructively or destructively with the power coming from the other RF modules [8]. The tests had revealed that this interference was close to maximum when the load was burnt. As suggested by the experts from SOLEIL [8] and shown in Fig. 12, this interference could be made destructive by inserting 170 mm long delay lines at all of the outputs of all of the first combiner stages. The measurements shown in Fig. 12(b) show that the load power of a switched-off module then remained below 1200 W when operating the SSA at the specified maximum mismatch and varying the phase of the reflected wave. To complement the modification shown in Fig. 12, the 800 W circulator loads were replaced with 1200 W loads, as already shown in Fig. 8.

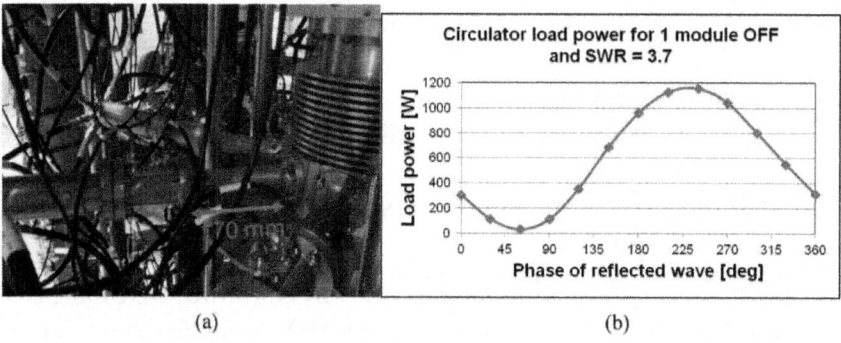

(a)   (b)

Figure 12: Minimization of the circulator load power of a switched-off RF module when the 150 kW SSA operates on a mismatch corresponding to 30% power reflection. (a) Insertion of a 170 mm delay line at the output of the first combiner stage; (b) measured load power is reduced from maximum 1700 W down to 1200 W.

The modifications described above were implemented for the three SSAs that are operated in continuous wave (CW) on the ESRF storage ring. This was not necessary for the ESRF booster, which is run in pulsed mode and where the relevant average load power remains well below 800 W.

Cavity Combiner Developed at ESRF

A compact single-stage cavity combiner is under development at ESRF. As shown in Fig. 13, its E010 resonance has a homogeneous azimuthal and longitudinal field distribution. For the 352.2 MHz ESRF application, the outer cylindrical wall is equipped with six vertical rows of 22 input loops distributed around the circumference. In total 132 transistor modules like that shown in Fig. 9, delivering up to 700 W each, will be connected to these coupling loops. The total output power is expected to be between 80 kW and 90 kW.

The coupling β_{module} must be the same for all of the input loops. It is easily adjusted by means of the loop size (typically a few square centimetres). The coupling of the output waveguide $\beta_{\text{waveguide}}$ is mainly determined by the size of the capacitive disc that couples to the electrical field of the E010 resonance. To obtain matched conditions and an optimum combining efficiency for a number n of RF modules, the coupling factors must be set according to:

$$\beta_{\text{waveguide}} \approx n \times \beta_{\text{module}} \gg 1 . \qquad (9)$$

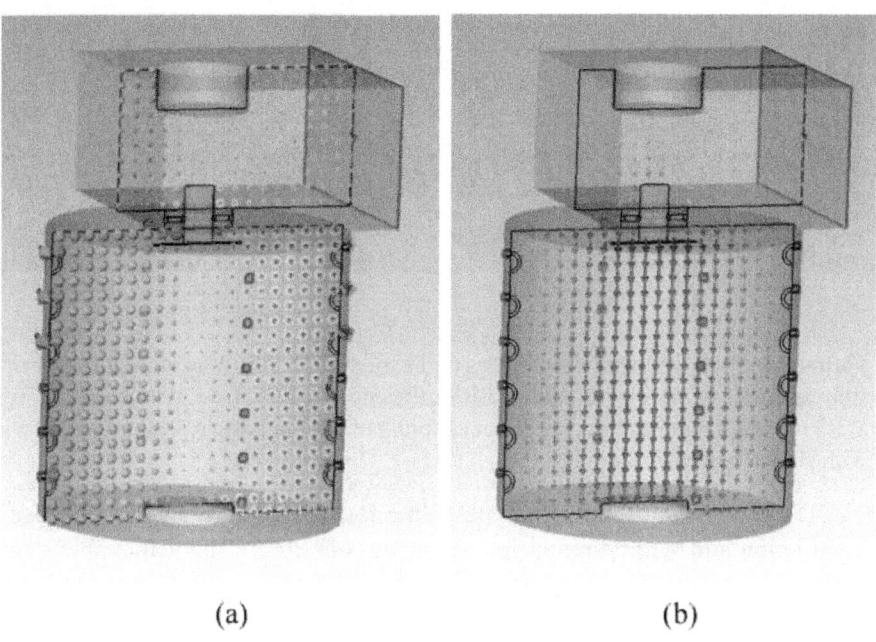

(a) (b)

Figure 13: Strongly loaded E010 resonance of a cylindrical resonator to combine the power from 132 RF modules in a single stage. (a) H field: homogeneous magnetic coupling to 132 input loops on the cylindrical wall; (b) E field: strong adjustable capacitive coupling to the output waveguide.

As shown in Fig. 13 and implemented on the ESRF prototype in Fig. 14, the piston attached to the top of the waveguide can be moved up and down, and the back short-circuit plane can be moved back and forth. This allows adjustment of the coupling factor β_{module} in a range of about 1:3 and thereby matching the SSA for a variable number of connected modules. The nominal power of the SSA can therefore be easily adapted to changing operating conditions by simply removing a number of RF modules and re-adapting the waveguide coupling accordingly. Figure 14 shows the prototype ESRF amplifier with cavity combiner. The water-cooled 'wings' that support six RF modules each constitute a section of the cavity wall with built-on coupling loops. The RF modules are hence directly flanged to the combiner and there is no need for coaxial RF power cables. The RF splitters for the distribution of the RF drive power as well as the DC power distribution are fixed onto the rear sides of the wings. The prototype in Fig. 14 with only three active wings and 19 blind flanges delivered as much as 12.4 kW of RF power with a DC-to-RF conversion efficiency of 63%.

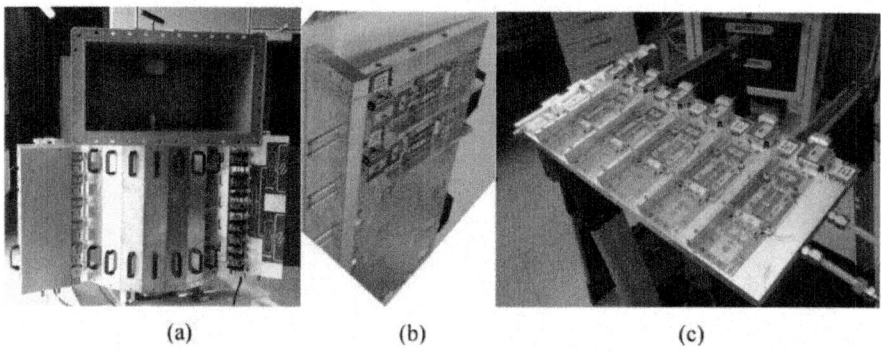

(a)   (b)   (c)

Figure 14: Prototype with three wings and a total of 18 700 W RF modules, successfully tested at 12.4 kW with a DC-to-RF efficiency of 63%. (a) Cavity combiner with WR2300 output waveguide; (b) direct coupling of RF modules; (c) water-cooled wing with six RF amplifier modules.

The extension of the prototype with a total of 22 active wings is close to completion and will be ready for power tests in 2015. The achievable output power is estimated to be between 80 kW and 90 kW, which is even slightly above the power obtained from one ELTA/SOLEIL coaxial tower. At the same time, the cavity combiner is more compact by far than an equivalent coaxial combiner.

Power Supply

As shown in Fig. 8 the ELTA/SOLEIL SSAs are equipped with two 280 V_{DC}-to-50 V_{DC} converters per RF module. Each of the three 150 kW SSAs in operation on the ESRF storage ring is fed with one 300 kW/400 V_{AC}-to-280 V_{DC} converter, which was developed by the ESRF Power Supply Group. As may be seen in Fig. 15, a 280 V_{DC}-to-RF conversion efficiency of 57% is obtained at 150 kW output power. Beyond nominal power, the efficiency approaches 60%.

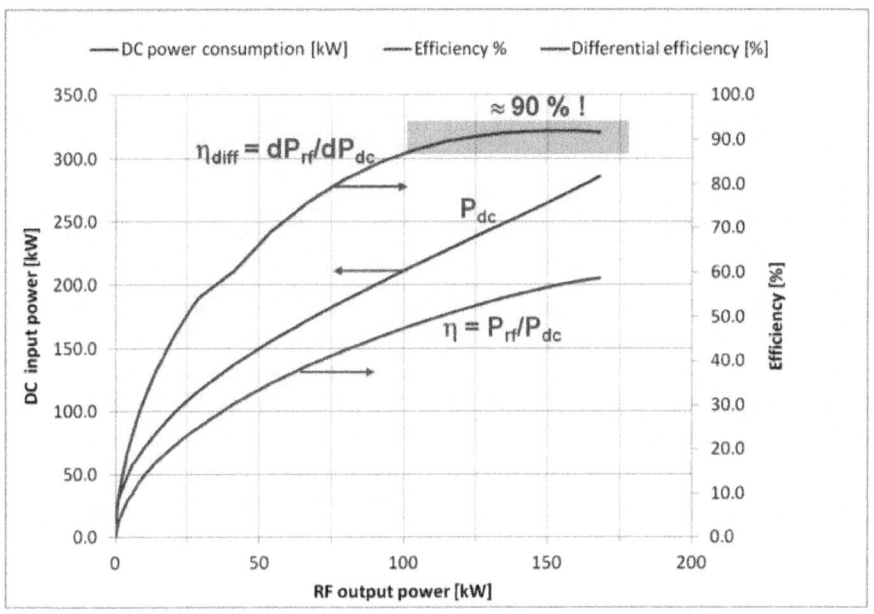

Figure 15: DC-to-RF efficiency and required DC power as a function of the output RF power of the ELTA/SOLEIL SSA implemented at ESRF.

It is worth noting that the efficiency drops fast when operating the amplifier at reduced power, and is only about 47% at 100 kW, i.e. at ⅔ of the nominal power. Note also that in the upper range 100– 150 kW, the differential efficiency is about 90%, meaning that, with 1 kW of additional DC power, one obtains as much as 0.9 kW of additional RF power. In other words: in order to make good use of the electrical power, it is important to correctly dimension the RF system and operate the SSAs close to their nominal power. Unfortunately, when designing an RF system, it is generally common practice to slightly over-dimension the available power in order to be able to overcome unforeseen transmission losses and to be prepared for possible power upgrades of the accelerator. A nice feature of the cavity combiner described in section 2.3.3 is that one can at any time remove a number of RF modules and easily re-adapt the output coupling. This way it is easy to bring down the nominal power of an over-dimensioned SSA close to required operation level and recover a good overall efficiency. Reference [9] describes another method for recovering good efficiency at reduced power by modulating the drain voltage of the transistors. However, it is important to check the stability of the amplifier modules over the entire range of the drain bias. Four 150 kW SSAs feed the accelerating RF cavities of the ESRF booster. Currently, a system of resonant AC and DC power supplies feed the booster magnets with a 10 Hz sine wave. The dipole magnet currents determine the modulation of particle energy E in the booster, as shown in Fig. 16. The RF voltage V_{tot} performs the bunching of the beam injected from the linac at 200 MeV and accelerates it to 6 GeV where it is extracted to the storage ring. In this process, the largest portion of RF voltage is needed to compensate for the synchrotron radiation losses that scale with E4 and lead to the red curve for the total accelerating voltage V_{tot} in Fig. 16. The total DC power $P_{dc\text{-}tot}$ absorbed by the SSAs to produce the RF power P_{rf} needed to obtain this voltage is oscillating at 10 Hz between about 100 kW and 1000 kW. An anti-flicker capacitor bank of 3.2 F has been installed to prevent this power fluctuation being transmitted to the mains. The residual flicker thus remains well below the legal limit of 0.29%. A good side effect is that an AC power of maximum 400 kW, slightly above the average DC power Pdcaver in Fig. 16, is drawn from the mains. This constitutes a reduction of nearly a factor 3 in power consumption with respect to the previous klystron transmitter, for which a 10 Hz filtering was not possible at 80 kV supply voltage. The klystron was operated with input RF modulation and constant beam power, most of this power being absorbed in the klystron collector.

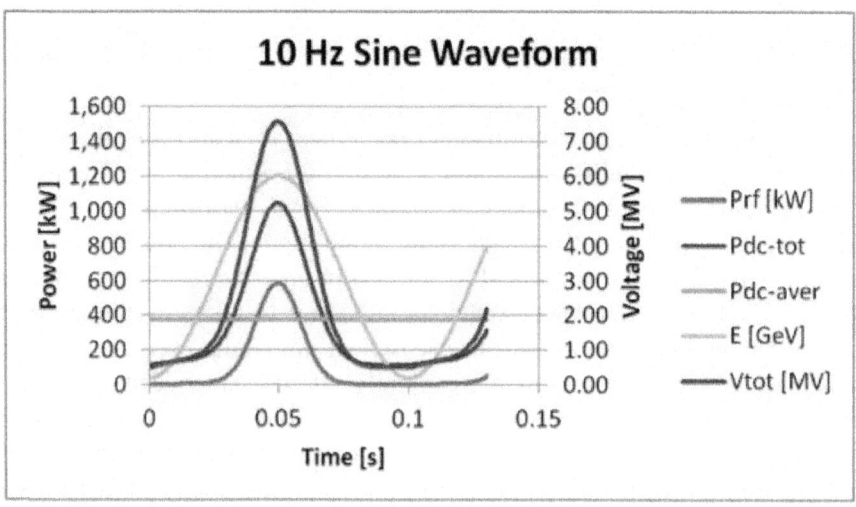

Figure 16: Pulsed RF for the ESRF booster.

For the SSA with the cavity combiner under development at ESRF, the RF modules will be fed from 22 power converters, one per wing, with a direct conversion from 400 V_{AC} to 50 V_{DC}. They will be installed close to the SSA in order not to carry the high currents over longer distances.

Specification

Having addressed the main ingredients of a high power RF SSA, typical performance data are now reviewed, however not exhaustively, based on the 352.2 MHz–150 kW SSAs delivered by ELTA to ESRF.

- The specified 280 V_{DC}-to-RF conversion efficiency was easily met:

✓ $\eta > 57\%$ at $P_{nom} = 150$ kW   (specified: $\eta > 55\%$);

✓ $\eta > 47\%$ at ⅔ $P_{nom} = 100$ kW   (specified: $\eta > 45\%$).

- Gain compression<1 d_B at $P_{nom} = 150$ kW.

    o Figure 17 shows the typical gain curve of one 150 kW SSA at ESRF. The P1 point, at which the gain compression reaches 1 dB, has been set above the nominal output power to limit distortion and safeguard the transistor against drain over-voltage. The P1 point is adjusted by means of the load resistance RL in Fig. 7.

- Avoid overdrive conditions

    o High peak drain voltage can damage the transistor. It is thus crucial to implement an overdrive protection interlock.

- Short pulses (20 µs).

    o RF cavities and their power couplers need to be slowly conditioned to high RF power, starting with cycles of 20 µs short pulses, then increasing the pulse length up to CW operation.

    o Up to 1.3 dB transient gain increase has been measured in pulsed operation, which bears a risk of overdrive. The overdrive protection therefore needs to be adjusted carefully.

- Requested redundancy for reliable operation.

    o All of the points from the specification in terms of power and operating conditions are achieved with up to 2.5% missing RF modules, i.e. with up to six RF modules in fault. As never more than one or two modules fail at the same time, the SSA never stops for output module failures. One can wait until the next programmed maintenance shutdown to exchange faulty modules.

o   The requested redundancy requires some power margin that is paid for with a slightly lower efficiency than that which is ultimately achievable. The power margin thus needs to be dimensioned carefully.

o   Note that if one of the RF power modules used in the drive chain fails, the output power can no longer be maintained and the SSA may trip the accelerator.

- Harmonics:

  ✓ H2 < −36 dBc   (36 dB below carrier at 704.4 MHz),
  ✓ H3 < −50 dBc   (50 dB below carrier at 1056.6 MHz).

- Spurious sidebands/phase noise:

✓ Sidebands < −68 dBc at 400 kHz   (DC/DC converter switching, harmless).

✓ A substantial improvement as compared with −50 dBc for klystrons from voltage ripples of the HV power supplies at 600 Hz, 900 Hz, 1200 Hz, etc., which are, moreover, much closer to the synchrotron frequency at which the stored beam is extremely sensitive.

- Operation on mismatched load

✓ 150 kW output power with ⅓, i.e. 50 kW, reflection at any phase,

✓ 80 kW output power with 100% reflection at any phase,

✓ 150 kW output power with 100% reflection at any phase for a duration of up to 20 μs.

o   Note that for high power reflections, due to the limited isolation of the circulators of the RF modules, the residual reverse power reaching the transistor output circuit provokes a gain modulation. Also, the power combiner itself contributes to some gain modulation in the presence of high reflected power. As a result, the specified output power cannot be reached for some phase values of the mismatch, due to the overdrive limitation.

- Reliability:

o   Not more than 0.7% of the installed RF power modules, including their individual DC/DC converters, are allowed to fail within one year. o With seven 150 kW SSAs gradually commissioned at ESRF between early 2012 and the end of 2013, not including early failures with less than 1000 hours of operation (debugging), the failure rate is in the range of the specified 0.7% per year. However, the cumulated operating time is still too low to give a statistically sound statement.

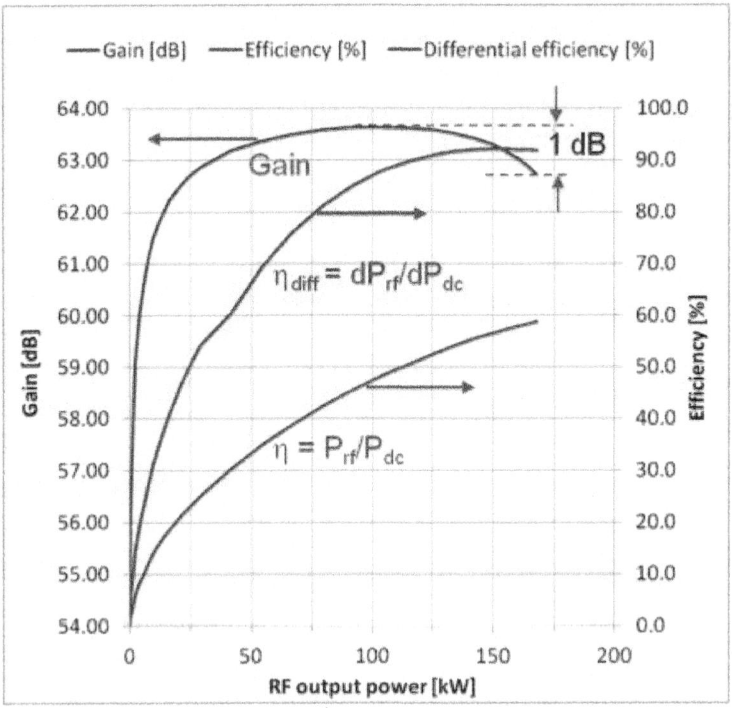

Figure 17: Gain and efficiency of an ESRF 352.2 MHz–150 kW SSA delivered by ELTA

Transient reflections for pulsed cavity conditioning

For the RF conditioning of accelerating cavities they are fed with short RF pulses. The physical explanation for the transient wave amplitudes in Fig. 18(a) is straightforward:

- the leading edge of the incoming RF pulse sees the cavity as a short circuit and is reflected with a factor of −1;

- after the filling time of the cavity ($Tf = Qo/(\pi frf (1 + \beta)) = 6.1 \ \mu s$), the reflection coefficient approaches the steady-state value $r = (\beta − 1)/(\beta + 1) = 0.58$;

- the negative falling edge of the incoming pulse again sees a short with $r = −1$ and gives rise to a positive transient reflection of amplitude 1 that adds to the 0.53 amplitude of the almost steady reflection: the total transient reflection therefore peaks at more than 1.5 times the incoming pulse amplitude, corresponding to more than twice the input power;

- afterthe cavity filling time of 6.1 μs, the reflected pulse vanishes; – The measurement of the incident and reflected power transients shown in Fig. 18(b) confirms this behaviour: the scope traces correspond to the squared amplitude plots shown Fig. 18(a).

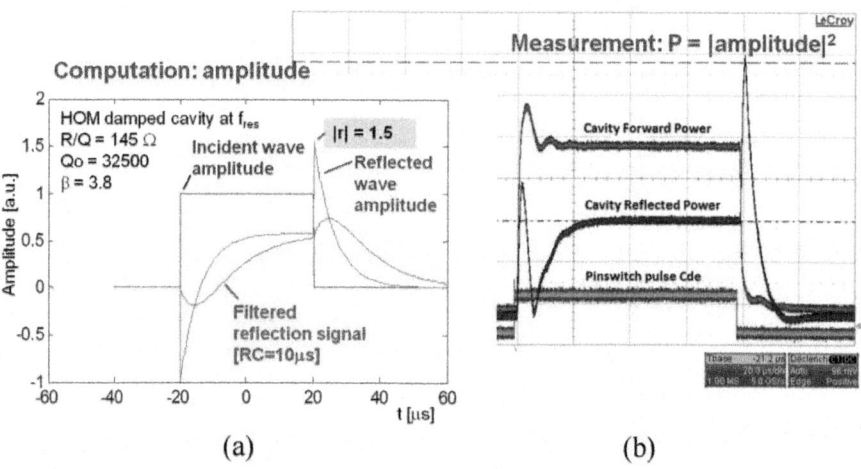

(a)   (b)

Figuer 18: Transient reflections when conditioning an accelerating cavity with short RF pulses. (a) Calculated transient wave amplitudes; (b) measured transient wave power.

The SSAs are protected against reflections exceeding the specified 50 kW as noted above. However, for a duration below 20 μs a reverse power of 150 kW is acceptable. Therefore, in order not to trip the SSAs for pulsed cavity conditioning, the signal triggering the 50 kW interlock was filtered as shown in the red curve in Fig. 18(a). This interlock then only triggers for power exceeding the maximum cavity conditioning power of 80 kW. The SSA is still protected by a second unfiltered fast interlock that triggers if the reflected power transient exceeds 150 kW. Taking into account the power reflection on the falling edge of the pulse, this interlock will trigger if the incident power exceeds the 80 kW needed for cavity conditioning.

CONCLUSIONS – A SHORT COMPARISON BETWEEN KLYSTRONS AND SOLID STATE AMPLIFIERS

To conclude this paper the pros (+) and cons (−) of implementing high power RF solid state amplifiers as an alternative to klystrons are tentatively addressed.

+ SSAs do not need a high voltage power supply (50 V instead of 100 kV):

+ consequently there is no need for X-ray shielding;

+ 20 dB less phase noise.

+ High modularity/redundancy providing a high reliability in operation:

  o the SSA is still fully operational if a few RF modules fail, except if a driver module fails.

• More space required per kW for a SSA than for a klystron:

  o however, thanks to the modularity, it is easier to precisely match the power of an SSA to the requirements;

  o the cavity combiner provides a substantial size reduction when compared to a coaxial combiner tree.

• Durability and obsolescence.

  o Klystrons: there is no problem with klystrons and tubes in general for as long as a particular model is still manufactured. But it can become very problematic in the event of obsolescence, as the development costs for new tubes are too high for medium-sized labs.

  o SSAs: transistors generally have a shorter product lifetime. However, there will always be comparable or even better transistors on the market. Nevertheless, operating SSAs commits the user to carefully follow the transistor market and react quickly enough to develop RF modules with new transistors that fit into the existing SSAs.

+ Easy maintenance of SSAs if sufficient spare parts are available.

• Investment costs:

− SSAs have still a higher price per kW than comparable tube solutions.

+ But SSA technology is progressing. A significant cost reduction is for instance expected from the possible mass production of fully planar RF modules as the ones developed at ESRF (see Section 2.1.2). Also the use of compact cavity combiners will reduce the fabrication costs as compared to building up large coaxial combiner trees.

+   Prices for SSA components should sink while prices for klystrons have strongly increased over the last decades.

+   Low cost of possession:

> o   With about 0.7% RF modules failing per year and relatively easy and inexpensive repair, the possession costs of SSAs are in principle very low and, in any case, substantially lower than for klystron transmitters, due to the high cost of spare klystrons

•   SSA and klystrons have comparable power efficiencies. However, this must be analysed case by case

+For the ESRF booster's pulsed RF system, for instance, a reduction of the power consumption by a factor of almost 3 was obtained, thanks to possible capacitive filtering of the DC supply voltage.

This list of arguments is far from being exhaustive. The question of the best selection of technology to provide high RF power for accelerator projects is a standing item on the agenda of RF workshops and conferences. Yet, more and more labs are considering SSAs as an adequate solution for their RF needs. For the moment, I would state that for multi-megawatt applications, klystrons remain advantageous. However, for accelerator applications of several hundreds of kilowatts, in particular for medium-sized projects with limited human resources, solid state amplifiers will become more and more attractive.

ACKNOWLEDGMENTS

This paper is in tribute to Ti Ruan, who passed away in March 2014. In the early 2000s Ti Ruan initiated the design and the implementation of high power SSAs combining hundreds of transistors for larger accelerators. He is the father of the big SSAs implemented at SOLEIL, ESRF, and many other places around the world. Many thanks are also due to the SOLEIL RF team: P. Marchand, R. Lopez, and F. Ribeiro; to the ELTA team: mainly J.-P. Abadie and A. Cauhepe; and to my RF colleagues at ESRF, in particular: J.- M. Mercier and M. Langlois. Their contributions constitute the backbone of this lecture.

REFERENCES

1. P. Marchand, T. Tuan, F. Ribeiro and R. Lopes, Phys. Rev. S.T. Accel. Beams 10 (2007) 112001.

2. P. Marchand, R. Lopes, F. Ribeiro and T. Ruan, Development of high RF power solid state amplifiers at SOLEIL, Proc. IPAC2011, San Sebastian, Spain, 4-9 September 2011, p. 376, http://accelconf.web.cern. ch/AccelConf/IPAC2011/papers/mopc127.pdf.

3. H. Büttig, A. Arnold, A. Buchner, M. Justus, M. Kuntzsch, U. Lehnert, P. Michel, R. Schurig, G. Staats and J. Teichert, RF power upgrade at the superconducting 1.3 GHz CW LINAC "ELBE" with solid state amplifiers, Nucl. Instrum. Meth. A 704 (2013) 7.

4. M. Getta, B. Aminov, A. Borisov, S. Kolesov, H. Piel and N. Pupeter, Modular high power solid state RF amplifiers for particle accelerators, Proc. PAC09, Vancouver, Canada, 4-8 May 2009, p. 1017, http:// accelconf.web.cern.ch/AccelConf/PAC2009/papers/tu5pfp081.pdf.

5. J. Jacob, J.-M. Mercier, M. Langlois and G. Gautier, 352.2 MHz – 150 kW solid state amplifiers at the ESRF, Proc. IPAC2011, San Sebastian, Spain, 4-9 September 2011, p. 71, http://accelconf.web.cern.ch/AccelConf/ IPAC2011/papers/mopc005.pdf.

6. J. Jacob, L. Farvacque, G. Gautier, M. Langlois and J.-M. Mercier, Commissioning of first 352.2 MHz – 150 kW solid state amplifiers at the ESRF and status of R&D, Proc. IPAC2013, Shanghai, China, 12-17 May 2013, p. 2708, http://accelconf.web.cern.ch/AccelConf/IPAC2013/ papers/wepfi004.pdf.

7. J. Jacob, New Developments on RF Power Sources, Proc. EPAC2006, Edinburgh, Scotland, 26- 30 June 2006, p. 1842, http://accelconf.web. cern.ch/AccelConf/e06/PAPERS/WEXPA02.pdf.

8. P. Marchand, private communication.

9. N. Pupeter, H. Piel, B. Aminov, A. Borisov, S. Kelesov and M. Nedos, A significant increase of the efficiency at the operating point of very high power solid state RF amplifiers (50 MHz to 1.3 GHz) by remote P1 control, 8th CWRF workshop, Elettra – Sincrotrone Trieste, Trieste, Italy, 13-16 May 2014. Slides are available at indico.cern.ch.

CITATION

CHAPTER 1

Antonio Agnesi and Federico Pirzio (2010). High Gain Solid-State Amplifiers for Picosecond Pulses, Advances in Solid State Lasers Development and Applications, Mikhail Grishin (Ed.), ISBN: 978-953-7619-80-0,

CHAPTER 2

Jiaxing Liu, Wei Wang, Zhaohua Wang, Zhiguo Lv, Zhiyuan Zhang and Zhiyi Wei; Diode-Pumped High Energy and High Average Power All-Solid-State Picosecond Amplifier Systems; doi:10.3390/app5041590

CHAPTER 3

Tianxiang Zhuge and Yulu Hu, "Design of a Novel High Power V-Band Helix-Folded Waveguide Cascaded Traveling Wave Tube Amplifier," Active and Passive Electronic Components, vol. 2015, Article ID 846425, 9 pages, 2015. doi:10.1155/2015/846425

CHAPTER 4

Mikhail Grishin and Andrejus Michailovas (2010). Dynamics of Continuously Pumped Solid-State Regenerative Amplifiers, Advances in Solid State Lasers Development and Applications, Mikhail Grishin (Ed.), ISBN: 978-953-7619-80-0,

CHAPTER 5

Tsung-Sum Lee (2010). Low-Voltage Fully Differential CMOS Switched-Capacitor Amplifiers, Advances in Solid State Circuit Technologies, Paul K Chu (Ed.), ISBN: 978-953-307-086-5, InTech, DOI: 10.5772/8621.

CHAPTER 6

Alessandro Cidronali, Iacopo Magrini and Gianfranco Manes (2010). Flexible Power Amplifier Architectures for Spectrum Efficient Wireless Applications, Advanced Microwave Circuits and Systems, Vitaliy Zhurbenko (Ed.), ISBN: 978-953-307-087-2

CHAPTER 7

Rajeev Kumar Ranjan, Surya Prasanna Yalla, Shubham Sorya, and Sajal K. Paul, "Active Comb Filter Using Operational Transconductance Amplifier," Active and Passive Electronic Components, vol. 2014, Article ID 587932, 6 pages, 2014. doi:10.1155/2014/587932

CHAPTER 8

Zhengyu Sun and Yuepeng Yan, "Design of a 2 GHz Linear-in-dB Variable-Gain Amplifier with 80-dB Gain Range," Active and Passive Electronic Components, vol. 2014, Article ID 434189, 7 pages, 2014. doi:10.1155/2014/434189

CHAPTER 9

Huiyong Li, Xun Li and Chen Wei, The analysis of the performance of multibeamforming in memory nonlinear power amplifier, doi:10.1186/1687-6180-2014-52

CHAPTER 10

Gregorio et al.: Power amplifier linearization technique with IQ imbalance and crosstalk compensation for broadband MIMO-OFDM transm.

CHAPTER 11

J. Jacob, Radio Frequency Solid State Amplifiers, DOI   10.5170/CERN-2015-003.197

INDEX